Lecture Notes in Computer Science 2383

Edited by G. Goos, J. Hartmanis, and J. van Leeuwen

T0241876

Springer
Berlin
Heidelberg
New York
Barcelona
Hong Kong
London
Milan
Paris
Tokyo

Michael S. Lew   Nicu Sebe   John P. Eakins (Eds.)

Image and Video Retrieval

International Conference, CIVR 2002
London, UK, July 18-19, 2002
Proceedings

 Springer

Series Editors

Gerhard Goos, Karlsruhe University, Germany
Juris Hartmanis, Cornell University, NY, USA
Jan van Leeuwen, Utrecht University, The Netherlands

Volume Editors

Michael S. Lew
Nicu Sebe
Leiden University, LIACS Media Lab
Niels Bohrweg 1, 2333 CA, Leiden, The Netherlands
E-mail: {mlew, nicu}@liacs.nl

John P. Eakins
University of Northumbria, Institute for Image Data Research
Newcastle NE1 8ST, UK
E-mail: john.eakins@unn.ac.uk

Cataloging-in-Publication Data applied for

Die Deutsche Bibliothek - CIP-Einheitsaufnahme

Image and video retrieval : international conference ; proceedings / CIVR
2002, London, UK, July 18 - 19, 2002. Michael S. Lew ... (ed.). - Berlin ;
Heidelberg ; New York ; Barcelona ; Hong Kong ; London ; Milan ; Paris ;
Tokyo : Springer, 2002
    (Lecture notes in computer science ; Vol. 2383)
    ISBN 3-540-43899-8

CR Subject Classification (1998):H.3, H.2, H.4, H.5.1, H.5.4-5

ISSN 0302-9743
ISBN 3-540-43899-8 Springer-Verlag Berlin Heidelberg New York

Springer-Verlag Berlin Heidelberg New York
a member of BertelsmannSpringer Science+Business Media GmbH

http://www.springer.de

© Springer-Verlag Berlin Heidelberg 2002
Printed in Germany

Typesetting: Camera-ready by author, data conversion by DA-TeX Gerd Blumenstein
Printed on acid-free paper   SPIN 10870499   06/3142   5 4 3 2 1 0

International Conference an Image and Video Retrieval 2002 Organization

Organizing Committee

Organizing Committee Chair:	John P. Eakins
	(University of Northumbria, UK)
Technical Program Chair:	Michael S. Lew
	(LIACS Media Lab, Leiden University, NL)
Practitioner Program Chair:	Margaret Graham
	(University of Northumbria, UK)
Publicity Chair:	Richard Harvey
	(University of East Anglia, UK)
Webmaster:	Paul Lewis (University of Southampton, UK)
Local Chair:	Chris Porter (Getty Images, London, UK)
	Peter Enser (University of Brighton, UK)
	Alan Smeaton (Dublin City University, IRL)

Program Committee

Jim Austin	University of York
Alberto Del Bimbo	University of Florence
Larry Chen	Kodak Research Labs
John P. Eakins	University of Northumbria
Peter Enser	University of Brighton
Graham Finlayson	University of East Anglia
David Forsyth	UC Berkeley
Theo Gevers	University of Amsterdam
Margaret Graham	University of Northumbria
Richard Harvey	University of East Anglia
Tom Huang	University of Illinois at Urbana-Champaign
Joemon Jose	University of Glasgow
Josef Kittler	University of Surrey
Clement Leung	University of Melbourne
Michael S. Lew	LIACS Media Lab, Leiden University
Paul Lewis	University of Southampton
Stephane Marchand-Maillet	University of Geneva
Jiří (George) Matas	CVUT Prague
Majid Mirmehdi	University of Bristol
Chahab Nastar	LTU technologies
Eric Pauwels Katholieke	University of Leuwen
Maria Petrou	University of Surrey
Chris Porter	Getty Images

Tony Rose	Reuters Limited
Yong Rui	Microsoft Research
Phillipe Salembier	University of Barcelona
Stan Sclaroff	Boston University
Nicu Sebe	LIACS Media Lab, Leiden University
Alan Smeaton	Dublin City University
Arnold Smeulders	University of Amsterdam
Barry Thomas	University of Bristol

Additional Reviewers

Erwin Bakker	Leiden University
Ira Cohen	University of Illinois at Urbana-Champaign
Ashutosh Garg	University of Illinois at Urbana-Champaign
Jan-Mark Geusebroek	University of Amsterdam
Alan Hanjalic	Delft University of Technology
Vicky Hodge	University of York
Nies Huijsmans	Leiden University
Thang Pham	University of Amsterdam
Cees Snoek	University of Amsterdam
Qi Tian	University of Illinois at Urbana-Champaign
Jeroen Vendrig	University of Amsterdam
Roy Wang	University of Illinois at Urbana-Champaign
Ziyou Xiong	University of Illinois at Urbana-Champaign

Sponsors

The British Computer Society, Information Retrieval Specialist Group
The British Machine Vision Association
The Institute for Image Data Research, University of Northumbria
The Institution of Electrical Engineers
The Leiden Institute of Advanced Computer Science, Leiden University

Table of Contents

Challenges of Image and Video Retrieval 1
Michael S. Lew, Nicu Sebe, and John P. Eakins

Image Retrieval I (Oral)

Visualization, Estimation and User-Modeling
for Interactive Browsing of Image Libraries 7
Qi Tian, Baback Moghaddam, and Thomas S. Huang

Robust Shape Matching 17
Nicu Sebe and Michael Lew

Semantics-Based Image Retrieval by Region Saliency 29
Wei Wang, Yuqing Song, and Aidong Zhang

The Truth about Corel – Evaluation in Image Retrieval 38
Henning Müller, Stephane Marchand-Maillet, and Thierry Pun

Non-retrieval: Blocking Pornographic Images 50
Alison Bosson, Gavin C. Cawley, Yi Chan, and Richard Harvey

Modelling I (Poster)

A Linear Image-Pair Model
and the Associated Hypothesis Test for Matching 61
Gregory Cox and Gerhard de Jager

On the Coupled Forward and Backward Anisotropic Diffusion Scheme
for Color Image Enhancement 70
Bogdan Smolka and Konstantinos N. Plataniotis

Multiple Regions and Their Spatial Relationship-Based Image Retrieval 81
ByoungChul Ko and Hyeran Byun

Feature Based Retrieval (Poster)

Query by Fax for Content-Based Image Retrieval 91
Mohammad F. A. Fauzi and Paul H. Lewis

Spectrally Layered Color Indexing 100
Guoping Qiu and Kin-Man Lam

Using an Image Retrieval System
for Vision-Based Mobile Robot Localization 108
Jürgen Wolf, Wolfram Burgard, and Hans Burkhardt

JPEG Image Retrieval Based on Features from DCT Domain 120
Guocan Feng and Jianmin Jiang

Image Retrieval Methods for a Database of Funeral Monuments 129
A. Jonathan Howell and David S. Young

Semantics/Learning I (Poster)

AtomsNet: Multimedia Peer2Peer File Sharing 138
Willem de Bruijn and Michael S. Lew

Visual Clustering of Trademarks Using the Self-Organizing Map 147
Mustaq Hussain, John Eakins, and Graham Sexton

FACERET: An Interactive Face Retrieval System Based
on Self-Organizing Maps 157
Javier Ruiz-del-Solar and Pablo Navarrete

Object-Based Image Retrieval Using Hierarchical Shape Descriptor 165
Man-Wai Leung and Kwok-Leung Chan

Video Retrieval (Oral)

Multimodal Person Identification in Movies 175
Jeroen Vendrig and Marcel Worring

Automated Scene Matching in Movies 186
F. Schaffalitzky and A. Zisserman

Content Based Analysis for Video from Snooker Broadcasts 198
H. Denman, N. Rea, and A. Kokaram

Retrieval of Archival Moving Imagery – CBIR Outside the Frame? 206
Peter G. B. Enser and Criss J. Sandom

Challenges for Content-Based Navigation of Digital Video
in the Físchlár Digital Library 215
Alan F. Smeaton

Image Retrieval II (Oral)

Spin Images and Neural Networks
for Efficient Content-Based Retrieval in 3D Object Databases 225
Pedro A. de Alarcón, Alberto D. Pascual-Montano, and José M. Carazo

Size Functions for Image Retrieval:
A Demonstrator on Randomly Generated Curves 235
A. Brucale, M. d'Amico, M. Ferri, L. Gualandri, and A. Lovato

An Efficient Coding of Three Dimensional Colour Distributions
for Image Retrieval 245
Jeff Berens and Graham D. Finlayson

Content-Based Retrieval of Historical Watermark Images: I-tracings 253
K. Jonathan Riley and John P. Eakins

Semantics/Learning II (Poster)

Object Recognition for Video Retrieval 262
Rene Visser, Nicu Sebe, and Erwin Bakker

Semantic Video Retrieval Using Audio Analysis 271
Erwin M. Bakker and Michael S. Lew

Extracting Semantic Information from Basketball Video Based
on Audio-Visual Features 278
Kyungsu Kim, Junho Choi, Namjung Kim, and Pankoo Kim

Video Indexing and Retrieval for Archeological Digital Library, CLIOH ... 289
Jeffrey Huang, Deepa Umamaheswaran, and Mathew Palakal

Fast *k*-NN Image Search with Self-Organizing Maps 299
Kun Seok Oh, Aghbari Zaher, and Pan Koo Kim

Video Retrieval by Feature Learning in Key Frames 309
Marcus J. Pickering, Stefan M. Rüger, and David Sinclair

Modelling II (Poster)

Local Affine Frames for Image Retrieval 318
Štěpán Obdržálek and Jiří Matas

A Ranking Algorithm Using Dynamic Clustering
for Content-Based Image Retrieval 328
Gunhan Park, Yunju Baek, and Heung-Kyu Lee

Online Bayesian Video Summarization and Linking 338
Xavier Orriols and Xavier Binefa

Face Detection for Video Summaries 348
Jean Emmanuel Viallet and Olivier Bernier

Evaluation/Benchmarking (Poster)

A Method for Evaluating the Performance
of Content-Based Image Retrieval Systems Based
on Subjectively Determined Similarity between Images 356
John A. Black, Jr., Gamal Fahmy, and Sethuraman Panchanathan

Evaluation of Salient Point Techniques 367
N. Sebe, Q. Tian, E. Loupias, M. Lew, and T. Huang

Personal Construct Theory as a Research Tool
for Analysing User Perceptions of Photographs 378
Mary A. Burke

Author Index 387

Challenges of Image and Video Retrieval

Michael S. Lew[1], Nicu Sebe[1], and John P. Eakins[2]

[1] LIACS Media Lab, Leiden University, The Netherlands
{mlew,nicu}@liacs.nl
[2] Institute for Image Data Research,
University of Northumbria at Newcastle, UK
john.eakins@unn.ac.uk

What use is the sum of human knowledge if nothing can be found? Although significant advances have been made in text searching, only preliminary work has been done in finding images and videos in large digital collections. In fact, if we examine the most frequently used image and video retrieval systems (i.e. www.google.com) we find that they are typically oriented around text searches where manual annotation was already performed.

Image and video retrieval is a young field which has its genealogy rooted in artificial intelligence, digital signal processing, statistics, natural language understanding, databases, psychology, computer vision, and pattern recognition. However, none of these parental fields alone has been able to directly solve the retrieval problem. Indeed, image and video retrieval lies at the intersections and crossroads between the parental fields. It is these curious intersections which appear to be the most promising.

What are the main challenges in image and video retrieval? We think the paramount challenge is bridging the semantic gap. By this we mean that low level features are easily measured and computed, but the starting point of the retrieval process is typically the high level query from a human. Translating or converting the question posed by a human to the low level features seen by the computer illustrates the problem in bridging the semantic gap.

However, the semantic gap is not merely translating high level features to low level features. The essence of a semantic query is understanding the meaning behind the query. This can involve understanding both the intellectual and emotional sides of the human, not merely the distilled logical portion of the query but also the personal preferences and emotional subtones of the query and the preferential form of the results.

In this proceedings, several papers [1,2,3,4,5,6,7,8] touch upon the semantic problem and give valuable insights into the current state of the art. Wang et al [1] propose the use of color-texture classification to generate a codebook which is used to segment images into regions. The content of a region is then characterize by its self-saliency which describes its perceptual importance. Bruijn and Lew [2] investigate multi-modal content-based browsing and searching methods for Peer2Peer retrieval systems. Their work targets the assumption that keyframes are more interesting when they contain people. Vendrig and Wor-

M. S. Lew, N. Sebe, and J. P. Eakins (Eds.): CIVR 2002, LNCS 2383, pp. 1–6, 2002.

ring [3] propose a system that allows character identification in movies. In order to achieve this, they relate visual content to names extracted from movie scripts. Denman et al [5] present the tools in a system for creating semantically meaningful summaries of broadcast Snooker footage. Their system parses the video sequence, identifies relevant camera views, and tracks ball movements. A similar approach presented by Kim et al [8] extracts semantic information from basketball videos based on audio-visual features. A semantic video retrieval approach using audio analysis is presented by Bakker and Lew [7] in which the audio can be automatically categorized into semantic categories such as explosions, music, speech, etc. A system for recognizing objects in video sequences is presented by Visser et al [6]. They use the Kalman filter to obtain segmented blobs from the video, classify the blobs using the probability ration test, and apply several different temporal methods, which results in sequential classification methods over the video sequence containing the blob. An automated scene matching algorithm is presented by Schaffalitzky and Zisserman [4]. Their goal is to match images of the same 3D scene in a movie. Ruiz-del-Solar and Navarrete [9] present a content-based face retrieval system that uses self-organizing maps (SOMs) and user feedback. SOMs were also employed by Oh et al [10], Hussain et al [11], and Huang et al [12] for visual clustering. A ranking algorithm using dynamic clustering for content-based image retrieval is proposed by Park et al [13]. A learning method using the AdaBoost algorithm and a k-nearest neighbor approach is proposed by Pickering et al [14] for video retrieval.

An overview of challenges for content-based navigation of digital video is presented by Smeaton [15]. The author presents the different ways in which video content can be used directly to support the navigation within large video libraries and lists the challenges that still remain to be addressed in this area. An insight into the problems and challenges of retrieval of archival moving imagery via the Internet is presented by Enser and Sandom [16]. The authors conclude that the combination of limited CBIR functionality and lack of adherence to cataloging standards seriously limits the Internet's potential for providing enhanced access to film and video-based cultural resources. Burke [17] describes a research project which applies Personal Construct Theory to individual user perceptions of photographs. This work presents a librarian viewpoint toward content-based image retrieval. A user-centric system for visualization and layout for content-based image retrieval and browsing is proposed by Tian et al [18].

An important segment of papers discusses the challenges of using different models of human image perception in visual information retrieval. Cox and de Jager [19] developed a statistical model for the image pair and used it to derive a minimum-error hypothesis test for matching. An online Bayesian formulation for video summarization and linking is proposed by Orriols and Binefa [20]. A model-based approach for detecting pornographic images is presented by Bosson et al [21]. Based on the idea that the faces of persons are the first information looked for in an image, Viallet and Bernier [22] propose a face detection system which automatically derives face summaries from a video sequence. A model for edge-preserving smoothing is presented by Smolka and Plataniotis [23]. Their algorithm is based on the combined forward and backward anisotropic diffusion with incorporated time dependent cooling process. The authors report that their method is able to efficiently remove image noise while preserving and enhancing image edges. Ko and Byun [24] present a model for content-based image retrieval based on regions-of-interest and their spatial relationship. A vision-based approach to mobile robot localization that integrates an image retrieval system with Monte-Carlo localization is presented by Wolf et al [25]. A layered-based model for image retrieval is proposed by Qiu [26].

Object recognition and detection is one of the most challenging problems in visual information retrieval. Several papers present the advances in this area [27,28,29,30]. Obdržálek and Matas [27] present a method that supports recognition of objects under a very wide range of viewing and illumination conditions and is robust to occlusion and background clutter. A hierarchical shape descriptor for object-based retrieval is proposed by Leung and Chan [29]. Brucale, et al [30] propose a class of topological-geometrical shape descriptors called size functions. Sebe and Lew [28] investigate the link between different metrics and the similarity noise model in an object-based retrieval application. They conclude that the prevalent Gaussian distribution assumption is often invalid and propose a Cauchy model. Furthermore, they explained how to create a maximum likelihood metric based on the similarity noise distribution and showed that it consistently outperformed all of the analytical metrics. Based on the idea that the distribution of colors in an image provides a useful cue for image retrieval and object recognition, Berens and Finlayson [31] propose an efficient coding of three dimensional color distribution for image retrieval.

In addition, new techniques are presented for a wide range of retrieval problems, including 3-D object matching [32] and compressed-domain image searching [33], as well as applications in areas as diverse as videos of snooker broadcasts [5], images of historical watermarks [34], funeral monuments [35], and low quality fax images [36].

In order for image and video retrieval to mature, we will need to understand how to evaluate and benchmark features, methods, and systems. Several papers which address these questions are [37,38], and [39]. While Sebe et al [39] perform an evaluation of different salient point extraction techniques to be used in content-based image retrieval, Black et al [38] propose a method for creation of a reference image set in which the similarity of each image pair is estimated using "visual content words" as a basis vector that allows the multidimensional content of each image to be represented with a content vector. The authors claim that the similarity measure computed with these content vectors correlates with the subjective judgment of human observers and provides a more objective method for evaluating and expressing the image content. Müller et al [37] compare different ways of evaluating the performance of content-based retrieval systems (CBIRSs) using a subset of the Corel images. Their aim is to show how easy it is to get differing results, even when the same image collection, CBIRS, and performance measures are used.

References

1. Wang, W., Song, Y., Zhang, A.: Semantics-based image retrieval by region saliency. In: International Conference on Image and Video Retrieval, Lecture Notes in Computer Science, vol. 2383, Springer (2002) 27–35 1
2. de Bruijn, W., Lew, M.: Atomsnet: Multimedia peer 2 peer for file sharing. In: International Conference on Image and Video Retrieval, Lecture Notes in Computer Science, vol. 2383, Springer (2002) 130–138 1
3. Vendrig, J., Worring, M.: Multimodal person identification in movies. In: International Conference on Image and Video Retrieval, Lecture Notes in Computer Science, vol. 2383, Springer (2002) 166–175 1, 2
4. Schaffalitzky, F., Zisserman, A.: Automated scene matching in movies. In: International Conference on Image and Video Retrieval, Lecture Notes in Computer Science, vol. 2383, Springer (2002) 176–185 1, 2
5. Denman, H., Rea, N., Kokaram, A.: Content based analysis for video from snooker broadcasts. In: International Conference on Image and Video Retrieval, Lecture Notes in Computer Science, vol. 2383, Springer (2002) 186–193 1, 2, 3
6. Visser, R., Sebe, N., Bakker, E.: Object recognition for video retrieval. In: International Conference on Image and Video Retrieval, Lecture Notes in Computer Science, vol. 2383, Springer (2002) 250–259 1, 2
7. Bakker, E., Lew, M.: Semantic video retrieval using audio analysis. In: International Conference on Image and Video Retrieval, Lecture Notes in Computer Science, vol. 2383, Springer (2002) 260–267 1, 2
8. Kim, K., Choi, J., Kim, N., Kim, P.: Extracting semantic information from basketball video based on audio-visual features. In: International Conference on Image and Video Retrieval, Lecture Notes in Computer Science, vol. 2383, Springer (2002) 268–277 1, 2
9. Ruiz-del-Solar, J., Navarrete, P.: FACERET: An interactive face retrieval system based on self-organizing maps. In: International Conference on Image and Video Retrieval, Lecture Notes in Computer Science, vol. 2383, Springer (2002) 149–156 2
10. Oh, K., Zaher, A., Kim, P.: Fast k-nn image search with self-organizing maps. In: International Conference on Image and Video Retrieval, Lecture Notes in Computer Science, vol. 2383, Springer (2002) 288–297 2
11. Hussain, M., Eakins, J., Sexton, G.: Visual clustering of trademarks using the self-organizing map. In: International Conference on Image and Video Retrieval, Lecture Notes in Computer Science, vol. 2383, Springer (2002) 139–148 2
12. Huang, J., Umamaheswaran, D., Palakal, M.: Video indexing and retrieval for archeological digital library, CLIOH. In: International Conference on Image and Video Retrieval, Lecture Notes in Computer Science, vol. 2383, Springer (2002) 278–287 2
13. Park, G., Baek, Y., Lee, H. K.: A ranking algorithm using dynamic clustering for content-based image retrieval. In: International Conference on Image and Video Retrieval, Lecture Notes in Computer Science, vol. 2383, Springer (2002) 316–324 2
14. Pickering, M., Rüger, S., Sinclair, D.: Video retrieval by feature learning in key frames. In: International Conference on Image and Video Retrieval, Lecture Notes in Computer Science, vol. 2383, Springer (2002) 298–306 2
15. Smeaton, A.: Challenges for content-based navigation of digital video in the Físchlár digital library. In: International Conference on Image and Video Retrieval, Lecture Notes in Computer Science, vol. 2383, Springer (2002) 203–215 2

16. Enser, P., Sandom, C.: Retrieval of archival moving imagery - CBIR outside the frame? In: International Conference on Image and Video Retrieval, Lecture Notes in Computer Science, vol. 2383, Springer (2002) 194–202 2
17. Burke, M.: Personal construct theory as a research tool for analysing user perceptions of photographs. In: International Conference on Image and Video Retrieval, Lecture Notes in Computer Science, vol. 2383, Springer (2002) 363–370 2
18. Tian, Q., Moghaddam, B., Huang, T.: Visualization, estimation, and user-modeling for interactive browsing of image libraries. In: International Conference on Image and Video Retrieval, Lecture Notes in Computer Science, vol. 2383, Springer (2002) 7–16 2
19. Cox, G., de Jager, G.: A linear image-pair model and the associated hypothesis test for matching. In: International Conference on Image and Video Retrieval, Lecture Notes in Computer Science, vol. 2383, Springer (2002) 56–63 2
20. Orriols, X., Binefa, X.: Online bayesian video summarization and linking. In: International Conference on Image and Video Retrieval, Lecture Notes in Computer Science, vol. 2383, Springer (2002) 325–334 2
21. Bosson, A., Cawley, G., Chan, Y., Harvey, R.: Non-retrieval: Blocking pornographic images. In: International Conference on Image and Video Retrieval, Lecture Notes in Computer Science, vol. 2383, Springer (2002) 46–55 2
22. Viallet, J., Bernier, O.: Face detection for video summaries. In: International Conference on Image and Video Retrieval, Lecture Notes in Computer Science, vol. 2383, Springer (2002) 335–342 2
23. Smolka, B., Plataniotis, K.: On the coupled forward and backward anisotropic diffusion scheme for color image enhancement. In: International Conference on Image and Video Retrieval, Lecture Notes in Computer Science, vol. 2383, Springer (2002) 64–73 2
24. Ko, B., Byun, H.: Multiple regions and their spatial relationship-based image retrieval. In: International Conference on Image and Video Retrieval, Lecture Notes in Computer Science, vol. 2383, Springer (2002) 74–83 3
25. Wolf, J., Burgard, W., Burkhardt, H.: Using an image retrieval system for vision-based mobile robot localization. In: International Conference on Image and Video Retrieval, Lecture Notes in Computer Science, vol. 2383, Springer (2002) 102–111 3
26. Qiu, G., Lam, K. M.: Spectrally layered color indexing. In: International Conference on Image and Video Retrieval, Lecture Notes in Computer Science, vol. 2383, Springer (2002) 93–101 3
27. Š. Obdržálek, Matas, J.: Local affine frames for image retrieval. In: International Conference on Image and Video Retrieval, Lecture Notes in Computer Science, vol. 2383, Springer (2002) 307–315 3
28. Sebe, N., Lew, M.: Robust shape matching. In: International Conference on Image and Video Retrieval, Lecture Notes in Computer Science, vol. 2383, Springer (2002) 17–26 3
29. Leung, M. W., Chan, K. L.: Object-based image retrieval using hierarchical shape descriptor. In: International Conference on Image and Video Retrieval, Lecture Notes in Computer Science, vol. 2383, Springer (2002) 157–165 3
30. Brucale, A., d'Amico, M., Ferri, M., Gualandri, M., Lovato, A.: Size functions for image retrieval: A demonstrator on randomly generated curves. In: International Conference on Image and Video Retrieval, Lecture Notes in Computer Science, vol. 2383, Springer (2002) 223–232 3

31. Berens, J., Finlayson, G.: An efficient coding of three dimensional colour distributions for image retrieval. In: International Conference on Image and Video Retrieval, Lecture Notes in Computer Science, vol. 2383, Springer (2002) 233–240 3

32. de Alarcón, P., Pascual-Montano, A., Carazo, J.: Spin images and neural networks for efficient content-based retrieval in 3D object databases. In: International Conference on Image and Video Retrieval, Lecture Notes in Computer Science, vol. 2383, Springer (2002) 216–222 3

33. Feng, G., Jiang, J.: JPEG image retrieval based on features from DCT domain. In: International Conference on Image and Video Retrieval, Lecture Notes in Computer Science, vol. 2383, Springer (2002) 112–120 3

34. Riley, K., Eakins, J.: Content-based retrieval of historical watermark images: I-Tracings. In: International Conference on Image and Video Retrieval, Lecture Notes in Computer Science, vol. 2383, Springer (2002) 241–249 3

35. Howell, A., Young, D.: Image retrieval methods for a database of funeral monuments. In: International Conference on Image and Video Retrieval, Lecture Notes in Computer Science, vol. 2383, Springer (2002) 121–129 3

36. Fauzi, M., Lewis, P.: Query by fax for content-based image retrieval. In: International Conference on Image and Video Retrieval, Lecture Notes in Computer Science, vol. 2383, Springer (2002) 84–92 3

37. Müller, H., Marchand-Maillet, S., Pun, T.: The truth about Corel - Evaluation in image retrieval. In: International Conference on Image and Video Retrieval, Lecture Notes in Computer Science, vol. 2383, Springer (2002) 36–45 3

38. Black, J., Fahmy, G., Panchanathan, S.: A method for evaluating the performance of content-based image retrieval systems based on subjectively determined similarity between images. In: International Conference on Image and Video Retrieval, Lecture Notes in Computer Science, vol. 2383, Springer (2002) 343–352 3

39. Sebe, N., Tian, Q., Loupias, E., Lew, M., Huang, T.: Evaluation of salient point techniques. In: International Conference on Image and Video Retrieval, Lecture Notes in Computer Science, vol. 2383, Springer (2002) 353–362 3

Visualization, Estimation and User-Modeling for Interactive Browsing of Image Libraries

Qi Tian[1], Baback Moghaddam[2], and Thomas S. Huang[1]

[1] Beckman Institute, University of Illinois, Urbana-Champaign, IL 61801, USA
{qitian,huang}@ifp.uiuc.edu
[2] Mitsubishi Electric Research Laboratory, Cambridge, MA 02139, USA
{baback}@merl.com

Abstract. We present a user-centric system for visualization and layout for content-based image retrieval and browsing. Image features (visual and/or semantic) are analyzed to display and group retrievals as thumbnails in a 2-D spatial layout which conveys mutual similarities. Moreover, a novel subspace feature weighting technique is proposed and used to modify 2-D layouts in a variety of context-dependent ways. An efficient computational technique for subspace weighting and re-estimation leads to a simple user-modeling framework whereby the system can learn to display query results based on layout examples (or relevance feedback) provided by the user. The resulting retrieval, browsing and visualization engine can adapt to the user's (time-varying) notions of content, context and preferences in style of interactive navigation. Monte Carlo simulations with synthetic "user-layouts" as well as pilot user studies have demonstrated the ability of this framework to accurately model or "mimic" users by automatically generating layouts according to their preferences.

1 Introduction

With the advances in technology to capture, generate, transmit and store large amounts of digital imagery and video, research in content-based image retrieval (CBIR) has gained increasing attention. In CBIR, images are indexed by their visual contents such as color, texture, etc. Many research efforts have addressed how to extract these low level features [1, 2, 3], evaluate distance metrics [4, 5] for similarity measures and look for efficient searching schemes [6, 7].

In this paper, we present designs for optimal (uncluttered) visualization and layout of images (or iconic data in general) in a 2-D display space. We further provide a mathematical framework for user-modeling, which adapts and mimics the user's (possibly changing) preferences and style for interaction, visualization and navigation.

Monte Carlo simulation results with machine-generated layouts as well as pilot user-preference studies with actual user-guided layouts have indicated the power of this approach to model (or "mimic") the user. This framework is currently being incorporated into a broader system for computer-human guided navigation, browsing, archiving and interactive story-telling with large photo libraries.

M. S. Lew, N. Sebe, and J. P. Eakins (Eds.): CIVR 2002, LNCS 2383, pp. 7-16, 2002.

1.1   Traditional Interfaces

The purpose of automatic content-based visualization is augmenting the user's under-standing of large information spaces that cannot be perceived by traditional sequential display (*e.g.* by rank order of visual similarities). The standard and commercially prevalent image management and browsing tools currently available primarily use tiled sequential displays – *i.e.,* essentially a simple 1-D similarity-based visualization.

However, the user quite often can benefit by having a global view of a working subset of retrieved images in a way that reflects the relations between *all pairs* of images – *i.e.,* N^2 measurements as opposed to only N. Moreover, even a narrow view of one's immediate surroundings defines "context" and can offer an indication on how to explore the dataset. The wider this "visible" horizon, the more efficient the new query will be formed. In [8], Rubner proposed a 2-D display technique based on multi-dimensional scaling (MDS) [9]. A global 2D view of the images is achieved that reflects the mutual similarities among the retrieved images. MDS is a nonlinear trans-formation that minimizes the stress between high dimensional feature space and low dimensional display space. However, MDS is rotation invariant, non-repeatable (non-unique), and often slow to implement. These drawbacks make MDS unattractive for real time browsing or visualization of high-dimensional data such as images.

1.2   Layout & Visualization

We propose an alternative 2-D display scheme based on Principal Component Analy-sis (PCA) [10]. Moreover, a novel window display optimization technique is pro-posed which provides a more perceptually intuitive, visually uncluttered and informa-tive visualization of the retrieved images.

Traditional image retrieval systems display the returned images as a list, sorted by decreasing similarity to the query. The traditional display has one major drawback. The images are ranked by similarity to the query, and relevant images (as for example used in a relevance feedback scenario) can appear at separate and distant locations in the list. We propose an alternative technique to MDS in [8] that displays mutual simi-larities on a 2-D screen based on visual features extracted from images. The retrieved images are displayed not only in ranked order of similarity from the query but also according to their mutual similarities, so that similar images are grouped together rather than being scattered along the entire returned 1-D list.

1.3   PCA Splats

To create such a 2-D layout, Principal Component Analysis (PCA) [10] is first per-formed on the retrieved images to project the images from the high dimensional fea-ture space to the 2-D screen. Image thumbnails are placed on the screen so that the screen distances reflect as closely as possible the similarities between the images. If the computed similarities from the high dimensional feature space agree with our perception, and if the resulting feature dimension reduction preserves these similarities reasonably well, then the resulting spatial display should be informative and useful.

For our image representation, we have used three visual features: color moments [1], wavelet-based texture [2], and water-filling edge-based structure feature [3]. We should note that the choice of visual representation is not important to the methodology of this paper. In our experiments, the 37 visual features (9 color moments, 10 wavelet moments and 18 water-filling features) are pre-extracted from the image database and stored off-line. The 37-dimensional feature vector for an image, when taken in context with other images, can be projected on to the 2-D $\{x, y\}$ screen based on the 1st two principal components normalized by the respective eigenvalues. Such a layout is denoted as a PCA Splat. We note that PCA has several advantages over nonlinear methods like MDS. It is a fast, efficient and unique linear transformation that achieves the maximum distance preservation from the original high dimensional feature space to 2-D space among all possible linear transformations [10]. The fact that it fails to model nonlinear mappings (which MDS succeeds at) is in our opinion a minor compromise given the advantages of real-time, repeatable and mathematically tractable linear projections.

Let us consider a scenario of a typical image-retrieval engine at work in which an actual user is providing relevance feedback for the purposes of query refinement. Figure 1 shows an example of the retrieved images by the system (which resembles most traditional browsers in its 1D tile-based layout). The database is a collection of 534 images. The 1st image (building) is the query. The other 9 relevant images are ranked in 2nd, 3rd, 4th, 5th, 9th, 10th, 17th, 19th and 20th places, respectively.

Fig. 1. Top 20 retrieved images (ranked in scan-line order; the query is first in the list)

Figure 2 shows an example of a PCA Splat for the top 20 retrieved images shown in Figure 1. In addition to visualization by layout, in this particular example, the sizes (alternatively contrast) of the images are determined by their visual similarity to the query. The higher the rank, the larger the size (or higher the contrast). There is also a number next to each image in Figure 2 indicating its corresponding rank in Figure 1.

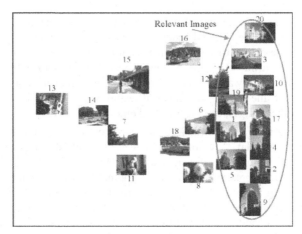

Fig. 2. PCA Splat of top 20 retrieved images in Figure 1

Clearly the relevant images are now better clustered in this new layout as opposed to being dispersed along the tiled 1-D display in Figure 1. Additionally, PCA Splats convey N^2 mutual distance measures relating all pair-wise similarities between images while the ranked 1-D display in Figure 1 provides only N.

One drawback of PCA Splats (or any low-dimensional mapping) is that inevitably some images are partially or totally overlapped (due to proximity or co-location in 2D space) which makes it difficult to view all the images at the same time. This image overlap will worsen when the number of retrieved images becomes larger. To address this problem, we have devised a unique layout optimization technique [13]. The goal is to minimize the total image overlap by finding an optimal set of image size and image positions with a little deviation as possible. A nonlinear optimization method was implemented by an iterative gradient descent method.

We note that Figure 2 is, in fact, just such an optimized PCA Splat. The overlap is clearly minimized while the relevant images are still close to each other to allow a global view. With such a display, the user can see the relations between the images, better understand how the query performed, and subsequently formulate future queries more naturally. Additionally, attributes such as contrast and brightness can be used to convey rank. We note that this additional visual aid is essentially a 3rd dimension of information display. A full discussion of the resulting enhanced layouts is deferred to our future work.

2   User-Modeling

Image content and "meaning" is ultimately based on semantics. The user's notion of content is a high-level concept, which is quite often removed by many layers of abstraction from simple low-level visual features. Even near-exhaustive semantic (keyword) annotations can never fully capture context-dependent notions of content. The same image can "mean" a number of different things depending on the particular circumstance.

By user-modeling or "context awareness" we mean that our system must be constantly aware of and adapting to the changing concepts and preferences of the user. A typical example of this human-computer synergy is having the system learn from a user-generated layout in order to visualize new examples based on identified relevant/irrelevant features. In other words, design smart browsers that "mimic" the user, and over-time, adapt to their style or preference for browsing and query display. Given information from the layout, *e.g.*, positions and mutual distances between images, a novel feature weight estimation scheme, noted as α-estimation is proposed, where α is a weighting vector for different features e.g., color, texture and structure (and possibly semantic keywords as well).

2.1   Subspace Estimation of Feature Weights

The weighting parameter vector is denoted as $\alpha = (\alpha_c, \alpha_t, \alpha_s)^T$, where α_c is the weight for color, α_t is the weight for texture, and α_s is the weight for structure. The number of images in the preferred clustering is N, and $\mathbf{X}_c$ is a $L_c \times N$ matrix where the ith column is the color feature vector of the ith image, $i = 1, \cdots, N$, $\mathbf{X}_t$ is the $L_t \times N$ matrix, the ith column is the texture feature vector of the ith image, $i = 1, \cdots, N$, and $\mathbf{X}_s$ is the $L_s \times N$ matrix, the ith column is the structure feature vector of the ith image, $i = 1, \cdots, N$. The lengths of color, texture and structure features are L_c, L_t, and L_s respectively. The distance, for example Euclidean,-based between the ith image and the jth image, for $i, j = 1, \cdots, N$, in the preferred clustering (distance in 2-D space) is d_{ij}. These weights α_c, α_t, α_s are constrained such that they always sum to 1.

We then define an energy term to minimize with an Lp norm (with $p = 2$). This cost function is defined in Equation (1). It is a nonnegative quantity that indicates how well mutual distances are preserved in going from the original high dimensional feature space to 2-D space. Note that this cost function is similar to MDS stress, but unlike MDS, the minimization is seeking the optimal feature weights α. Moreover, the low-dimensional projections in this case are already known. The optimal weighting parameter recovered is then used to weight original feature-vectors prior to a PCA Splat, resulting in the desired layout.

$$J = \sum_{i=1}^{N}\sum_{j=1}^{N}\{d_{ij}^{~p} - \sum_{k=1}^{L_c}\alpha_c^{~p} \mid \mathbf{X}_{c(i)}^{(k)} - \mathbf{X}_{c(j)}^{(k)} \mid^p - \sum_{k=1}^{L_t}\alpha_t^{~p} \mid \mathbf{X}_{t(i)}^{(k)} - \mathbf{X}_{t(j)}^{(k)} \mid^p$$

$$- \sum_{k=1}^{L_s}\alpha_s^{~p} \mid \mathbf{X}_{s(i)}^{(k)} - \mathbf{X}_{s(j)}^{(k)} \mid^p\}^2 \qquad\qquad (1)$$

The parameter α is then estimated by a constrained non-negative least-squares optimization procedure. Defining the following Lp-based deviations for each of the 3 subspaces:

$$V_{(ij)}^c = \sum_{k=1}^{L_c} |\mathbf{X}_{c(i)}^{(k)} - \mathbf{X}_{t(j)}^{(k)}|^P \ , \ V_{(ij)}^t = \sum_{k=1}^{L_t} |\mathbf{X}_{t(i)}^{(k)} - \mathbf{X}_{t(j)}^{(k)}|^P \ , \ V_{(ij)}^s = \sum_{k=1}^{L_s} |\mathbf{X}_{s(i)}^{(k)} - \mathbf{X}_{s(j)}^{(k)}|^P$$

Equation (1) can be re-written as the following cost function:

$$J = \sum_{i=1}^{N} \sum_{j=1}^{N} (d_{ij}^{\ P} - \alpha_c^{\ P} V_{(ij)}^c - \alpha_t^{\ P} V_{(ij)}^t - \alpha_s^{\ P} V_{(ij)}^s)^2 \tag{2}$$

which we differentiate and set to zero to obtain a linear system in the p-th power of the desired subspace coefficients, α^P. This system is easily solved using *constrained* linear least-squares, since the coefficients must be *non-negative*: $\alpha^P > 0$. Subsequently, the subspace parameters we set out to estimate are simply the p-th root of the solution. In our experiments we used $p = 2$.

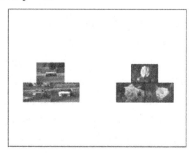

Fig. 3. An example of a user-guided layout

Figure 3 shows a simple user layout where 3 car images are clustered together despite their different colors. The same is performed with 3 flower images (despite their texture/structure). These two clusters maintain a sizeable separation thus suggesting two separate concept classes implicit by the user's placement. Specifically, in this layout the user is clearly concerned with the distinction between *car* and *flower* regardless of color or other possible visual attributes.

Applying the α-estimation algorithm to Figure 3, the feature weights learned from this 2-D layout are $\alpha_c = 0.3729$, $\alpha_t = 0.5269$ and $\alpha_s = 0.1002$. This shows that the most important feature in this case is texture and not color, which is in accord with the concepts of car *vs.* flower as graphically indicated by the user in Figure 3.

Now that we have the learned feature weights (or modeled the user) what can we do with them? Figure 4 shows an example of a typical application: automatic layout of a larger (more complete data set) in the style indicated by the user. Fig. 4(a) shows the PCA splat using the learned feature weight for 18 cars and 19 flowers. It is obvious that the PCA splat using the estimated weights captures the essence of the configuration layout in Figure 3. Figure 4(b) shows a PCA splat of the same images but with a randomly generated α, denoting an arbitrary but coherent 2-D layout, which in this case, favors color ($\alpha_c = 0.7629$). This comparison reveals that proper feature weighting is an important factor in generating the desired layouts.

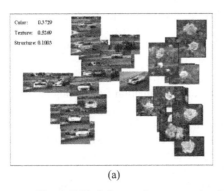 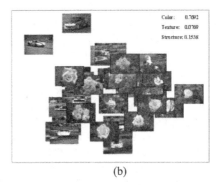

(a)   (b)

Fig. 4. PCA Splat on a larger set using (a) estimated weights (b) arbitrary weights

3   Performance Analysis

Given the lack of sufficiently numerous (and willing) human subjects to test our system with, we undertook a Monte Carlo approach to evaluating our user-modeling estimation algorithm. Thus, we generated 1000 synthetic "user-layouts" (with random values of α's representing the "ground-truth") to sample the space of all possible *consistent* user-layouts that could be conceivably generated by a human subject (an *inconsistent* layout would correspond to randomly "splattered" thumbnails in the 2-D display). In each case, the α-estimation method was used to estimate (recover) the original ("ground-truth") values. We should note that the parameter recovery is non-trivial due to the information lost whilst projecting from the high-dimensional feature space down to the 2-D display space. As a control, 1000 randomly generated feature weights were used to see how well they could match the synthetic user layouts (*i.e.*, by chance alone). Note that these controls were also *consistent* layouts.

Our primary test database consists of 142 images from the COREL database. It has 7 categories of car, bird, tiger, mountain, flower, church and airplane. Each class has about 20 images. Feature extraction based on color, texture and structure has been done off-line and pre-stored. Although we will be reporting on this test data set -- due to its common use and familiarity to the CBIR community -- we should emphasize that we have also successfully tested our methodology on larger and much more heterogeneous image libraries. For example: real personal photo collections of 500+ images (including family, friends, vacations, etc.).

The following is the Monte Carlo procedure was used for testing the significance and validity of user-modeling with α-estimation:

```
Simulation: Randomly select M images from the database.
Generate arbitrary (random) feature weights α in order
to simulate a "user-layout". Do a PCA Splat using this
"ground truth" α. From the resulting 2-D layout, esti-
mate α. Select a new distinct (non-overlapping) set of
M images from the database. Do PCA Splats on the second
```

```
set using the original α, the estimated α and a third
random α (as control). Calculate the resulting stress
and layout deviation (2-D position error) for the
original, estimated and random (control) values of α.
Repeat 1000 times
```

The scatter-plot of 1000 Monte Carlo trials of α-estimation from synthetic "user-generated" layouts is shown in Figure 5. Clearly there is a direct linear relationship between the original weights and the estimated weights. Note that when the original weight is very small (<0.1) or very large (>0.9), the estimated weight is zero or one correspondingly. This means that when one particular feature weight is very large (or very small), the corresponding feature will become the most dominant (or least dominant) feature in the PCA, therefore the estimated weight for this feature will be either 1 or 0.

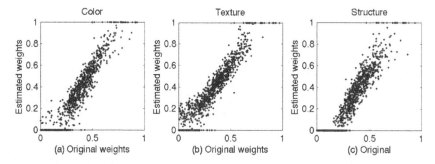

Fig. 5. Scatter-plot of α-estimation: Estimated weights vs. original weights

In terms of actual measures of stress and deviation we found that the α-estimation scheme yielded the smaller deviation 78.4% of the time and smaller stress 72.9%. The main reason these values are less than 100% is due to the nature of the Monte Carlo testing and the fact that in low-dimensional (2-D) spaces, random weights can become close to the original weights and hence yield similar "user" layouts (in this case apparently ~ 25% of the time).

Another control other than random weights is to compare the deviation of an α-estimation layout generator to a simple scheme which assigns each new image to the 2-D location of its (unweighted) 37-dimensional nearest-neighbor from the set previously laid out by the "user". This control essentially operates on the principle that new images should be displayed on screen at the same location as their nearest neighbors in the original 37-dimensional feature space and thus ignores the subspace defined by the "user" in a 2-D layout.

The results of this Monte Carlo simulation are shown in Figure 6 where we see that the layout deviation using α-estimation (red: mean=0.9691 std=0.7776) was consistently lower -- by almost an order of magnitude -- than nearest neighbor (blue: mean=7.5921, std=2.6410).

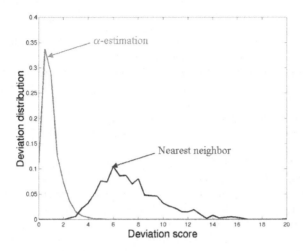

Fig. 6. Comparison of the distribution of α-estimation vs. nearest neighbor deviation scores (note that in every one of the random trials the α-estimation deviation was smaller)

4   Discussion

There are several areas of future work. First, more extensive user-layout studies are needed to replace Monte Carlo simulations. It is critical to have real users with different notions of content do layouts and to see if our system can model them accurately. Moreover, we have already integrated hybrid visual/semantic feature weighting that requires further testing with human subjects. We have designed our system with general CBIR in mind but more specifically for personalized photo collections. The final visualization and retrieval interface can be displayed on a computer screen, large panel projection screens, or -- for example -- on embedded tabletop devices [12], designed specifically for purposes of story-telling or multi-person collaborative exploration of large image libraries.

Finally, we note that although the focus of this paper has been on visual content analysis, the same framework for visualization and user-modeling would apply to other data entities such as video clips, audio files, specialized documents (legal, medical, etc), or web pages. The main difference would be the choice of features used, their representation in high-dimensional spaces and the appropriate metrics.

Acknowledgements

This work was supported in part by Mitsubishi Electric Research Laboratories (MERL), Cambridge, MA, and National Science Foundation Grant CDA 96-24396 and EIA 99-75019.

References

1. M. Stricker, M. Orengo, "Similarity of Color Images", Proc. SPIE Storage and Retrieval for Image and Video Databases, 1995
2. J. R. Smith, S. F. Chang, "Transform Features for Texture Classification and Discrimination in Large Image Database", Proc. IEEE Intl. Conf. on Image Proc., 1994
3. S. X. Zhou, Y. Rui and T. S. Huang, "Water-filling algorithm: A novel way for image feature extraction based on edge maps", in Proc. IEEE Intl. Conf. On Image Proc., Japan, 1999
4. S. Santini, R. Jain, "Similarity measures", IEEE PAMI, vol. 21, no. 9, 1999
5. M. Popescu, P. Gader, "Image Content Retrieval From Image Databases Using Feature Integration by Choquet Integral", in SPIE Conference Storage and Retrieval for Image and Video Databases VII, San Jose, CA, 1998
6. D. M. Squire, H. Müller, and W. Müller, "Improving Response Time by Search Pruning in a Content-Based Image Retrieval System, Using Inverted File Techniques", Proc. of IEEE workshop on CBAIVL, June 1999
7. D. Swets, J. Weng, "Hierarchical Discriminant Analysis for Image Retrieval", IEEE PAMI, vol. 21, no.5, 1999
8. Y. Rubner, "Perceptual metrics for image database navigation", Ph.D. dissertation, Stanford University, 1999
9. W. S. Torgeson, Theory and methods of scaling, John Wiley & Sons, New York, NY, 1958
10. Jolliffe, I. T., Principal Component Analysis, Springer-Verlag, New-York, 1986
11. S. Santini, Ramesh Jain, "Integrated browsing and querying for image databases", July-September Issue, IEEE Multimedia Magazine, pp.26-39, 2000
12. B. Moghaddam et al; "Visualization and Layout for Personal Photo Libraries," International Workshop on Content-Based Multimedia Indexing (CBMI'01), September, 2001
13. Q. Tian, B. Moghaddam, T. S. Huang, "Display Optimization for Image Browsing," International Workshop on Multimedia Databases and Image Communication (MDIC'01), September, 2001

Robust Shape Matching

Nicu Sebe and Michael Lew

Leiden Institute of Advanced Computer Science, Leiden University
NielsBohrweg 1, 2333CA Leiden, The Netherlands
{nicu,mlew}@liacs.nl

Abstract. Many visual matching algorithms can be described in terms
of the features and the inter-feature distance or metric. The most commonly used metric is the sum of squared differences (SSD), which is valid
from a maximum likelihood perspective when the real noise distribution
is Gaussian. However, we have found experimentally that the Gaussian
noise distribution assumption is often invalid. This implies that other
metrics, which have distributions closer to the real noise distribution,
should be used. In this paper we considered a shape matching application. We implemented two algorithms from the research literature and
for each algorithm we compared the efficacy of the SSD metric, the SAD
(sum of the absolute differences) metric, and the Cauchy metric. Furthermore, in the case where sufficient training data is available, we discussed
and experimentally tested a metric based directly on the real noise distribution, which we denoted the maximum likelihood metric.

1 Introduction

We are interested in using shape descriptors in content-based retrieval. Assume
that we have a large number of images in the database. Given a query image, we
would like to obtain a list of images from the database which are most similar
(here we consider the shape aspect) to the query image. For solving this problem,
we need two things - first, a measure which represents the shape information of
the image and second a similarity measure to compute the similarity between
corresponding features of two images.

The similarity measure is a matching function and gives the degree of similarity for a given pair of images (represented by shape measures). The desirable
property of a similarity measure is that it should be a metric (that is, it has the
properties of symmetry, transitivity, and linearity). The SSD (L_2) and the SAD
(L_1) are the most commonly used metrics. This brings to mind several questions. First, under what conditions should one use the SSD versus the SAD?
From a maximum likelihood perspective, it is well known that the SSD is justified when the additive noise distribution is Gaussian. The SAD is justified when
the additive noise distribution is Exponential (double or two-sided exponential).
Therefore, one can determine which metric to use by checking if the real noise
distribution is closer to the Gaussian or the Exponential. The common assumption is that the real noise distribution should fit either the Gaussian or the

M. S. Lew, N. Sebe, and J. P. Eakins (Eds.): CIVR 2002, LNCS 2383, pp. 17–28, 2002.
© Springer-Verlag Berlin Heidelberg 2002

Exponential, but what if there is another distribution which fits the real noise distribution better? Toward answering this question, we have endeavored to use international test sets and promising algorithms from the research literature.

In this paper, the problem of image retrieval using shape was approached by active contours for segmentation and invariant moments for shape measure. Active contours were first introduced by Kass et al. [1] and were termed snakes by the nature of their movement. Active contours are a sophisticated approach to contour extraction and image interpretation. They are based on the idea of minimizing energy of a continuous spline contour subject to constraints on both its autonomous shape and external forces derived from a superposed image that pull the active contour toward image features such as lines and edges.

Moments describe shape in terms of its area, position, orientation, and other parameters. The set of invariant moments [2] makes a useful feature vector for the recognition of objects which must be detected regardless of position, size, or orientation. Matching of the invariant moments feature vectors is computationally inexpensive and is a promising candidate for interactive applications.

2   Active Contours and Invariant Moments

Active contours challenge the widely held view of bottom-up vision processes. The principal disadvantage with the bottom-up approach is its serial nature; errors generated at a low-level are passed on through the system without the possibility of correction. The principal advantage of active contours is that the image data, the initial estimate, the desired contour properties, and the knowledge-based constraints are integrated into a single extraction process.

In the literature, del Bimbo et al. [3] deformed active contours over a shape in an image and measured the similarity between the two based on the degree of overlap and on how much energy the active contour has to spend in the deformation. Jain et al. [4] used a matching scheme with deformable templates. Our work is different in that we use a Gradient Vector Flow (GVF) based method [5] to improve the automatic fit of the snakes to the object contours.

Active contours are defined as energy-minimizing splines under the influence of internal and external forces. The internal forces of the active contour serve as a smoothness constraint designed to hold the active contour together (elasticity forces) and to keep it from bending too much (bending forces). The external forces guide the active contour towards image features such as high intensity gradients. The optimal contour position is computed such that the total energy is minimized. The contour can hence be viewed as a reasonable balance between geometrical smoothness properties and local correspondence with the intensity function of the reference image.

Let the active contour be given by a parametric representation $v(s) = (x(s), y(s))$, with s the normalized arc length of the contour. The expression for the total energy can then be decomposed as follows:

$$E_{total} = \int_0^1 [E_{int}(v(s)) + E_{image}(v(s)) + E_{con}(v(s))] \, ds \qquad (1)$$

where E_{int} represents the internal forces (or energy) which encourage smooth curves, E_{image} represents the local correspondence with the image function, and E_{con} represents a constraint force that can be included to attract the contour to specific points in the image plane. In the following discussions E_{con} will be ignored. E_{image} is typically defined such that locations with high image gradients or short distances to image gradients are assigned low energy values.

2.1 Internal Energy

E_{int} is the internal energy term which controls the natural behavior of the active contour. It is designed to minimize the curvature of the active contour and to make the active contour behave in an elastic manner. According to Kass et al. [1] the internal energy is defined as

$$E_{int}(v(s)) = \alpha(s) \left| \frac{dv(s)}{ds} \right|^2 + \beta(s) \left| \frac{d^2 v(s)}{ds^2} \right|^2 \qquad (2)$$

The first order continuity term, weighted by $\alpha(s)$, makes the contour behave elastically, while the second order curvature term, weighted by $\beta(s)$, makes it resistant to bending. Setting $\beta(s) = 0$ at a point s allows the active contour to become second order discontinuous at that point and to develop a corner. Setting $\alpha(s) = 0$ at a point s allows the active contour to become discontinuous. Active contours can interpolate gaps in edges phenomena known as subjective contours due to the use of the internal energy. It should be noted that $\alpha(s)$ and $\beta(s)$ are defined to be functions of the curve parameter s, and hence segments of the active contour may have different natural behavior. Minimizing the energy of the derivatives gives a smooth function.

2.2 Image Energy

E_{image} is the image energy term derived from the image data over which the active contour lies and is constructed to attract the active contour to desired feature points in the image, such as edges and lines. The edge based functional attracts the active contour to contours with large image gradients - that is, to locations of strong edges.

$$E_{edge} = -|\nabla I(x, y)| \qquad (3)$$

2.3   Problems with Active Contours

There are a number of fundamental problems with the active contours and solutions to these problems sometimes create problems in other components of the active contour model.

Initialization. The final extracted contour is highly dependent on the position and shape of the initial contour due to the presence of many local minima in the energy function. The initial contour must be placed near the required feature otherwise the contour can become obstructed by unwanted features like JPEG compression artifacts, closeness of a nearby object, etc.

Non-convex shapes. How do we extract non-convex shapes without compensating the importance of the internal forces or without a corruption of the image data? For example, pressure forces [6] (addition to the external force) can push an active contour into boundary concavities, but cannot be too strong or otherwise weak edges will be ignored. Pressure forces must also be initialized to push out or push in, a condition that mandates careful initialization.

The original method of Kass et al. [1] suffered from three main problems: dependence on the initial contour, numerical instability, and lack of guaranteed convergence to the global energy minimum. Amini et al. [7] improved the numerical instability by minimizing the energy functional using dynamic programming, which allows inclusion of hard constraints into the energy functional. However, memory requirements are large, being $O(nm^2)$, and the method is slow, being $O(nm^3)$ where n is the number of contour points and m is the neighborhood size to which a contour point is allowed to move in a single iteration. Seeing the difficulties with both previous methods, Williams and Shah [8] developed the *greedy algorithm* which combines speed, flexibility, and simplicity. The greedy algorithm is faster $O(nm)$ than the dynamic programming and is more stable and flexible for including constraints than the variational approach of Kass et al. [1]. During each iteration, a neighborhood of each point is examined and a point in the neighborhood with the smallest energy value provides the new location of the point. Iterations continue till the number of points in the active contour that moved to a new location in one iteration is below a specified threshold.

2.4   Gradient Vector Flow

Since the greedy algorithm easily accommodates new changes, there are three things we would like to add to it: (1) the ability to inflate the contour as well as to deflate it, (2) the ability to deform to concavities, and (3) the ability to increase the capture range of the external forces. These three additions reduce the sensitivity to initialization of the active contour and allow deformation inside concavities. This can be done by replacing the existing external force (image term) with the gradient vector flow (GVF) [5]. The GVF is an external force computed as a diffusion of the gradient vectors of an image, without blurring the edges.

Xu and Prince [5] define the gradient vector flow (GVF) field to be the vector field $\mathbf{v}(i,j) = (u(i,j), v(i,j))$ which is updated with every iteration of the diffusion equations:

$$u_{i,j}^{n+1} = (1 - b_{i,j})u_{i,j}^n + (u_{i+1,j}^n + u_{i,j+1}^n + u_{i-1,j}^n + u_{i,j-1}^n - 4u_{i,j}^n) + c_{i,j}^1 \quad (4)$$

$$v_{i,j}^{n+1} = (1 - b_{i,j})v_{i,j}^n + (v_{i+1,j}^n + v_{i,j+1}^n + v_{i-1,j}^n + v_{i,j-1}^n - 4v_{i,j}^n) + c_{i,j}^2 \quad (5)$$

where $b_{i,j} = G_i(i,j)^2 + G_j(i,j)^2$, $c_{i,j}^1 = b_{i,j}G_i(i,j)$, and $c_{i,j}^2 = b_{i,j}G_j(i,j)$ with G_i and G_j the first and the second elements of the gradient vector.

The second term in (4) and (5) is the Laplacian operator. The intuition behind the diffusion equations is that in homogeneous regions, the first and third terms are 0 since the gradient is 0, and within those regions, u and v are each determined by Laplace equation. This results in a type of "filling-in" of information taken from the boundaries of the region. In regions of high gradient v is kept nearly equal to the gradient.

Creating GVF field yields streamlines to a strong edge. In the presence of these streamlines, blobs and thin lines in the way to strong edges do not form any impediments to the movement of the active contour. It can be considered as an advantage if the blobs are in front of the shape, nevertheless it can be considered as a disadvantage if the active contour enters the silhouette of the shape.

2.5   Invariant Moments

Perhaps the most popular method for shape description is the use of invariant moments [2] which are invariant to affine transformations. In the case of a digital image, the moments are approximated by

$$m_{pq} = \sum_x \sum_y x^p y^q f(x,y) \quad (6)$$

where the order of the moment is $(p+q)$, x and y are the pixel coordinates relative to some arbitrary standard origin, and $f(x,y)$ represents the pixel brightness.

To have moments that are invariant to translation, scale, and rotation, first the central moments μ are calculated

$$\mu_{pq} = \sum_x \sum_y (x - \overline{x})^p (y - \overline{y})^q f(x,y), \quad \overline{x} = \frac{m_{10}}{m_{00}}, \quad \overline{y} = \frac{m_{01}}{m_{00}} \quad (7)$$

Further, the normalized central moments η are calculated

$$\eta_{pq} = \frac{\mu_{pq}}{\mu_{00}^\lambda}, \quad \lambda = \frac{(p+q)}{2}, \quad p + q \geq 2 \quad (8)$$

From these normalized parameters a set of invariant moments $\{\phi\}$ found by Hu [2], can be calculated. The 7 equations of the invariant moments contain terms up to order 3:

$$\phi_1 = \eta_{20} + \eta_{02}$$
$$\phi_2 = (\eta_{20} - \eta_{02})^2 + 4\eta_{11}^2$$
$$\phi_3 = (\eta_{30} - 3\eta_{12})^2 + (3\eta_{21} - \eta_{03})^2$$
$$\phi_4 = (\eta_{30} - \eta_{12})^2 + (\eta_{21} - \eta_{03})^2$$
$$\phi_5 = (\eta_{30} - 3\eta_{12})(\eta_{30} + \eta_{12})\left((\eta_{30} + \eta_{12})^2 - 3(\eta_{21} + \eta_{03})^2\right) +$$
$$(3\eta_{21} - \eta_{03})(\eta_{21} + \eta_{03})\left(3(\eta_{30} + \eta_{12})^2 - (\eta_{21} + \eta_{03})^2\right)$$
$$\phi_6 = (\eta_{20} - \eta_{02})\left((\eta_{30} + \eta_{12})^2 - (\eta_{21} + \eta_{03})^2\right) + 4\eta_{11}(\eta_{30} + \eta_{12})(\eta_{21} + \eta_{03})$$
$$\phi_7 = (3\eta_{21} - \eta_{30})(\eta_{30} + \eta_{12})\left((\eta_{30} + \eta_{12})^2 - 3(\eta_{21} + \eta_{03})^2\right) +$$
$$(3\eta_{12} - \eta_{03})(\eta_{21} + \eta_{03})\left(3(\eta_{30} + \eta_{12})^2 - (\eta_{21} + \eta_{03})^2\right) \tag{9}$$

Global (region) properties provide a firm common base for similarity measure between shapes silhouettes where gross structural features can be characterized by these moments. Since we do not deal with occlusion, the invariance to position, size, and orientation, and the low dimensionality of the feature vector represent good reasons for using the invariant moments in matching shapes. The logarithm of the invariant moments is taken to reduce the dynamic range.

3   Maximum Likelihood Approach

In the previous sections we were discussing about extracting the shape information in a feature vector. In order to implement a content-based retrieval application we still need to provide a framework for selecting the similarity measure to be used when the feature vectors are compared.

In our previous work [9], we showed that the maximum likelihood theory allows us to relate a noise distribution to a metric. Specifically, if we are given the noise distribution then the metric which maximizes the similarity probability is

$$\sum_{i=1}^{M} \rho(n_i) \tag{10}$$

where n_i represents the ith bin of the discretized noise distribution and ρ is the maximum likelihood estimate of the negative logarithm of the probability density of the noise. Typically, the noise distribution is represented by the difference between the corresponding elements given by the ground truth.

To analyze the behavior of the estimate we take the approach described in [10] and based on the influence function. The influence function characterizes the bias that a particular measurement has on the solution and is proportional to the derivative, ψ, of the estimate

$$\psi(z) \equiv \frac{d\rho(z)}{dz} \tag{11}$$

In the case where the noise is Gaussian distributed:

$$P(n_i) \sim \exp(-n_i{}^2) \tag{12}$$

then,

$$\rho(z) = z^2 \qquad \text{and} \qquad \psi(z) = z \tag{13}$$

If the errors are distributed as a double or two-sided exponential, namely,

$$P(n_i) \sim \exp(-|n_i|) \tag{14}$$

then,

$$\rho(z) = |z| \qquad \text{and} \qquad \psi(z) = \text{sgn}(z) \tag{15}$$

In this case, using (10), we minimize the mean absolute deviation, rather than the mean square deviation. Here the tails of the distribution, although exponentially decreasing, are asymptotically much larger than any corresponding Gaussian.

A distribution with even more extensive tails is the Cauchy distribution,

$$P(n_i) \sim \frac{a}{a^2 + n_i{}^2} \tag{16}$$

where the *scale* parameter a determines the height and the tails of the distribution.

This implies

$$\rho(z) = \log\left(1 + \left(\frac{z}{a}\right)^2\right) \qquad \text{and} \qquad \psi(z) = \frac{z}{a^2 + z^2} \tag{17}$$

For normally distributed errors, (13) says that the more deviant the points, the greater the weight. By contrast, when tails are somewhat more prominent, as in (14), then (15) says that all deviant points get the same relative weight, with only the sign information used. Finally, when the tails are even larger, (17) says that ψ increases with deviation, then starts decreasing, so that very deviant points - the true outliers - are not counted at all.

Maximum likelihood gives a direct connection between the noise distributions and the comparison metrics. Considering ρ as the negative logarithm of the probability density of the noise, then the corresponding metric is given by Eq. (10).

Consider the Minkowski-form distance L_p between two vectors x and y:

$$L_p(x, y) = \left(\sum_i |x_i - y_i|^p\right)^{\frac{1}{p}} \tag{18}$$

If the noise is Gaussian distributed, so $\rho(z) = z^2$, then (10) is equivalent to (18) with $p = 2$. Therefore, in this case the corresponding metric is L_2. Equivalently, if the noise is Exponential, so $\rho(z) = |z|$, then the corresponding metric is L_1 (Eq. (18) with $p = 1$). In the case the noise is distributed as a

Cauchy distribution with scale parameter a, then the corresponding metric is no longer a Minkovski metric. However, for convenience we denote it as L_c:

$$L_c(x, y) = \sum_i \log \left(1 + \left(\frac{x_i - y_i}{a} \right)^2 \right) \tag{19}$$

In practice, the probability density of the noise can be approximated as the normalized histogram of the differences between the corresponding feature vectors elements. For convenience, the histogram is made symmetric around zero by considering pairs of differences (e.g., $x - y$ and $y - x$). Using this normalized histogram, we extract a metric, called *maximum likelihood (ML) metric*. The ML metric is given by Eq. (10) where $\rho(n_i)$ is the negative logarithm of $P(n_i)$:

$$\rho(n_i) = -\log(P(n_i)). \tag{20}$$

The ML metric is a discrete metric extracted from a discrete normalized histogram having a finite number of bins. When n_i does not exactly match any of the bins, for calculating $P(n_i)$ we perform linear interpolation between $P(n_{inf})$ (the histogram value at bin n_{inf}) and $P(n_{sup})$ (the histogram value at bin n_{sup}), where n_{inf} and n_{sup} are the closest inferior and closest superior bins to n_i, respectively:

$$P(n_i) = \frac{(n_{sup} - n_i)P(n_{inf}) + (n_i - n_{inf})P(n_{sup})}{n_{sup} - n_{inf}} \tag{21}$$

4   Experiments

We assume that representative ground truth is provided. The ground truth is split into two non-overlapping sets: the training set and the test set. First, for each image in the training set a feature vector is extracted. Second, the real noise distribution is computed as the normalized histogram of differences from the corresponding elements in feature vectors taken from similar images according to the ground truth. The Gaussian, Exponential, and Cauchy distributions are fitted to the real distribution. The Chi-square test is used to find the fit between each of the model distributions and the real distribution. We select the model distribution which has the best fit and its corresponding metric L_k is used in ranking. The ranking is done using only the test set.

It is important to note that for real applications, the parameter in the Cauchy distribution is found when fitting this distribution to the real distribution from the training set. This parameter setting would be used for the test set and any future comparisons in that application.

As noted in the previous section, it is also possible to create a metric based on the real noise distribution using maximum likelihood theory. Consequently, we denote the maximum likelihood (ML) metric as (10) where ρ is the negative logarithm of the normalized histogram of the absolute differences from the training set. Note that the histogram of the absolute differences is normalized

Fig. 1. Example of images of one object rotated with 60°

to have area equal to one by dividing the histogram by the total number of examples in the training set. This normalized histogram is our approximation for the probability density function.

For the performance evaluation let $\mathcal{Q}_1, \cdots, \mathcal{Q}_n$ be the query images and for the i-th query $\mathcal{Q}_i$, $\mathcal{I}_1^{(i)}, \cdots, \mathcal{I}_m^{(i)}$ be the images similar with $\mathcal{Q}_i$ according to the ground truth. The retrieval method will return this set of answers with various ranks. As an evaluation measure of the performance of the retrieval method we used recall vs. precision at different scopes: For a query $\mathcal{Q}_i$ and a scope $s > 0$, the recall r is defined as $|\{\mathcal{I}_j^{(i)}|rank(\mathcal{I}_j^{(i)}) \leq s\}|/m$, and the precision p is defined as $|\{\mathcal{I}_j^{(i)}|rank(\mathcal{I}_j^{(i)}) \leq s\}|/s$.

In our experiments we used a database of 1,440 images of 20 common house hold objects from the COIL-20 database [11]. Each object was placed on a turntable and photographed every 5° for a total of 72 views per object. Examples are shown in Fig. 1.

In creating the ground truth we had to take into account the fact that the images of one object may look very different when an important rotation is considered. Therefore, for a particular instance (image) of an object we consider as similar the images taken for the same object when it was rotated within $\pm r \times 5°$. In this context, we consider two images to be r-similar if the rotation angle of the object depicted in the images is smaller than $r \times 5°$. In our experiments we used $r = 3$ so that one particular image is considered to be similar with 6 other images of the same object rotated within $\pm 15°$. We prepared our training set by selecting 18 equally spaced views for each object and using the remaining views for testing.

The first question we asked was, "Which distribution is a good approximation for the similarity noise distribution?" To answer this we needed to measure the similarity noise caused by the object rotation and depending on the feature extraction algorithm (greedy or GVF). The real noise distribution was obtained as the normalized histogram of differences between the elements of feature vectors corresponding to similar images from the training set.

Fig. 2 presents the real noise distribution obtained for the greedy algorithm. The best fit Exponential had a better fit to the noise distribution than the Gaussian. Consequently, this implies that L_1 should provide better retrieval results than L_2. The Cauchy distribution is the best fit overall, and the results obtained with L_c should reflect this. However, when the maximum likelihood metric (ML) extracted directly from the similarity noise distribution is used we expect to obtain the best retrieval results.

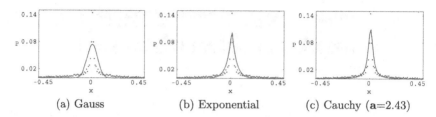

(a) Gauss   (b) Exponential   (c) Cauchy (a=2.43)

Fig. 2. Similarity noise distribution for the greedy algorithm compared with (a) the best fit Gaussian (approximation error is 0.156), (b) the best fit Exponential (approximation error is 0.102), and (c) the best fit Cauchy (approximation error is 0.073)

In the case of GVF algorithm the approximation errors for matching the similarity noise distribution with a model distribution are given in Table 1. Note that the Gaussian is the worst approximation. Moreover, the difference between the Gaussian fit and the fit obtained with the other two distributions is larger than in the previous case and therefore the results obtained with L_2 will be much worse. Again the best fit by far is provided by the Cauchy distribution.

Table 1. The approximation error for matching the similarity noise distribution with one of the model distributions in the case of GVF algorithm (for Cauchy a=3.27)

Gauss	Exponential	Cauchy
0.0486	0.0286	0.0146

The results are presented in Fig. 3 and Table 2. In the precision-recall graphs the curves corresponding to L_c are above the curves corresponding to L_1 and L_2 showing that the method using L_c is more effective. Note that the choice of the noise model significantly affects the retrieval results. The Cauchy distribution was the best match for the measured similarity noise distribution and the results in Table 2 show that the Cauchy model is more appropriate for the similarity noise than the Gaussian and Exponential models. However, the best results are obtained when the metric extracted directly from the noise distribution is used. One can also note that the results obtained with the GVF method are significantly better than the ones obtained with the greedy method.

In summary, L_c performed better than the analytic distance measures, and the ML metric performed best overall.

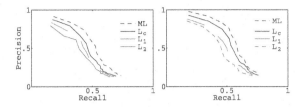

Fig. 3. Precision/Recall for COIL-20 database using the greedy algorithm (for L_c **a**=2.43) (left) and the GVF algorithm (for L_c **a**=3.27) (right)

Table 2. Precision and Recall for different Scope values

		Precision			Recall		
Scope		6	10	25	5	10	25
Greedy	L_2	0.425	0.258	0.128	0.425	0.517	0.642
	L_1	0.45	0.271	0.135	0.45	0.542	0.675
	L_c **a**=2.43	0.466	0.279	0.138	0.466	0.558	0.692
	ML	0.525	0.296	0.146	0.525	0.592	0.733
GVF	L_2	0.46	0.280	0.143	0.46	0.561	0.707
	L_1	0.5	0.291	0.145	0.5	0.576	0.725
	L_c **a**=3.27	0.533	0.304	0.149	0.533	0.618	0.758
	ML	0.566	0.324	0.167	0.566	0.635	0.777

5   Conclusions

In this paper we showed that the GVF based snakes give better retrieval results than the traditional snakes. In particular, the GVF snakes have the advantage in that it is not necessary to know apriori whether the snake must be expanded or contracted to fit the object contour. Furthermore, the GVF snakes have the ability to fit into concavities of the object which traditional snakes cannot do. Both of these factors resulted in significant improvement in the retrieval results.

We also addressed the problem of finding the appropriate metric to use for computer vision applications in shape based retrieval. From our experiments, L_2 is typically not justified because the similarity noise distribution is not Gaussian. We showed that better accuracy was obtained when the Cauchy metric was substituted for the L_2 and L_1. Minimizing the Cauchy metric is optimal with respect to maximizing the likelihood of the difference between image elements when the real noise distribution is equivalent to a Cauchy distribution. Therefore, the breaking points occur when there is no ground truth, the ground truth is not representative, or when the real noise distribution is not a Cauchy distribution. We also make the assumption that one can measure the fit between the real distribution and a model distribution, and that the model distribution which has the best fit should be selected. We used the Chi-square test as the measure of fit between the distributions, and found in our experiments that it served as a reliable indicator for distribution selection.

In conclusion, we showed that the prevalent Gaussian distribution assumption is often invalid, and we proposed the Cauchy metric as an alternative to both L_1 and L_2. In the case where representative ground truth can be obtained for an application, we provided a method for selecting the appropriate metric. Furthermore, we explained how to create a maximum likelihood metric based on the real noise distribution, and in our experiments we found that it consistently outperformed all of the analytic metrics.

References

1. Kass, M., Witkin, A., Terzopoulos, D.: Snakes: Active contour models. IJCV **1** (1988) 321–331 18, 19, 20
2. Hu, M.: Visual pattern recognition by moment invariants. IRA Trans. on Information Theory **17-8** (1962) 179–187 18, 21
3. Del Bimbo, A., Pala, P.: Visual image retrieval by elastic matching of user sketches. IEEE Trans. Pattern Analysis and Machine Intelligence **19** (1997) 121–132 18
4. Jain, A., Zhong, Y., Lakshmanan, S.: Object matching using deformable template. IEEE Trans. Pattern Analysis and Machine Intelligence **18** (1996) 267–278 18
5. Xu, C., Prince, J.: Gradient vector flow: A new external force for snakes. CVPR (1997) 66–71 18, 20, 21
6. Cohen, L.: On active contour models and balloons. CVGIP: Image Understanding **53** (1991) 211–218 20
7. Amini, A., Tehrani, S., Weymouth, T.: Using dynamic programming for minimizing the energy of active contours in the presence of hard constraints. ICCV (1988) 95–99 20
8. Williams, D., Shah, M.: A fast algorithm for active contours and curvature estimation. CVGIP: Image Understanding **55** (1992) 14–26 20
9. Sebe, N., Lew, M., Huijsmans, D.: Toward improved ranking metrics. IEEE Transactions on Pattern Analysis and Machine Intelligence **22** (2000) 1132–1141 22
10. Hampel, F., Ronchetti, E., Rousseeuw, P., Stahel, W.: Robust Statistic: The Approach Based on Influence Functions. John Wiley and Sons, New York (1986) 22
11. Murase, H., Nayar, S.: Visual learning and recognition of 3D objects from appearance. IJCV **14** (1995) 5–24 25

Semantics-Based Image Retrieval by Region Saliency

Wei Wang, Yuqing Song, and Aidong Zhang

Department of Computer Science and Engineering
State University of New York at Buffalo, Buffalo, NY 14260, USA
{wwang3,ys2,azhang}@cse.buffalo.edu

Abstract. We propose a new approach for semantics-based image retrieval. We use color-texture classification to generate the codebook which is used to segment images into regions. The content of a region is characterized by its self-saliency and the lower-level features of the region, including color and texture. The context of regions in an image describes their relationships, which are related to their relative-saliencies. High-level (semantics-based) querying and query-by-example are supported on the basis of the content and context of image regions. The experimental results demonstrate the effectiveness of our approach.

1   Introduction

Although the content-based image retrieval (CBIR) techniques based on low-level features such as color, texture, and shape have been extensively explored, their effectiveness and efficiency are not satisfactory. The ultimate goal of image retrieval is to provide the users with the facility to manage large image databases in an automatic, flexible and efficient way. Therefore, image retrieval systems should be armed to support high-level (semantics-based) querying and browsing of images.

The basic elements to carry semantic information are the image regions which correspond to semantic objects, if the image segmentation is effective. Most approaches proposed for region-based image retrieval, although successful in some aspects, could not integrate the semantic descriptions into the regions, therefore cannot support the high-level querying of images. After the regions are obtained, proper representation of the content and context remains a challenge.

In our previous work [7], we used color-texture classification to generate the semantic codebook which is used to segment images into regions. The content and context of regions were then extracted in a probabilistic manner and used to perform the retrieval. The computation of the content and context needed some heuristic weight functions, which can be improved to combine visual features. In addition, when retrieving images, users may be interested in both semantics and some specific visual features of images. Thus the content and context of image regions should be refined to incorporate different visual features. The saliencies of the image regions [2], [3], [6] represent the important visual cues of regions, therefore can be used to improve our previous method. In this paper, we will introduce an improved image-retrieval method, which incorporates region saliency to the definition of content and context of image regions.

The saliencies of regions have been used to detect the *region of interest* (ROI). In [6], saliency was further categorized as *self-saliency* and *relative-saliency*. Self-saliency was defined as "what determines how conspicuous a region is on its own",

M. S. Lew, N. Sebe, and J. P. Eakins (Eds.): CIVR 2002, LNCS 2383, pp. 29–37, 2002.
© Springer-Verlag Berlin Heidelberg 2002

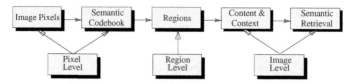

Fig. 1. System design of our approach

while relative-saliency was used to measure how distinctive the region appears among other regions. Apparently, saliencies of regions represent the intrinsic properties and relationships for image regions.

Our approach consists of three levels. At the pixel level, color-texture classification is used to form the semantic codebook. At the region level, the semantic codebook is used to segment the images into regions. At the image level, content and context of image regions are defined and represented on the basis of their saliencies to support the semantics retrieval from images. The three levels are illustrated in Figure 1.

The remainder of the paper is organized as follows. From Section 2 to Section 5, each step of our approach is elaborated. In Section 6 experimental results will be presented, and we will conclude in Section 7.

2   Image Segmentation

2.1   Generation of the Semantic Codebook

We generate the semantic codebook by using color-texture classification to classify the color-texture feature vectors for pixels in the images into cells in the color-texture space. About the color-texture feature vectors for each pixel, the three color features we choose are the averaged *RGB* values in a neighborhood and the three texture features are *anisotropy, contrast and flowOrEdge. Anisotropy* measures the energy comparison between the dominant orientation and its orthogonal direction. *Contrast* reflects the contrast, or harshness of the neighborhood. *FlowOrEdge* can distinguish whether an 1D texture is a flow or an edge. (See [7] for detailed description of these values). The color and texture feature vectors are denoted as $Color_{FV}$ and $Texture_{FV}$, respectively. Therefore for each pixel, we have a six-dimensional feature vector where three dimensions are for color, and three for texture. The color-texture classification is performed in the following way: by $Color_{FV}$, the color space is uniformly quantized into $4 \times 4 \times 4 = 64$ classes; by $Texture_{FV}$, texture space is classified into 7 classes: one class for low contrast and edge, respectively; two classes for flow and three for 2D texture. Formal definition of these classes can be found in [7]. Because edges cannot reflect the color-texture characteristics of image regions semantically, we classify all the pixels, except those pixels corresponding to edges, to 6 texture classes $LowContrast, 2D_0, 2D_1, 2D_2, Flow_0, Flow_1$. Corresponding to $T (T = 6)$ texture classes and $C (C = 64)$ color classes, we now split the color-texture space (denoted as CTS, excluding the pixels on the edges) to $C \times T$ cells.

We then define the major semantic categories based on the semantic constraints of the images in the database: $SC = \{sc_1, ..., sc_M\}$. For each semantic category SC_i, certain number of images are chosen to be the training images such that the semantic codebook generated from them can represent as much as possible the color-texture characteristics of all images in the database belonging to the SC. The set of all pixels in the training images, except those in class $Edge$, is denoted as $TrainingPixels$. For each pixel in $TrainingPixels$, its feature vector will fall into one cell of CTS. After all the pixels in $TrainingPixels$ are classified, for each cell in CTS we count the number of pixels in it. Only those cells whose number of pixels exceeds a threshold will be one entry of the semantic codebook of the that database. Therefore the size of semantic codebook will be less or equal to $C \times T$. In the following discussion, we use $SCDB$ to denote the semantic codebook for color-texture space, and suppose its size is N.

2.2   Image Segmentation Based on Semantic Codebook

We segment images by $SCDB$ in this way: for each image I, we extract the color-texture feature vector for each pixel q of I. For each feature vector, we find the cell in CTS where the pixel belongs to. If the cell is one of $SCDB$'s entries, q is replaced with the $index$ of that entry; otherwise its value is set to $C \times T + 1$ to distinguish it from any valid entry of the semantic codebook. After the process, I becomes a matrix of indices corresponding to entries in $SCDB$. Because the number of valuable regions in an image is usually less than 5 in our experiments, we only choose 5 most dominant indices (referred as $DOMINANT$) and use $DOMINANT$ to re-index the pixels with indices not present in $DOMINANT$ [1]. Finally we run the encoded images through a connected-components algorithm and remove the small regions (with area less than 200 pixels).

3   Representation of Saliency for Regions and Semantic Categories

As stated before, saliency features of the regions can be either self-saliency or relative-saliency. Self-saliency features are computed by the visual features of the regions. Relative-saliency features are computed by taking the value of a feature for the current region and comparing it with the average value of that feature in the neighboring regions. Gaussian and Sigmoid functions are used to map the feature values to the interval [0, 1] for the purpose of normalization and integration of different saliency features. Self and relative-saliency features we are using include:

(1) *Self-saliency features of image regions*:
  - *Size (S_{size})*: $S_{size}(R_i) = max(\frac{A(R_i)}{A_{max}}, 1.0)$. Here $A(R_i)$ is the area of region R_i and A_{max} is a constant used prevent excessive saliency being given to very large regions. It was shown in [3] that larger regions are more likely to have larger saliency. However a saturation points exists, after which the size saliency levels off. A_{max} is set to the 5% of the total image area. The value is in the interval [0, 1].

---

[1] re-index means to find the closest codebook entry in $DOMINANT$, not in the whole $SCDB$.

- *Eccentricity (S_{ecce}, for shape):* $S_{ecce}(R_i) = \frac{major\_axis(R_i)}{minor\_axis(R_i)}$. A Gaussian function maps the value to the interval [0, 1].
- *Circularity (S_{circ}, for shape):* $S_{circ}(R_i) = \frac{Peri(R_i)^2}{A(R_i)}$. Here $Peri(R_i)$ is the perimeter of the region R_i and a Sigmoid function maps the value to the interval [0, 1].
- *Perpendicularity (S_{perp}, for shape):* $S_{perp}(R_i) = \frac{angle(major\_axis(R_i))}{\pi/2}$. A Sigmoid function maps the value to the interval [0, 1].
- *Location (S_{loca}):* $S_{loca}(R_i) = \frac{center(R_i)}{A(R_i)}$. Here $center(R_i)$ is the number of pixels in region R_i which are also in the center 25% of the image.

(2) *Relative-saliency features of image regions:* similar to [2], relative-saliency features are computed as:

$$(Relative-saliency)^{feature}_{R_i} = \sum_{R_i \in NR, i \neq 0} \frac{||feature_{R_0} - feature_{R_i}||}{feature_{R_i}}$$

Here *NR* refers to the neighboring regions. We use *Brightness* and *Location* as the features and compute *Relative brightness* and *relative location* as the relative-saliency features in the system. The values are mapped to interval [0, 1] using Sigmoid functions.

(3) *Combining self and relative-saliency to generate the saliency of the regions and semantic categories:* It is not necessary that all the saliency features will be used for each semantic categories. We choose particular saliency feature(s) to represent the most dominant visual features of each semantic categories, if possible. For instance, assume we have a category "water falls". We will use *Eccentricity* and *Perpendicularity* as well as *Relative brightness* for the saliency features because usually "water falls" will be long and narrow, and can be approximately described as perpendicular to the ground. Another example is the category "flower", the *Circularity* and *Location* as well as *Relative brightness* are used since usually flower will be round and in the middle of the images and brighter compared with surrounding scenes. A table in Section 6 lists the selection of the saliency features for all the semantic categories we have in the system.

Because all the self and relative-saliency values are normalized to the interval [0, 1], we can simply add them together as the *Saliency of the region* with regard to a certain semantic category, denoted as $Sal(R_i, SC_j)$, for the region R_i and the semantic category SC_j. For all the regions in training images that represent the semantic category SC_j, we calculate the mean and variance of their saliency and store them as the *Saliency of the semantic category*, denoted as μ_j and σ_j for the SC_j.

4   Representation of Content and Context of Regions

By collecting the statistical data from the training images, we derive the logic to represent the content and context for all the images in the database for the semantics retrieval. We first generate the statistical data from the training images. For each entry e_i in the semantic codebook *SCDB*, and each semantic category SC_j, we count the number of regions R in the training images such that (i) the pixels in R belong to e_i (e_i is a cell in the

CTS); and (ii) the region R represents an object belong to the category SC_j. The result is stored in a table called *cell-category statistics table*. In addition, we count the times that two or three different categories present in the same training images. The result is stored in a table, called *category-category statistics table*.

Based on the cell-category statistics table, for each codebook entry of $SCDB$, we can calculate its probability of representing a certain semantic category. Let N be the size of $SCDB$, M be the number of SC, $i \rightarrow SC_j, i \in [0...N-1], j \in [0...M-1]$ denote the event that index of $SCDB$ i represents semantics described in SC_j. The probability of the event $i \rightarrow SC_j$ is $P(i \rightarrow SC_j) = \frac{T(i \rightarrow SC_j)}{T(i)}$, where $T(e)$ represents event e's presence times in the training images.

Based on the category-category statistics table, we define the Bi-correlation factor and Tri-correlation factor, for representing the context of regions.

Definition 1 For any two different semantic categories SC_i and SC_j, we define the **Bi-correlation factor** $B_{i,j}$ as: $B_{i,j} = \frac{n_{i,j}}{n_i + n_j - n_{i,j}}$.

Definition 2 For any three different semantic categories SC_i, SC_j and SC_k, we define the **Tri-correlation factor** $T_{i,j,k}$ as: $T_{i,j,k} = \frac{n_{i,j,k}}{n_i + n_j + n_k - n_{i,j} - n_{i,k} - n_{j,k} + n_{i,j,k}}$. Here n_i, $n_{i,j}$ and $n_{i,j,k}$ are entries in the category-category statistics table. Bi and Tri-correlation factors reflect the correlation between two or three different categories within the scope of training images. If the training images are selected suitably, they reflect the relationship of pre-defined semantic categories we are interested in.

Armed with the statistical data we have generated from training images, we can define and extract content and context of regions for all the images in the database. Assume we have an image I with number of Q regions. The $SCDB$'s codebook indices of regions are $C_i, i \in [0...Q-1]$, and regions are represented by $R_i, i \in [0...Q-1]$, and each C_i is associated with i_N possible semantic categories $SC_{i_{j(i)}}, i_{j(i)} \in [0...N-1], j(i) \in [0...i_N-1]^2$. For I, let P_{all} be the set of all possible combinations of *indices* of semantic categories represented by regions $R_i, i \in [0...Q-1]$. We have

$$P_{all} = \{(0_{j(0)}, ..., i_{j(i)}, ..., Q-1_{j(Q-1)}) \mid \forall i, P(C_i \rightarrow SC_{i_{j(i)}}) > 0, i \in [0...Q-1],$$
$$i_{j(i)} \in [0...N-1], j(i) \in [0...i_N-1]\}$$

Note there are totally $\prod_{i=0}^{Q-1} i_N$ possible combinations for P_{all}, therefore P_{all} has $\prod_{i=0}^{Q-1} i_N$ tuples of semantic categories indices, with each tuple having Q fields.

Corresponding to each tuple $\kappa \in P_{all}$, for $Q >= 2$ we have $B(\kappa) = \sum_{p,q \in \kappa, p} {}_q B_{p,q}$, $\kappa \in P_{all}$, and for $Q >= 3$ we have $T(\kappa) = \sum_{p,q,r \in \kappa} {}_q {}_r T_{p,q,r}$, $\kappa \in P_{all}$. Here p, q, r are the indices belonging to tuple κ, $B(\kappa)$ represents the sum of Bi-correlation factors of different semantic categories with regard to tuple κ, and $T(\kappa)$ represents the sum of Tri-correlation factors of different semantic categories with regard to tuple κ.

Definition 3 We define **Context** $C_{Score}(\kappa)$ of I as: $C_{Score}(\kappa) = \frac{1}{Norm(\kappa)}(B(\kappa) + \beta T(\kappa))$, $\kappa \in P_{all}$ here $Norm(\kappa)$ is the normalization function with tuple κ, β is the weight for

---

[2] N is the size of $SCDB$

$T(\kappa)$, since $T(\kappa)$ will be more effective in distinguishing contexts of images than $B(\kappa)$[3]. We normalize the C_{Score} because several region indices may point to the same semantic category, we need to guarantee that removal the redundant semantic category will not influence the effectiveness of C_{Score}.

Definition 4 We define **ProbScore** $P_{Score}(\kappa)$ of I as: $P_{Score}(\kappa) = \sum_{i=0}^{Q-1} w(R_i, SC_{i_j})$ $P(C_i \rightarrow SC_{i_j}), i_j \in \kappa, \kappa \in P_{all}$ $P_{Score}(\kappa)$ represents the probability score corresponding to tuple κ, here $w(R_i, SC_{i_j})$ is the weight function with regard to region R_i and semantic category SC_{i_j}, it can be determined by using the saliency of the region and the semantic category:

$$w(R_i, SC_{i_j}) = \frac{1}{\sigma_{i_j}\sqrt{2\pi}}e^{-(Sal(R_i, SC_{i_j})-\mu_{i_j})^2/2\sigma_{i_j}^2}$$

where $Sal(R_i, SC_{i_j})$ is the saliency of region R_i with regard to the semantic category SC_{i_j}, μ_{i_j} and σ_{i_j} are the saliency of the semantic category SC_{i_j}.

Definition 5 We define the **TotalScore** T_{Score} of image I as $T_{Score} = Max\{P_{Score}(\kappa) + \gamma C_{Score}(\kappa) \mid \kappa \in P_{all}\}$ where $Max\{t\}$ represents the maximum value of value t.

Definition 6 We define the **Content** of image I as the semantic categories corresponding to T_{Score}. By computing the maximum value of $P_{Score}(\kappa) + \gamma C_{Score}(\kappa)$ over all tuples in P_{all}, we find the semantic categories that best interpret the semantics of the regions in image I as the *Content* of I. We store the *Content*, T_{Score} and the each region's *SCDB* codebook index and saliency as the final features for I. Note that for those regions whose codebook indices are invalid corresponding to semantic codebook, we will mark its semantic category as "UNKNOWN".

5   Semantics Retrieval

Both semantic keyword query and query-by-example are supported in our approach. According to the submitted semantic keywords (corresponding to semantic categories), the system will first find out those images that contain all of the categories (denoted as set RI), then rank the documents by sorting the T_{Score} stored for each image in the database. For those images belonging to RI that has "UNKNOWN" category, its T_{Score} will be multiplied by a diminishing factor. When user submits the query-by-example, if the query image is in the database, its *Content* will be used as query keywords to perform the retrieval, otherwise it will first go through the above steps to obtain the *Content* and T_{Score}. When users are interested in retrieving images with not only same semantics, but also the similar visual features as the query, Euclidean distance of saliency value will be calculated between the query image Q and image I in RI:

$$d(I,Q) = \sqrt{\sum_{i=0}^{W-1}(Sal(I_{Ri}, SC_j) - Sal(Q_{Ri}, SC_j))^2}$$

---

[3] In our experiments, we select $\beta = 10$.

Here assume Q has W regions and I's region I_{Ri} has same semantic category SC_j with Q's region Q_{Ri}. $d(I, Q)$ indicates the distance of saliency between I and Q and therefore can be used to fine tune T_{Score}.

6   Experiments

To test the effectiveness of our approach in retrieving the images, we conduct experiments to compare the performance between our approach (with and without using the saliency features) and the traditional CBIR techniques including Keyblock [1], color histogram [5], color coherent vector [4]. The comparison is made by the precisions and recalls of each method on all the semantic categories. In the following table, we define 10 semantic categories and list the saliency features adopted for each semantic category as "Y", otherwise as "N". Abbreviations represent the saliency features: S–Size, E–Eccentricity, C–Circularity, P–Perpendicularity, L-Location, RB–Relative brightness, RL–Relative location.

Category	Explanation	No. in db	S	E	C	P	L	RB	RL
SKY	Sky (no sunset)	1323	N	N	N	N	Y	N	Y
WATER	Water	474	N	N	N	N	Y	N	Y
TREE_GRASS	Tree or Grass	778	Y	N	N	N	N	Y	N
FALLS_RIVER	Falls or Rivers	144	N	Y	N	Y	N	Y	N
FLOWER	Flower	107	N	N	Y	N	Y	Y	N
EARTH_ROCK _MOUNTAIN	Earth or Rocks or Mountain composed of	998	Y	N	N	N	N	N	Y
ICE_SNOW _MOUNTAIN	Ice or Snow or Mountain composed of	204	Y	N	N	N	N	N	Y
SUNSET	Sunset scene	619	N	N	N	N	N	Y	N
NIGHT	Night scene	171	Y	N	N	N	Y	N	N
SHADING	Shading	1709	N	N	N	N	N	N	N

We use an image database with name *COREL* and size 3865. The *COREL* images can be considered as scenery and non-scenery. Scenery part has 2639 images consisting images containing the defined semantic categories, while non-scenery part has 1226 images including different kinds of textures, indoor, animals, molecules, etc. We choose 251 training images from scenery images as training images and form the semantic codebook with size 149. For each semantic category SC_i, we calculate and plot the precision-recall of our approach in the following way. Let $RETRIEVELIST$ denote the images retrieved with SC_i. Suppose $RETRIEVELIST$ has n images. We calculate the precisions and recalls of first $\frac{n}{30}$, $\frac{2n}{30}$, ... , and n images in $RETRIEVELIST$, respectively.

Since traditional CBIR approaches accepts only query-by-example, we have to solve the problem of comparing the approach of query-by-semantics with query-by-example. Let us take Keyblock as the example of traditional CBIR to show how we choose query sets and calculate the precision-recall for these methods. Suppose user submits a semantic keyword query of semantic category of *Sky*. There are total of 1323 images in *COREL* containing *Sky*. For each image containing *Sky*, we use it as a query on *COREL* to select top 100 images by Keyblock, and count the number of images containing *Sky* in the retrieved set. Then we sort the 1323 images descendingly by the numbers of *Sky* images in their corresponding retrieved sets. Let the sorted list be *SKYLIST*. Then we select the first 5% of *SKYLIST* as query set, denoted as *QUERYSET*. Then for each *COREL* image *I*, we calculate shortest distance to $QUERYSET - \{I\}$ by Keyblock [4]. The *COREL* images are sorted ascendantly by this distance. Top 1323 *COREL* images are retrieved and we calculate and plot the precision-recall of Keyblock on *Sky*, as we did for our approach.

The average precision-recall on all semantic categories is shown in Figure 2. We can see our method outperforms traditional approaches and retrieval performance improves when saliency features are used.

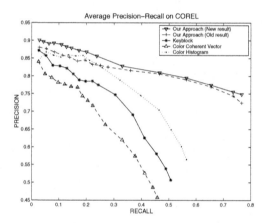

Fig. 2. Average Precision-recall on all the *SC*s

7 Conclusion

The saliency of image regions, which describes the perceptual importance of the regions, is used for the semantics-based image retrieval. It helps refine the content and context of regions to represent the semantics of regions more precisely. The experimental results show that our approach outperforms the traditional CBIR approaches we compared with.

---

[4] images in the *QUERYSET* will have distance zero to *QUERYSET*. Thus the query images will be automatically be retrieved as the top 5% images, which is unfair when making comparison.

Acknowledgment

This research is supported by the NSF Digital Government Grant EIA-9983430. The authors thank the anonymous reviewers for their valuable comments to the paper.

References

1. L. Zhu, A. Zhang, A. Rao and R. Srihari. Keyblock: An approach for content-based image retrieval. In *Proceedings of ACM Multimedia 2000*, pages 157–166, Los Angeles, California, USA, Oct 30 - Nov 3 2000. 35
2. J. Luo and A. Singhal. On measuring low-level saliency in photographic images. In *Proc. IEEE Comp. Vision and Pattern Recognition*, pages Vol. 1 pp 84–89, 2000. 29, 32
3. W. Osberger and A. J. Maeder. Automatic identification of perceptually important regions in an image. In *Proc. IEEE Int. Conf. Pattern Recognition*, 1998. 29, 31
4. Greg Pass, Ramin Zabih, and Justin Miller. Comparing images using color coherence vectors. In *Proceedings of ACM Multimedia 96*, pages 65–73, Boston MA USA, 1996. 35
5. M. J. Swain and D. Ballard. Color Indexing. *Int Journal of Computer Vision*, 7(1):11–32, 1991. 35
6. T. F. Syeda-Mahmood. Data and model-driven selection using color regions. In *Int. J. Comp. Vision*, pages Vol. 21 No. 1. pp 9–36, 1997. 29
7. W. Wang, Y. Song, and A. Zhang. Semantics retrieval by content and context of image regions. In *Proc. of the 15th International Conference on Vision Interface (VI'2002), Calgary, Canada*, May 27-29, 2002. 29, 30

The Truth about Corel - Evaluation in Image Retrieval

Henning Müller, Stephane Marchand-Maillet, and Thierry Pun

Computer Vision Group, University of Geneva
Henning.Mueller@cui.unige.ch

Abstract. To demonstrate the performance of content-based image retrieval systems (CBIRSs), there is not yet any standard data set that is widely used. The only dataset used by a large number of research groups are the Corel Photo CDs. There are more than 800 of those CDs, each containing 100 pictures roughly similar in theme. Unfortunately, basically every evaluation is done on a different subset of the image sets thus making comparison impossible.

In this article, we compare different ways of evaluating the performance using a subset of the Corel images with the same CBIRS and the same set of evaluation measures. The aim is to show how easy it is to get differing results, even when using the same image collection, the same CBIRS and the same performance measures. This pinpoints the fact that we need a standard database of images with a query set and corresponding relevance judgments (RJs) to really compare systems.

The techniques used in this article to "enhance" the apparent performance of a CBIRS are commonly used, sometimes described, sometimes not. They all have a justification and seem to change the performance of a CBIRS but they do actually not. With a larger subset of images it is of course much easier to generate even bigger differences in performance. The goal of this article is not to be a guide of how to make the "apparent" performance of systems look good, but rather to make readers aware of CBIRS evaluations and the importance of standardized image databases, queries and RJ.

1 Introduction

The Corel Photo CDs have been used in many publications to demonstrate the performance of content-based image retrieval systems (CBIRSs) and has become a de-facto standard in the field [10,19,28,26,25,3,27,18,2,12]. Sometimes the Corel images are mixed with other images [6]. Unfortunately, Corel (http://www.corel.com/) does not only offer one set of images but a collection of more than 800 Photo CDs containing 100 images each of a certain theme but also differently compiled collections of smaller-sized images with sets of several thousand images grouped into same-subject groups [28]. Even people of the same research lab presenting at the same conference often use completely different sets of images from the Corel collection to evaluate their systems [13,10,19]. This makes it very hard to compare the performance of systems as different

M. S. Lew, N. Sebe, and J. P. Eakins (Eds.): CIVR 2002, LNCS 2383, pp. 38–49, 2002.

groups of images can be either very easy or hard to separate from one another, depending on the images they contain. When having a large number of image sets, it does make sense to compile a collection of images with groups where each one contains a distinct feature, so they can be separated easily and the performance of a system looks better (sunsets, planes on a blue sky, divers, deserts images, ...).

Once a collection is compiled, a decision has to be taken as to which images to chose as query images. Here, it is important to chose images that represent a group well and not images that differ from all other images within the group. Sometimes, the first image is taken [15] but mostly nothing is said about the process of selecting a query image [26,18] or it is selected by hand [5,12] or randomly [10]. Both these options leave a lot of room for choosing the images as query images that perform best for a given system, which, as we will show, can improve the performance enormously (Section 3.3). Often, more than one image of a class is chosen as query image [12,5].

Although the Corel sets contain images of the same subject, many groups contain a few images that are visually dissimilar. It does make sense to keep these images out of the relevance sets for a query as they will make the performance look bad, although it was only a few visually dissimilar pictures that were not found. Thus, often a subset of the images is chosen to be in the relevance group [5,26]. Sometimes the entire grouping of Corel is discarded and a new grouping is done [18,28] which basically means that the developers can decide upon the performance of the system under evaluation. This can also be done by choosing easy-to-separate groups. Basically all researchers use a small subset of the images. Choosing the easiest subsets can drastically improve the performance as shown in Sections 3.5 and 3.6. In [27], the system developers judge the results themselves after every query step, which makes all results irreproducible.

One of the problems with using the groupings of the Corel collection as relevance judgments (RJs) is that the grouping does not always comply with what a human might chose visually as a good result for this group. What we really would like to evaluate from a CBIRS, is how well its response fits with what a human would expect to be returned as a result. Humans are subjective and have different opinions [24]. Thus, it might be sensible to use real user judgments as a basis for evaluation, as done in [23].

The problem induced by the fact that retrieving images from small sets is easier than from large collections has already been mentioned in [24] where a correction for discounting the chance factor was proposed. Similarly, the use of *generality* in [8] leads to appropriate measures to compare the retrieval quality when the database and relevance sets have a varying size. The *normalized averaged rank* proposed in [16] shows to be effective for comparing the performance when removing unused images (Section 3.6).

In Section 2, we shortly describe the CBIRS we used for our experiments. Section 3 explains the infrastructure of the system evaluation setup and shows the results of the different techniques to prune the performance of a system. As shown beforehand, all these techniques are common practices in CBIR evaluation. This highlights the need for a standard testbed for benchmarking because otherwise results are incomparable and researchers can basically choose the performance of their system by compiling a set of nice images plus friendly query images and corresponding groundtruth.

There is a large need for a common, freely-available image database plus standard queries and RJs for these queries. The Benchathlon (http://www.benchathlon.net/) is an iniative that tries to compile an image database and a testbed for evaluation. It is very important to support such an iniative and help to build databases and ground truth to get serious with benchmarking in CBIR. In other areas, such as in text retrieval, standard databases exist since the 1960s [4] and the TREC (Text REtrieval Conference) is doing a standard evaluation every year since 1992 [7].

2   The *Viper* System

For the validation of our approach, we use the *Viper* (http://viper.unige.ch/) system, which has been developed at the University of Geneva. A stable version called *GIFT* (GNU Image Finding Tool, http://www.gnu.org/software/gift/) can be downloaded free of charge. The system is described in more detail in [23,17].

The main difference between *GIFT* and other systems is the usage of more than $85,000$ possible features. Most images contain between $1,000$ and $2,000$ of theses features. The access method to the features is the *inverted file*, which is the most common access method used in text retrieval (TR). The system implements four different groups of image features:

- A global color histogram based on the HSV color space corresponding roughly to the human color perception [21];
- local color blocks at different scales for fixed-size regions by using the mode color for each of the fixed blocks; the image is successively partitioned into four equally sized blocks and each block is partitioned again four times;
- global texture characteristics are represented by the histograms of the response to Gabor filters of different frequencies and directions; Gabor filters are known to be a good model for the human perception of edges [14];
- local Gabor filters at different scales and in different regions by using the smallest blocks of the local color features and applying Gabor filters with different directions and scales to these blocks.

3   Results

3.1   Technical Infrastructure and Images Used

Unfortunately, we only have access to a small subset of images from the Corel Photo CDs, containing 61 different groups of 100 images each. They do not contain the groups most often used by other systems and also relatively easy to query such as "sunsets", "tigers", "eagles on the sky". We are thus limited with the upper performance we can reach, but in the following sections it will become clear that even with a small subset the apparent performance can be boosted.

For evaluation, we use the automated benchmark described in [15]. Such a benchmark makes it easy to perform a large number of example evaluations to test and optimize the apparent system performance. The performance measures we use are described in [16] which also gives an overview about different methods used in CBIR. These measures are:

- N_{rel}, number of relevant images and t, the time it takes to execute the query;
- rel_1, $\overline{Rank}$ and $\widetilde{Rank}$: rank at which the first relevant image is retrieved, average rank and normalized average rank of relevant images, respectively;
- $P(20)$, $P(50)$ and $P(N_{rel})$: *precision* after 20, 50 and N_{rel} images are retrieved;
- $R_P(.5)$ and $R(100)$: *recall at precision* .5 and after 100 images are retrieved;
- *Precision vs. recall (PR)-graph.*

A simple *average rank* is difficult to interpret, since it depends on both the collection size N and the number of relevant images N_{rel} for a given query. Consequently, we normalize by these numbers and use the *normalized average rank*, $\widetilde{Rank}$:

$$\widetilde{Rank} = \frac{1}{NN_{rel}} \left(\sum_{i=1}^{N_{rel}} R_i - \frac{N_{rel}(N_{rel} + 1)}{2} \right) \qquad (1)$$

where R_i is the rank at which the ith relevant image is retrieved. This measure is 0 for perfect performance, approaches 1 as performance worsens, and is 0.5 for random retrieval. It is basically a complement of the normalized recall proposed by Salton [20].

3.2 Evaluation of the Corel Database

Our initial evaluation procedure using databases with fixed grouping consists in using the first image of a group as a query image and the entire set as RJ for the query.

The left side of Fig. 1 and Table 1 show that the results do not look very good, although the use of RF does improve sensibly the *precision* in the first $n = 20$ images. Having an average of 5 relevant images in the first 20 returned does not sound too bad. However, we can see that rel_1 is well above 1, which means that there are image sets where no relevant image was retrieved in the top 20. In this special case $P(Nrel)$ and $R(100)$ have the same values as $N_{rel} = 100$ and thus the precision and recall after 100 images are retrieved are the same.

When taking a look at the sets with the worst results we can see that sometimes the first image is visually very different from other images in the same group. An example is a query for images of "Paris" where the first image of the group was the Eiffel tower at night, which perfectly helped retrieving images of other buildings at night, but only few of them were from "Paris". Therefore, it sounds logical to chose a query image that represents the class in a better way.

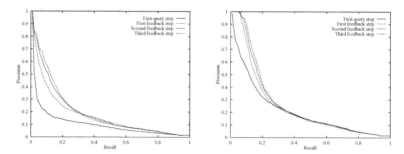

Fig. 1. *PR-graph* using the first image as query image (left) and an optimized query image (right)

Table 1. Performance using the first image (left) and an optimized image (right) as query

	1st step	RF 1	RF 2	RF 3		1st step	RF 1	RF 2	RF 3
N_{rel}	100	100	100	100	N_{rel}	100	100	100	100
time t	8.53	11.69	12.62	13.19	*time t*	8.35	11.60	13.08	12.89
rel_1	28.79	21.13	24.30	23.56	rel_1	1.49	1	1	1
$R(P(.5))$	.0552	.1336	.1487	.1525	$R(P(.5))$	.1590	.1904	.2074	.2136
$\overline{Rank}$	1713.4	1602.8	1561.0	1543.5	$\overline{Rank}$	1223.9	1205.7	1185.9	1176.1
$\widetilde{Rank}$	.2535	.2366	.2301	.2275	$\widetilde{Rank}$	.1787	.1759	.1729	0.1714
$P(20)$	.2508	.4082	.4795	.5164	$P(20)$	.5508	.6352	.6959	.7344
$P(50)$	.1711	.2603	.2846	.2895	$P(50)$	.3531	.3750	.3892	.4006
$P(N_{rel})$	.1267	.1821	.1857	.19	$P(N_{rel})$	.2452	.2449	.2566	0.2549
$R(100)$	.1267	.1821	.1857	.19	$R(100)$	.2452	.2449	.2566	0.2549

3.3 Optimizing the Query Image

For not having to chose a "best" query image by hand, we wanted to find the image that represents the class best from our system's point of view. We thus evaluated a query for every image of every class and choose the image with the highest $P(20)$. If there were several images with the same value, then $P(50)$ gave the decision. These two measures correspond to what a normal user would look at on the screen and therefore we regard these measures as the most important ones.

Fig. 1 and Table 1 illustrate how strongly the performance of our system improved, although the same system with the same relevance sets and the same database was used. For every group, we have a relevant image in the top 20 after one step of feedback, we even have an average of 11 relevant images in the top 20 and after three steps of RF even 15 relevant images in the top 20.

Comparing the two data sets in Table 1, we can see clearly that all the performance measures improve strongly. $P(20)$ more than doubles and the rank-based measures improve substantially.

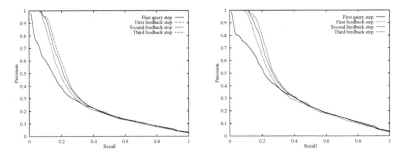

Fig. 2. *PR-graph* when removing bad images from the RJ (left) and removing bad groups (right)

3.4 Removing Bad Images

We said earlier that there are images in each group that are visually dissimilar from several images in the same group. Although these images are not used as query images anymore, they still worsen the results of retrieval. We do not want to cut too many images and set an upper limit of 20 images to be removed from each set of the database. We assume that an image is really bad for the query and can never be retrieved in one of the top ranks if the image is ranked below one third of the entire database (we used $Rank < 2000$).

As changing this does not at all change the ranking of the images and neither the *precision* after the first images are retrieved, we can only hope for an improvement in the *recall* measure. Fig. 2 shows that the entire *PR-graph* looks improved as the curve is slightly lifted up. This makes the performance looks quite a bit better although the *precision* measures in Table 2 show that the performance in the first 2000 result images did not change at all. As expected, the average rank measures and the $R(100)$ do improve significantly.

Table 2. Performance when removing bad images from the RJ (left) and bad groups (right)

	1st step	RF 1	RF 2	RF 3		1st step	RF 1	RF 2	RF 3
N_{rel}	82.41	82.41	82.41	82.41	N_{rel}	83.68	83.68	83.68	83.68
time t	8.39	11.56	12.33	12.77	*time t*	8.66	11.90	12.43	12.46
rel_1	1.49	1	1	1	rel_1	1.43	1	1	1
$R(P(.5))$	.1871	.2268	.2465	.2544	$R(P(.5))$	.2519	.2839	.3058	.3124
$\overline{Rank}$	850.3	883.3	860.0	850.39	$\overline{Rank}$	600.5	651.3	628.2	616.2
$\widetilde{Rank}$	.1325	.1379	.1341	.1325	$\widetilde{Rank}$	.0915	.0997	.0960	.0940
$P(20)$	.5508	.6352	.6959	.7344	$P(20)$	.6463	.7412	.8088	.8463
$P(50)$	.3531	.3751	.3891	.400	$P(50)$	.4365	.4495	.4685	.4845
$P(N_{rel})$	.2717	.2729	.2829	.2882	$P(N_{rel})$	.3368	.3290	.3395	.3456
$R(100)$	.2931	.2933	.3077	.3055	$R(100)$	.3609	.3532	.3648	.3650

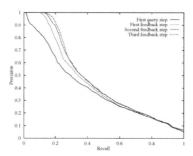

Fig. 3. *PR-graph* when creating an inverted file with only the good images and groups

3.5   Removing Bad Image Groups

The easiest way of improving the performance is of course to only have a set of easily distinguishable groups of images. With the Corel Photo CDs containing more than 80,000 images it is easy to choose a small subset of groups that work well. Unfortunately, we do not have access to many of the image groups commonly mentioned in citations, but we can already see that removing the 21 worst relevance sets from the evaluation (leaving us with 40 query images) still leads to a massive improvement in performance. And this with still using the exact same system, exact same database with the exactly same measures for evaluation.

Fig. 2 and Table 2 show that basically all measured results are massively improved by simply removing a few bad groups from the evaluation setup. We now have an average of 13 relevant images in the top 20 and 22 in the top 50. This is 13 images more than what we had with our first evaluation. All values basically doubled in result.

3.6   Creating a Smaller Database

Until now, we always used the exact same database for the queries. Just like we remove image sets or single images from the query process, it makes sense to remove all these unused images completely from the database. Like this, we create a database where every image is part of exactly one relevance group. This database now contains 40 image groups and a total of 3500 images.

We can see clearly in Fig. 3 and Table 3 that this again improves the performance strongly, especially the *average rank* drops to roughly half the value. The query time is strongly reduced and all *precision* and *recall* values go up. Interestingly, the *normalized average rank* keeps almost exactly the same value as before, which shows that the basic performance of the system did not improve and that it simply looks better because of the smaller database size. As the *normalized average rank* is normalized by the database size, it characterizes well the ranks of relevant images in dependent of the database size.

Table 3. Performance when creating an inverted file with only the good images and groups

	1st step	RF 1	RF 2	RF 3
N_{rel}	83.68	83.68	83.68	83.68
time t	4.71	5.80	6.51	6.52
rel_1	1.18	1	1	1
$R(P(.5))$	.3228	.3654	.3863	.3923
$\overline{Rank}$	350.3	364.3	347.88	351.89
$\widetilde{Rank}$	.0907	.0948	.0900	.0911
$P(20)$	.7212	.8250	.8813	.9163
$P(50)$	.504	.5345	.5585	0.5650
$P(N_{rel})$	.3848	.4013	.4094	.4092
$R(100)$	.4162	.4268	0.4387	.4391

3.7   Global Comparison

As it is very hard to compare graphical values over several graphs, Fig. 4 shows a comparison of the different evaluations in the first query step. The graph shows how strongly the performance of the system may be improved by simply choosing well the image sets, query images and RJs.

Fig. 4 also shows the improvement in performance obtained for the first RF step. The increase in performance between the best and worst graph is often more than 100%.

The same can be seen when we take a look at the other performance measures shown in Tables 1–3. $P(20)$ improved from .25 to .72, $P(50)$ from .17 to .50 and $R(100)$ from .13 to .42. Still, we have to recall that it is exactly the same algorithm with exactly the same results, only the method of evaluation is changing between the different steps.

With having a larger number of Corel image sets we could surely increase the performance even more, and with not only removing the worst 20 images but a larger number, we can as well get a much higher improvement. If we even create

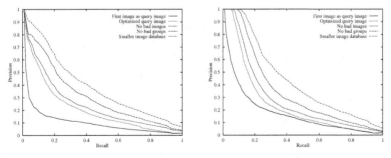

Fig. 4. Comparing the initial query (a) and the first feedback step (b) of all evaluation runs

Table 4. $P(50)$ for the easiest image groups

$P(50)$	group number	content	$P(50)$	group number	content
0.92	211000	oil paintings	0.66	153000	swimming Canada
0.9	156000	divers and diving	0.6	524000	works of art: engravings
0.9	99000	religious stained glass	0.54	161000	land of the pyramids
0.72	373000	everyday objects	0.54	502000	works of art: landscapes
0.7	125000	coins and currency	0.54	96000	candy backgrounds

the relevance sets on our own such as [18], we can basically select the apparent performance of our system even when using the same metrics and features as before.

3.8   Easy Query Groups

Finally, to have an idea about the performance differences of single images sets, we list in Table 4 the images groups with the highest $P(50)$, characterizing an easy group. We can see how much easier the very easy image sets are to retrieve. By having the entire set of more than 800 groups it should be possible to have an almost perfect performance at least for the first 50 results. Based on this study, it would be interesting to create a difficulty measure for every Corel image set, but unfortunately the difficulty of a group depends on the other groups present in an image database and the CBIRS used. Not only the intra-group similarity but also inter-group similarities are important to measure the difficulty of an image collection.

4   Conclusions and Future Work

This article shows how easy it is to change the apparent performance of a CBIRS without changing the system and even by using the same image set or a subset. This demonstrates that it is impossible to objectively compare CBIRS performances unless it is clearly stated which images where used, which images were used as queries and what had been used as RJs. Most of the times, this is not stated in articles on CBIR and only the database used plus a *precision* measure or a *PR-Graph* of the system is shown. This makes it completely impossible to compare any two CBIRSs.

The introduction of measures based on *generality* in [8] and the correction of retrieval by chance in [24] do help in some way, but many of the problems still persist when the other factors in evaluation are not clear.

The aim of this article is to highlight the need for standardized evaluation in CBIR and especially the need for a standard image database. In text retrieval, this was already realized many years ago [22] and large efforts were made to create such databases. In CBIR, an effort to create a database exists (http://www.cs.washington.edu/research/imagedatabase/groundtruth/) and there is also an iniative to create a benchmark for CBIR, the Benchathlon. It is very important to support these efforts if we want to be able to really compare CBIRSs.

Not only the creation of a standard database for CBIR is important but also the creation of proper RJs. Predefined groups of the same subject such as the Corel Photo CDs are interesting because they are easy to use without any effort to create RJs. However, the fact that groups contain very dissimilar images gives room and reasons for manipulations. Although real user tests might be the best option to gain RJs, they are costly and have to be repeated once the database gets bigger. Thus, it is important to develop alternatives such as generating RJs based on annotations [11].

References

1. *Proceedings of the ACM Multimedia Workshop on Multimedia Information Retrieval (ACM MIR 2001)*, Ottawa, Canada, October 2001. The Association for Computing Machinery. 47, 48
2. C. Carson, S. Belongie, H. Greenspan, and J. Malik. Color- and texture-based segmentation using em and its application to image querying and classification. *IEEE Transactions on Pattern Analysis and Machine Intelligence*, 2002 (to appear). 38
3. K. Chakrabarti, K. Porkaew, and S. Mehrotra. Efficient query refinement in multimedia databases. In *Proceedings of the 16th International Conference on Data Engineering (ICDE2000)*, San Diego, CA, USA, March 1–3 2000. IEEE Computer Society. 38
4. C. W. Cleverdon, L. Mills, and M. Keen. Factors determining the performance of indexing systems. Technical report, ASLIB Cranfield Research Project, Cranfield, 1966. 40
5. L. Duan, W. Gao, and J. Ma. A rich get richer strategy for content-based image retrieval. In R. Laurini, editor, *Fourth International Conference On Visual Information Systems (VISUAL'2000)*, number 1929 in Lecture Notes in Computer Science, pages 290–299, Lyon, France, November 2000. Springer-Verlag. 39
6. K.-S. Goh, E. Chang, and K.-T. Cheng. Support vector machine pairwise classifiers with error reduction for image classification. In ACMMIR2001 [1], pages 32–37. 38
7. D. Harman. Overview of the first Text REtrieval Conference (TREC-1). In *Proceedings of the first Text REtrieval Conference (TREC-1)*, pages 1–20, Washington DC, USA, 1992. 40
8. D. P. Huijsmans and N. Sebe. Extended performance graphs for cluster retrieval. In *Proceedings of the 2000 IEEE Conference on Computer Vision and Pattern Recognition (CVPR'2001)*, pages 26–31, Kauai, Hawaii, USA, December 9–14 2001. IEEE Computer Society. 39, 46
9. IEEE. *Proceedings of the second International Conference on Multimedia and Exposition (ICME'2001)*, Tokyo, Japan, August 2001. IEEE. 48, 49
10. F. Jing, B. Zhang, F. Lin, W.-Y. Ma, and H.-J. Zhang. A novel region-based image retrieval method using relevance feedback. In ACMMIR2001 [1], pages 28–31. 38, 39
11. C. Jörgensen and P. Jörgensen. Testing a vocabulary for image indexing and ground truthing. In G. Beretta and R. Schettini, editors, *Internet Imaging III*, volume 4672

of *SPIE Proceedings*, pages 207–215, San Jose, California, USA, January 21–22 2002. (SPIE Photonics West Conference). 47

12. C. S. Lee, W.-Y. Ma, and H. Zhang. Information embedding based on user's relevance feedback in image retrieval. In S. Panchanathan, S.-F. Chang, and C.-C. J. Kuo, editors, *Multimedia Storage and Archiving Systems IV (VV02)*, volume 3846 of *SPIE Proceedings*, pages 294–304, Boston, Massachusetts, USA, September 20–22 1999. (SPIE Symposium on Voice, Video and Data Communications). 38, 39

13. M. Li, Z. Chen, L. Wenyin, and H.-J. Zhang. A statistical correlation model for image retrieval. In ACMMIR2001 [1], pages 42–45. 38

14. W. Y. Ma, Y. Deng, and B. S. Manjunath. Tools for texture- and color-based search of images. In B. E. Rogowitz and T. N. Pappas, editors, *Human Vision and Electronic Imaging II*, volume 3016 of *SPIE Proceedings*, pages 496–507, San Jose, CA, February 1997. 40

15. H. Müller, W. Müller, S. Marchand-Maillet, D. M. Squire, and T. Pun. Automated benchmarking in content-based image retrieval. In ICME'2001 [9], pages 321–324. 39, 41

16. H. Müller, W. Müller, D. M. Squire, S. Marchand-Maillet, and T. Pun. Performance evaluation in content-based image retrieval: Overview and proposals. *Pattern Recognition Letters*, 22(5):593–601, April 2001. 39, 41

17. H. Müller, W. Müller, D. M. Squire, Z. Pečenović, S. Marchand-Maillet, and T. Pun. An open framework for distributed multimedia retrieval. In *Recherche d'Informations Assistée par Ordinateur (RIAO'2000) Computer-Assisted Information Retrieval*, volume 1, pages 701–712., Paris, France, apr 12-14 2000. 40

18. M. Ortega, Y. Rui, K. Chakrabarti, K. Porkaew, S. Mehrotra, and T. S. Huang. Supporting ranked boolean similarity queries in MARS. *IEEE Transactions on Knowledge and Data Engineering*, 10(6):905–925, December 1998. 38, 39, 46

19. F. Qian, M. Li, W.-Y. Ma, F. Ling, and B. Zhang. Alternating features spaces in relevance feedback. In ACMMIR2001 [1], pages 14–17. 38

20. G. Salton. *The SMART Retrieval System, Experiments in Automatic Document Processing*. Prentice Hall, Englewood Cliffs, New Jersey, USA, 1971. 41

21. J. R. Smith and S.-F. Chang. VisualSEEk: a fully automated content-based image query system. In *The Fourth ACM International Multimedia Conference and Exhibition*, Boston, MA, USA, November 1996. 40

22. K. Sparck Jones and C. van Rijsbergen. Report on the need for and provision of an ideal information retrieval test collection. British Library Research and Development Report 5266, Computer Laboratory, University of Cambridge, 1975. 46

23. D. M. Squire, W. Müller, H. Müller, and J. Raki. Content-based query of image databases, inspirations from text retrieval: inverted files, frequency-based weights and relevance feedback. In *The 11th Scandinavian Conference on Image Analysis (SCIA'99)*, pages 143–149, Kangerlussuaq, Greenland, June 7–11 1999. 39, 40

24. D. M. Squire and T. Pun. A comparison of human and machine assessments of image similarity for the organization of image databases. In M. Frydrych, J. Parkkinen, and A. Visa, editors, *The 10th Scandinavian Conference on Image Analysis (SCIA'97)*, pages 51–58, Lappeenranta, Finland, June 1997. Pattern Recognition Society of Finland. 39, 46

25. N. Vasconcelos and A. Lippman. Learning over multiple temporal scales in image databases. In D. Vernon, editor, *6th European Conference on Computer Vision (ECCV2000)*, number 1842 in Lecture Notes in Computer Science, pages 33–47, Dublin, Ireland, June 26–30 2000. Springer-Verlag. 38

26. N. Vasconcelos and A. Lippman. A propabilistic architecture for content-based image retrieval. In *Proceedings of the 2000 IEEE Conference on Computer Vision and Pattern Recognition (CVPR'2000)*, pages 216–221, Hilton Head Island, South Carolina, USA, June 13–15 2000. IEEE Computer Society. 38, 39

27. J. Z. Wand, J. Li, and G. Wiederhold. SIMPLIcity: Semantics-sensitive integrated matching for picture libraries. *IEEE Transactions on Pattern Analysis and Machine Intelligence*, 23 No 9:1–17, 2001. 38, 39

28. L. Zhu, C. Tang, A. Rao, and A. Zhang. Using thesaurus to model keyblock-based image retrieval. In ICME'2001 [9], pages 237–240. 38, 39

Non-retrieval: Blocking Pornographic Images

Alison Bosson[1], Gavin C. Cawley[2], Yi Chan[2], and Richard Harvey[2]

[1] Clearswift Corporation, 1310 Waterside
Arlington Business Park, Theale, Berkshire, RG7 4SA, UK
`alison.bosson@clearswift.com`
[2] School of Information Systems, University of East Anglia
Norwich, NR4 7TJ, UK
`{gcc,yc,rwh}@sys.uea.ac.uk`

Abstract. We extend earlier work on detecting pornographic images. Our focus is on the classification stage and we give new results for a variety of classical and modern classifiers. We find the artificial neural network offers a statistically significant improvement. In all cases the error rate is too high unless deployed sensitively so we show how such a system may be built into a commercial environment.

1 Introduction

Dealing with pornography in the workplace is a serious challenge for many large organisations but employing a block-all-images email policy no longer provides a viable solution. Email is more media-based than ever before, and it is common for business mail to contain images such as logos, publicity shots etc. In a commercial environment, an image analysis is required to automatically classify embedded or attached images as acceptable or inappropriate. The problem therefore is the *non*-retrieval of certain types of image.

Although the identification of human skin is commonplace in vision systems, the detection of pictures containing nudity and pornography is a fairly specialised area (some relevant systems include [1,2,3] and [4,5]). These systems contain a skin filter which is usually based on colour sometimes with texture as a secondary feature. Skin filters are now fairly standard so we give only a brief explanation Section 2. Here we wish to focus on the classification and deployment of such systems which we describe in Section 3 and subsequent sections.

2 Image Processing

For skin filters based on colour we note that the choice of colour feature usually leads to some discussion of the correct colour space (see [4] for discussions of alternative colour spaces). In practice we [4,5], and others [2], find that provided there is enough training data and a histogram-based representation of the colour distribution is used then the choice of colour space is not critical. We compute the likelihood ratio $L(c|\text{skin}) = \Pr\{c|\text{skin}\}/\Pr\{c|\text{not skin}\}$ for a quantized

M. S. Lew, N. Sebe, and J. P. Eakins (Eds.): CIVR 2002, LNCS 2383, pp. 50–60, 2002.
© Springer-Verlag Berlin Heidelberg 2002

Fig. 1. Original image (left) and associated log-likelihood image (second from left) displayed so that the lowest non-zero likelihood ($\log L = -7.84$) is black and the maximum likelihood, ($\log L = 4.99$) is white; seed points for region growing algorithm (third from left) and final mask (right)

colour space. Figure 1 (second from left) shows the likelihood of pixel colours for an example image using a likelihood histogram with 25^3 bins in RGB space. Likelihood images such as the one shown on the right of Figure 1 may be used to produce segments that represent regions of skin by thresholding the likelihood image at the odds set by the ratio of the priors. Care is needed to avoid two common problems: firstly that an image may contain isolated pixels that have the same colour as skin but are associated with the background (examples of such pixels can be seen on the bottom right of the second image in Figure 1) and secondly the likelihood distribution for a particular image is not guaranteed to contain the mode of the training set likelihood distribution which can cause low likelihood values. In the image in Figure 1 for example, part of the skin segment associated with the woman's face appears to have a lower likelihood that those of the bus in the background. However a legitimate assumption is that skin regions are of reasonable area compared to the total image area and contain a locally maximum likelihood value. We therefore use a region-growing algorithm that uses as its seed points likelihood local maxima above a certain threshold. The regions are then grown out to a lower likelihood threshold. A typical sequence of operations is shown in Figure 1.

This likelihood segmentation approach has been tested using a database consisting of 1000 training images and 1000 test images manually segmented to provide the ground truth. The manually generated skin segments are polygonal and include interior regions such as eyes, mouths, hair and shadows that may not be skin coloured. For each putative colour space we compute ROC curves for varying upper and lower thresholds [5]. Along the curves thresholds vary in the interval $(0.1, 0.9)$ of the peak likelihood for that image. Doing this confirms the conclusions in [4] that the HSV colour space gives the best performance. A

Fig. 2. Example segmentation from [4] and [5]

typical operating point for the HSV system is:

$$P = \begin{bmatrix} p(\bar{s}|\bar{s}) \ p(\bar{s}|s) \\ p(s|\bar{s}) \ p(s|s) \end{bmatrix} = \begin{bmatrix} 0.82 \ 0.18 \\ 0.17 \ 0.83 \end{bmatrix} \tag{1}$$

where, for example, $p(\bar{s}|s)$ denotes the probability that a pixel from a skin region is classified as one from a non-skin region. It is useful to compare these results with [2] in which the authors also conclude that a histogram-based approach is superior to parametric representations of colour distributions. The ROC curves and operating point in [2] are similar to the ones reported here but the definition of skin in [2] is narrower because here shadows, mouths and some hair are contained in the skin masks. The labelled skin set in this paper also includes pornography unlike the public database in [2].

Figure 2 shows an example segmentation for an image drawn from the test set. High resolution images such as this one usually give qualitatively better results than the low resolution images but, provided the test images contain skin colours found in the training set the automatic segmentations are close to those obtained manually. Having identified areas of skin it is necessary to extract higher level features on which to distinguish the classes of image. For this task a larger data set is needed.

These data consist of 11,005 images collected from email and web traffic in a commercial environment. The manually segmented images are a subset of the this set. The data are hand-classified into five categories: 1994 pornographic images (nude pictures that show genitalia or sexual acts); 1973 images of nudity; 1626 images of people (showing people in all poses not covered in other categories showing people); 1803 images of portraiture (which is restricted to head and shoulders portraits of a type prevalent on the web); 1767 graphics images (containing computer generated web graphics, buttons and so on) and 1842 miscellaneous images that could not be classified into one of the previous classes. There is considerable overlap between classes which are subjective. Additionally we define two meta-classes consisting of the unacceptable images (nude plus pornography) and the acceptable (all other images). The proportions of images were chosen to be broadly representative of a range of commercial environments but we know there is considerable variation in these priors between sites. This issue in discussed further later.

There are suggestions for high-level features based on grouping of skin segments [1] that might distinguish these classes but here we have a requirement to process the images speedily so, along with [2] and [3], are interested to try simpler features. For each blob in the image we have computed: area; centroid; the length of the major axis of an ellipse with the same second-order moments as the blob; the minor axis length; eccentricity and orientation of the same ellipse; the area of a convex hull fitted to the blob; the diameter of a circle with the same area as the blob; the solidity (the proportion of the convex hull area accounted for by the blob); the extent (the proportion of the area of a rectangular bounding box accounted for by the blob); the number of colours in the image (graphics are often associated with few colours) and the area of any faces located in the image (we use a commercial face finder to detect and localise faces). These features are ranked using the mutual information of the class given the single feature. Doing this gives the subset of five features that we use: the fractional area of the largest skin blob; the number of skin segments; the fractional area of the largest skin segment; the number of colours in the image and the fractional area of skin that is accounted for by a face.

3   Pattern Recognition

The image processing and feature extraction steps, described in the previous section, produce a vector of features for each image, that we hope will serve to distinguish pornographic from non-pornographic images. The task is then to find the decision rule that optimally separates acceptable from unacceptable images, given a set of labelled examples, $\mathcal{D} = \{(\boldsymbol{x}_i, t_i)\}_{i=1}^{n}$, $\boldsymbol{x}_i \in \mathcal{X} \subset \mathbb{R}^d$, $t_i \in \{0, +1\}$, where $\boldsymbol{x}_i$ represents the feature vector for pattern i, and t_i indicates whether pattern i is considered dubious ($t_i = 1$) or acceptable ($t_i = 0$). In the remainder of this section, we briefly describe the four statistical pattern recognition methods compared in this paper.

The output of a generalised linear model [6] is given by $y = g(\boldsymbol{w} \cdot \boldsymbol{x} + b)$, where, in this case, the *link* function, $g(a)$, is the logistic function, $g(a) = 1/(1 + e^{-a})$. The link function constrains the output of the linear model to lie within the range $[0, 1]$, such that it can be regarded as an estimate of conditional probability, $y_i \approx P(t_i \mid \boldsymbol{x}_i)$. Assuming the target patterns, t_i, are an independent identically distributed (i.i.d) sample drawn from a Bernoulli distribution conditioned on the corresponding input vectors, $\boldsymbol{x}_i$, the negative log-likelihood of the data, known as the cross-entropy, is given by

$$E_{\mathcal{D}} = -\sum_{i=1}^{n} \{t_i \log y_i + (1 - t_i) \log(1 - y_i)\}. \tag{2}$$

The vector of optimal model parameters $(\boldsymbol{w}, b)$ is given by the minimum of (2), which may be found via the iterative reweighted least squares algorithm. For multi-class problems, a 1-of-c coding scheme is normally adopted in which the model has c output units, one for each class, and the target for the kth output unit, for a pattern belonging to class $\mathcal{C}_l$, is $t_k = \delta_{kl}$, where δ_{kl} is the Kronecker delta function. The cross-entropy then becomes $E_\mathcal{D} = -\sum_{i=1}^{n} \sum_{k=1}^{c} t_i^k \log y_i^k$. The *softmax* link function, $y_k = \exp(a_k)/\sum_{k'} \exp(a_{k'}))$, is then used to constrain the outputs of the model to lie within the range $[0, 1]$ and to sum to one.

The k-nearest neighbour classifier [7] assigns a test pattern $\boldsymbol{x}$ to the class most strongly represented by the k most similar patterns contained in the training set, according to some distance metric, D, in this case the Euclidean distance, $D_{\text{Euclid}}(\boldsymbol{x}, \boldsymbol{x}') = \|(\boldsymbol{x} - \boldsymbol{x}')\|_2$. The fraction of nearest neighbours belonging to class $\mathcal{C}_a$, provides a simple estimate of *a-posteriori* probability, i.e. $P(\mathcal{C}_a \mid \boldsymbol{x}) \approx k_a/k$. As k tends to infinity this estimate is equal to the true *a-posteriori* probability. The distance metric and k can be chosen so as to minimise the leave-one-out error rate (for two-class problems, k is normally odd in order to prevent ties).

A multi-layer perceptron classifier (see e.g. Bishop [8]), consists of a network of simple neurons (each having a structure similar to a generalised linear model) arranged in layers with strictly feed-forward connections. The parameters of this model, $\boldsymbol{w}$, are determined by minimising a functional, $M = E_\mathcal{D} + \alpha E_\mathcal{W}$, consisting of a data misfit term, $E_\mathcal{D}$, in this case the cross-entropy (2), and a regularisation term, $E_\mathcal{W}$, penalising overly complex models. In this study we adopt the regularisation term $E_\mathcal{W} = \sum_{i=1}^{W} |w_i|$ (which corresponds to a Laplacian prior over model parameters), where W is the number of parameters. This regularisation term provides both formal regularisation and structural stabilisation as redundant weights are set exactly to zero and can be pruned from the network [9]. The regularisation parameter α, which controls the bias-variance trade-off (e.g. [8]), is integrated out analytically as described by Williams [9].

The support vector machine (e.g. [10]) constructs a maximal margin linear classifier in a high dimensional feature space, $\mathcal{F}(\boldsymbol{\Phi} : \mathcal{X} \to \mathcal{F})$, defined by a positive definite kernel function, $\mathcal{K}(\boldsymbol{x}, \boldsymbol{x}')$, giving the inner product $\mathcal{K}(\boldsymbol{x}, \boldsymbol{x}') = \boldsymbol{\Phi}(\boldsymbol{x}) \cdot \boldsymbol{\Phi}(\boldsymbol{x}')$. For this study, we use the anisotropic Gaussian radial basis function (RBF) kernel $\mathcal{K}(\boldsymbol{x}, \boldsymbol{x}') = \exp\left\{-(\boldsymbol{x} - \boldsymbol{x}')^T \text{diag}(\boldsymbol{\gamma})(\boldsymbol{x} - \boldsymbol{x}')\right\}$, where $\boldsymbol{\gamma}$ is a vector of scaling factors for each attribute. The output of a support vector machine is given by the expansion $f(\boldsymbol{x}) = \sum_{i=1}^{n} \alpha_i t_i \mathcal{K}(\boldsymbol{x}_i, \boldsymbol{x}) - b$. The optimal coefficients, $\boldsymbol{\alpha}$, of this expansion are given by the maximiser of $W(\boldsymbol{\alpha}) = \sum_{i=1}^{n} \alpha_i - \frac{1}{2} \sum_{i,j=1}^{n} t_i t_j \alpha_i \alpha_j k(\boldsymbol{x}_i, \boldsymbol{x}_j)$, subject to $0 \leq \alpha_i \leq C$, $i = 1, \ldots, n$, and $\sum_{i=1}^{n} \alpha_i t_i = 0$. C is a regularisation parameter controlling a compromise between maximising the margin and minimising the number of training set errors. The bias parameter, b, is chosen in order to satisfy the second Karush-Kuhn-Tucker (KKT) condition, $0 < \alpha_i < C \Rightarrow t_i f(\boldsymbol{x}_i) = 1$. Fortunately many of the coefficients will assume non-zero values, so the kernel expansion will generally be sparse. Estimates of *a-posteriori* probabilities can be obtained via logistic regression on $f(\boldsymbol{x})$ [11]. The regularisation parameter, C, and kernel parameters,

Table 1. Confusion matrices for generalised linear model (a), k-nearest neighbour (b), multilayer perceptron (c) and support vector machine (d) classification of acceptable and unacceptable images

		Observed					Observed	
		T	F				T	F
(a)	Predicted T	2787	880		(b)	Predicted T	3355	814
	Predicted F	1180	6158			Predicted F	612	6224

		Observed					Observed	
		T	F				T	F
(c)	Predicted T	3327	764		(d)	Predicted T	3219	705
	Predicted F	640	6274			Predicted F	748	6333

such as γ, are selected so as to minimise an upper-bound on the leave-one-out error [12].

We adopt a 10-fold cross-validation strategy to obtain an almost unbiased estimate of generalisation performance [13]. Table 1 shows the composite confusion matrices for the four classifiers compared, compiled over the test partitions resulting from 10-fold cross-validation. The optimal value of k, for the k-nearest neighbour classifier, was selected in each cross-validation trial to minimise the leave-one-out cross-validation error over the training partition. The mean value of k was 47.8 (std. error 3.71). For the MLP classifier, a single layer of hidden units was used, initially consisting of 32 neurons, giving 225 free parameters. The Bayesian regularisation and pruning algorithm reduced this to a mean of 9.4 units (std. error 0.476) and 43.7 parameters (std. error 1.57) over 10 cross-validation trials. The mean number of support vectors used in the SVM classifier is 4555.2 (std. error of 8.38).

Table 2 summarises the mean classification accuracy of each classifier over the test partitions resulting from 10-fold cross-validation. The k-NN, MLP and SVM classifiers are all superior to the GLM approach, justifying the use of non-linear methods. The relative performance of classifiers systems can be assessed via tests of statistical significance. McNemar's test [14] is used to determine whether the difference in the accuracies of a pair of classifiers is statistically significant. In conducting the necessary set of 6 tests the probability of falsely rejecting the null hypothesis (that there is no significant difference) in at least one test at the 0.05 level of statistical significance is $1-(1-0.05)^6 \approx 0.265$ (assuming that the results of the tests are independent). As we are more concerned in this study with type I error than type II error (accepting a null hypothesis that is false), we should use the *Bonferroni* adjustment [15]; to obtain a statistical significance at the

0.05 level across all 6 tests, $\alpha = 1 - \sqrt[n]{(1 - 0.05)} \approx 0.0085$. Table 3 summarises the results of McNemar's test of statistical significance. The non-linear methods (k-NN, MLP and SVM) are found to be significantly better than linear methods and the MLP and k-NN significantly superior to the SVM, but the difference in performance between the k-NN and MLP is not statistically significant.

Plotting the true-positive rate of a classifier, which is defined as the proportion of positive patterns correctly classified as positive, versus the false-positive rate, which the proportion of negative patterns incorrectly classified as positive, gives the receiver operating characteristic (ROC). The ROC curve then provides a graphical assessment of the performance of a classifier under different misclassification costs, by showing the increasing rate of false-positive errors that must be tolerated in order to improve the true-positive rate. The best classification rules appear toward the upper-left hand corner of the ROC plot. Figure 3 shows the receiver operating characteristic for the four classifiers evaluated in this study. If nothing is known about the true operational *a-priori* probabilities or equivalently misclassification costs, the area under the ROC curve provides a reasonable performance statistic for comparing classifier systems [16]. Table 2 gives the mean area under the ROC curve of each classifier over the test partitions resulting from 10-fold cross-validation. Fitting a convex hull to individual ROC curves gives an area of 0.943, indicating that a combination of classifiers is preferred in uncertain environments.

Multi-layer perceptron networks were also used to solve the six-class pattern recognition task. A further classification stage into the meta-classes gives similar results to those reported with the two-class classifier with that advantage of being able to more sensitively adjust for new class priors. Table 4 shows a composite confusion matrix compiled over the test partitions resulting from 10-fold cross-validation. The MLP classifier achieves a mean test-partition accuracy of 0.520940 with a standard error of 0.005724. The six-class multi-layer perceptron network can also be used to implement the 2-class detector, designating an image as unacceptable if the sum of the *a-posteriori* probabilities for classes "pornography" and "nude" exceeds 0.5. Table 5 shows a composite confusion matrix compiled over the test partitions resulting from 10-fold cross-validation. The MLP classifier achieves a mean test-partition accuracy of 0.872331 with a standard error of 0.003481. As expected, this is almost identical to the accuracy achieved by the two-class multi-layer perceptron classifier.

Table 2. Mean test-partition accuracy by classification method and also area under ROC curves by classification method

Method	Mean accuracy	Std. err.	Mean area	Std. err.
GLM	0.813	0.004	0.889	0.004
k-NN	0.870	0.004	0.931	0.003
MLP	0.872	0.004	0.937	0.002
SVM	0.868	0.005	0.915	0.004

Table 3. Statistical significance of classifier system performance. The upper triangle gives the superior classifier in a pair-wise comparison, statistically superior victors are shown underlined, the lower triangle gives the corresponding level of statistical significance. For example, the entry in the fourth column of the third row indicates that the MLP is superior to the SVM, the third column of the fourth row indicates that the difference in performance is statistically significant according to McNemar's test

Classifier	GLM	k-NN	MLP	SVM
GLM	-	k-NN	MLP	SVM
k-NN	< 0.001	-	MLP	k-NN
MLP	< 0.001	0.252	-	MLP
SVM	< 0.001	< 0.01	< 0.01	-

Table 4. Confusion matrix for multi-layer perceptron classification of images into six categories, under 10-fold cross-validation. The true class runs horizontally and predicted class runs vertically

	porn	nude	people	portrait	misc	graphics
porn	1357	765	50	112	124	14
nude	457	809	64	251	175	28
people	5	23	487	62	248	398
portrait	65	190	194	1126	230	116
misc	101	162	332	187	920	177
graphics	9	24	499	65	145	1034

Table 5. Confusion matrix for multi-layer perceptron classification of pornographic images, under 10-fold cross-validation

Observed

		T	F
Predicted	**T**	3327	765
	F	640	6273

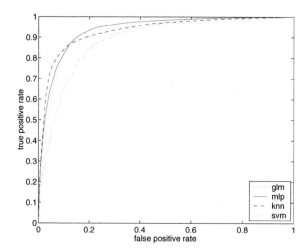

Fig. 3. ROC curves for GLM, k-NN, MLP and SVM classifiers

4   Discussion

The image classifier described in this paper is integrated into a mail-based security product MAILsweeper[TM][1] which is a *content security* solution that sits at an SMTP gateway, assessing email traffic entering and leaving a company and protecting the organisation from mail-borne threats such as viruses, breaches of confidentiality, offensive email content, legal liability and copyright infringement etc. MAILsweeper disassembles emails into their components, for example, zipped email attachments will be unzipped. These are then analysed according to user-defined policies which may be company-wide, department-wide or unique to an individual employee.

The outcome for a particular mail message is determined by its classification. Mails that are *clean* are allowed to pass to the intended recipient but for mails that, for example, contain large attachments, unknown file-types, offensive or confidential material, delivery may be delayed until a user-defined time; the item may be copied; returned to the sender; quarantined or deleted. Notifications and alerts to administrators/senders/recipients may accompany these final message classifications.

The image analyser add-on for MAILsweeper is called PORNsweeper[TM]. As emails are disassembled into their components, any images are passed to PORNsweeper for classification. It first tries to match the incoming image to any of the images in its *exception list*. These are common images, stored as an MD5 hash, that may be pre-classified by an administrator as pornographic or safe. Any incoming image not in the exceptions list is passed to the analyser. If an image is classified as safe the email will be delivered as usual. If, however

---

[1] All trademarks are the property of their respective owners

it is found to be unacceptable, the MAILsweeper system will quarantine the image for the administrators inspection. Any false positives that are blocked may be released from quarantine and may be added to the clean exceptions list to prevent future incorrect classifications. From an administrative perspective, PORNsweeper may be used to constantly monitor all images in emails entering and/or leaving an organisation or it may be used in short bursts, providing a snapshot of email activity.

This paper has provided evidence of a successful skin segmentation algorithm and suggested how this might form part of an automated pornography detector. The results of the classification experiments show that a non-linear classifier is essential. The choice of classifier depends on implementation issues such as speed and memory usage. The MLP performs well on both these counts and also has the best classification performance. The performance of the SVM is disappointing given the strong theoretical justification of this approach. A possible explanation might be that the model selection criterion unduly favours hyperparameters specifying highly regularised classifiers.

References

1. Fleck, M. M., Forsyth, D. A., Bregler, C.: Finding naked people. In: European Conference on Computer Vision. Volume II., Springer-Verlag (1996) 593–602   50, 53
2. Jones, M. J., Rehg, J. M.: Statistical color models with application to skin detection. Technical Report CRL 98/11, Compaq Cambridge Research Laboratory (1998)   50, 52, 53
3. Wang, J., Wiederhold, G., Firschein, O.: System for screening objectionable images using Daubechies' wavelets and color histograms. In Stenmetz, R., Wolf, L. C., eds.: Proc. IDMS'97. Volume 1309., Springer-Verlag LNCS (1997) 20–30   50, 53
4. Chan, Y., Harvey, R., Smith, D.: Building systems to block pornography. In Eakins, J., Harper, D., eds.: Challenge of Image Retrieval, BCS Electronic Workshops in Computing series (1999) 34–40   50, 51, 52
5. Chan, Y., Harvey, R., Bagham, J.: Using colour features to block dubious images,. In: Proc. Eusipco 2000. (2000)   50, 51, 52
6. McCullagh, P., Nelder, J. A.: Generalized Linear Models. 2nd edn. Volume 37 of Monographs on Statistics and Applied Probability. Chapman & Hall (1989)   53
7. Dasarathy, B. V., ed.: Nearest neighbour (NN) norms: NN pattern classification techniques. IEEE Computer Society, Washingtion, DC (1991)   54
8. Bishop, C. M.: Neural networks for pattern recognition. OUP (1995)   54
9. Williams, P. M.: Bayesian regularisation and pruning using a Laplace prior. Neural Computation **7** (1995) 117–143   54
10. Cristianini, N., Shawe-Taylor, J.: An Introduction to Support Vector Machines (and other kernel-based learning methods). CUP (2000)   54
11. Platt, J. C.: Probabilities for SV machines. In Smola, A. J., Bartlett, P. J., Schölkopf, B., Schuurmans, D., eds.: Advances in Large Margin Classifiers. MIT Press, Cambridge, Massachusetts (2000) 61–73   54
12. Joachims, T.: Estimating the generalization performance of a SVM efficiently. Technical Report LS-8 No. 25, Univerität Dortmund, Fachbereich Informatik (1999)   55

13. Stone, M.: Cross-validatory choice and assessment of statistical predictions. Journal of the Royal Statistical Society **B 36** (1974) 111–147   55
14. Gillick, L., Cox, S.: Some statistical issues in the comparison of speech recognition algorithms. In: Proceedings, ICASSP. Volume 1. (1989) 532–535   55
15. Zalzberg, S. L.: On comparing classifiers: pitfalls to avoid and a recommended approach. Data Mining and Knowledge Discovery **1** (1997) 317–327   55
16. Bradley, A. P.: The use of the area under the ROC curve in the evaluation of machine learning algorithms. Pattern Recognition **30** (1997) 1145–1159   56

A Linear Image-Pair Model
and the Associated Hypothesis Test
for Matching

Gregory Cox[1] and Gerhard de Jager[2]

[1] Machine Intelligence Group, DebTech Research, De Beers Consolidated Mines
Johannesburg 2013, South Africa
gregory.cox@debeersgroup.com
[2] Digital Image Processing Laboratory, Department of Electrical Engineering
University of Cape Town, Rondebosch 7701, South Africa
gdj@eng.uct.ac.za

Abstract. A statistical model is developed for the image pair and used
to derive a minimum-error hypothesis test for matching. For reasons of
tractability a multivariate normal image model and linear dependence
between images are assumed. As one would expect, the optimal test outperforms the standard approaches when the assumed model is in force,
but the extent of the optimal test's superiority suggests that there is significant potential for improvement on the standard methods of assessing
image similarity.

1 Introduction

Direct image matching, where image correspondence is obtained directly from
pixel values or simple statistics of those values, is still an important early stage
in many sophisticated image analysis algorithms. It is normally accomplished by
computing a similarity measure between the two images. *Correlation-based measures* are estimators of linear correlation between corresponding pixels in the
image pair. Variations include normalized cross-correlation for scale independence, zero-mean correlation for intensity offset invariance and the correlation
coefficient for both scale and offset invariance. *Difference-based measures* view
similarity in terms of the distance between two images in image space. The
standard measure is the sum of squared (or absolute) pixel differences. Robust
estimators of correlation- and difference-based measures are less sensitive to outliers in the image data [1,2]. *Non-parametric measures*, like the stochastic sign
change (based on sign changes in the difference image) [3] and ordinal measures
(based on pixel value ordering) [4], represent a more drastic departure from the
classical measures and can tolerate phenomena such as occlusion. *Histogram-
based measures*, based either on joint histograms of the pixel values in both
images (eg. mutual information [5]) or histograms of the pixel differences [6],
have proved effective in some applications, but are only feasible for images large
enough to create a meaningful histogram estimate.

M. S. Lew, N. Sebe, and J. P. Eakins (Eds.): CIVR 2002, LNCS 2383, pp. 61–69, 2002.

The comparative effectiveness of similarity measures has been explored in some detail for specific applications [7,8], or experimentally for a chosen set of images [9]. General statements, however, were not made because the results were only applicable to the images chosen for the experiments. A model could be the basis of a more generic investigation, and although much research has been devoted to models of individual images, very little has explicitly addressed their interdependence.

In this paper we develop a model for the image pair (essentially a joint probability density function (pdf) of all pixels in both images) which is then used to derive an optimal test of whether a match hypothesis H_1 or mismatch hypothesis H_0 should be accepted for an observed image pair. Consider the space, Ψ, of all possible image pairs in a particular application, and a particular image-pair observation, $\mathbf{w} \in \Psi$. The optimal test creates a partition of Ψ into image pairs that match, $\mathbf{w} \in \Psi_1$, and image pairs that don't, $\mathbf{w} \in \Psi_0 = \Psi - \Psi_1$, using the decision rule

$$\text{Accept } \begin{cases} H_1 \text{ if } \mathbf{w} \in \Psi_1 \\ H_0 \text{ if } \mathbf{w} \in \Psi_0 \end{cases}, \tag{1}$$

which optimizes the matching performance criteria over all potential decision rules and satisfies any other requirements of a matching algorithm.

Hypothesis testing based on an image-pair model is not covered in the literature — previous work has either reformulated the matching problem as object detection and applied existing techniques, projected the image-pair into a subspace (eg. the difference image) where modelling is simpler, or considered only the marginal pixel pdfs as the basis of comparison (eg. mutual information).

2   Stochastic Model for the Image Pair

For the purpose of this research, specific information about the content of the image is assumed to be unknown and therefore the option of deterministic modelling is rejected. Further assumptions simplify mathematical analysis based on the model: first, it is assumed that the images are samples of a stationary and multivariate normal (MVN) random field. Analysis with non-MVN distributions is rarely tractable. Spatial stationarity allows convenient analysis and leads to more efficient algorithms. Although a cursory analysis of typical images with natural or man-made scene content reveals that these assumptions are often unrealistic, authors have proposed techniques for transforming image data so that it better represents samples of a stationary MVN process (eg. [10,11]).

Second, it is assumed that a linear model adequately represents the match-mismatch relationship between the images. Given $n \times n$ images[1] $\mathbf{a}$ and $\mathbf{b}$ the pdf for the image pair $\mathbf{w}^T = \begin{bmatrix} \mathbf{a}^T, \mathbf{b}^T \end{bmatrix}$ is given by

$$p_{\mathbf{w}}(\mathbf{w}) = \frac{\exp\left[-\frac{1}{2}(\mathbf{w} - \mathbf{m_w})^T \mathbf{K_w}^{-1}(\mathbf{w} - \mathbf{m_w})\right]}{\sqrt{(2\pi)^{2n^2} |\mathbf{K_w}|}}, \tag{2}$$

---

[1] $n \times n$ images are written as n^2-vectors with pixels in row-column order.

where $\mathbf{m_w^T} = [\mathbf{m_a^T}, \mathbf{m_b^T}]$ is simply a concatenation of the mean vectors for the individual images. $\mathbf{K_w}$ is the joint image-pair covariance matrix, which can be written as

$$\mathbf{K_w} = \begin{bmatrix} \sigma_a^2 \mathbf{R_a} & \sigma_a \sigma_b \mathbf{R_{ab}} \\ \sigma_a \sigma_b \mathbf{R_{ab}} & \sigma_b^2 \mathbf{R_b} \end{bmatrix}, \tag{3}$$

where $\mathbf{K_a} = \sigma_a^2 \mathbf{R_a}$ and $\mathbf{K_b} = \sigma_b^2 \mathbf{R_b}$ are the covariance matrices of the individual images, and $\sigma_a \sigma_b \mathbf{R_{ab}}$ is their cross-covariance matrix.

Third, it is assumed that the images $\mathbf{a}$ and $\mathbf{b}$ share the same intra-image correlation structure, and therefore $\mathbf{R_a} = \mathbf{R_b} = \mathbf{R}$. In a matching application the two images will probably contain the same sort of subject matter, making this a reasonable assumption in most cases. Applications that require multimodal matching are possible exceptions, although it should be noted that the model still allows the images to differ by a systematic offset (mean vectors $\mathbf{m_a}$ and $\mathbf{m_b}$) and overall scale (variances σ_a^2 and σ_b^2).

Finally, it is assumed that the individual images are corrupted by additive white noise, implying the image-pair covariance matrix,

$$\mathbf{K_w} = \begin{bmatrix} \sigma_a^2 \mathbf{R} + \sigma_\mu^2 \mathbf{I} & \sigma_a \sigma_b \mathbf{R_{ab}} \\ \sigma_a \sigma_b \mathbf{R_{ab}} & \sigma_b^2 \mathbf{R} + \sigma_\nu^2 \mathbf{I} \end{bmatrix}, \tag{4}$$

for the noisy image pair $\mathbf{w}^T = [\mathbf{u}^T, \mathbf{v}^T]$, where $\mathbf{u} = \mathbf{a} + \boldsymbol{\mu}$ and $\mathbf{v} = \mathbf{b} + \boldsymbol{\eta}$.

3   Models for Match and Mismatch

The normalized cross-covariance matrix $\mathbf{R_{ab}}$ in equation (4) determines the relationship between images $\mathbf{u}$ and $\mathbf{v}$, and will therefore be instrumental in defining the image-pair model for match and mismatch hypotheses. Meaningful structure for $\mathbf{R_{ab}}$ is now developed for two separate models based on inter-image correlation and inter-image differences, respectively.

Correlation-Based Model Here we make the assumption that all corresponding pixel-pairs share a correlation coefficient ρ_{ab}, the value of which is specified differently in the cases of match and mismatch. For images that only differ by the systematic offset and scaling factor determined by the respective image mean vectors and variances, $\rho_{ab} = 1$, for statistically independent images $\rho_{ab} = 0$, and for values in-between the images exhibit varying degrees of correlation between corresponding pixels. The match and mismatch hypotheses can be defined as $H_1 \iff \rho_{ab} = \rho_1$ and $H_0 \iff \rho_{ab} = \rho_0$, respectively, where $0 \leq \rho_0 < \rho_1 \leq 1$.

In this case the normalized cross-covariance matrix has ρ_{ab} as diagonal elements, but the off-diagonal elements are unspecified. Two constraints limit the form of the matrix: (1) For $\rho_0 = 0$, mismatching images are statistically independent and $\mathbf{R_{ab}}|_{\rho_{ab}=0} = \mathbf{0}$. (2) For $\rho_1 = 1$, matching images are identical within a scaling factor and $\mathbf{R_{ab}}|_{\rho_{ab}=1} = \mathbf{R}$. One valid structure for $\mathbf{R_{ab}}$ with these constraints is $\mathbf{R_{ab}} = \rho_{ab}\mathbf{R}$. This is not a unique solution (another has elements

$\rho_k \mathbf{R}\,[i,j]^{\rho_k}$, where $\mathbf{R}\,[i,j]$ is the element of $\mathbf{R}$ at (i,j)), but its simplicity is appealing. In this case the joint covariance matrix for the correlation-based model becomes

$$\mathbf{K_w} = \begin{bmatrix} \sigma_a^2\mathbf{R} + \sigma_\mu^2\mathbf{I} & \sigma_a\sigma_b\rho_{ab}\mathbf{R} \\ \sigma_a\sigma_b\rho_{ab}\mathbf{R} & \sigma_b^2\mathbf{R} + \sigma_\nu^2\mathbf{I} \end{bmatrix}. \tag{5}$$

Difference-Based Model Here match and mismatch are defined in terms of differences between the images. Figure 1 shows a hypothetical process that uses two independent random images to generate an image pair with marginal covariance matrices $\mathbf{K_a}$ and $\mathbf{K_b}$, mean vectors $\mathbf{m_a}$ and $\mathbf{m_b}$, and with control over an image difference parameter, δ_{ab}. To guarantee identical images (aside from scaling and offset differences), $\delta_{ab} = 0$, for statistically independent images, $\delta_{ab} = 1$, and for values in-between the images have differences of varying magnitudes. For the matching problem there will be two processes: one that generates matching images, where $H_1 \iff \delta_{ab} = \delta_1$ and one that generates mismatching images, where $H_0 \iff \delta_{ab} = \delta_0$. Note that $0 \le \delta_1 < \delta_0 \le 1$.

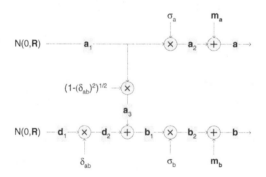

Fig. 1. Process for generating an image pair with control over a difference parameter

It can be shown that the process in Figure 1 does indeed produce images with the required marginal covariance matrices according to equation (3). The cross covariance matrix implied by Figure 1 is given by

$$\mathrm{Cov}\,[\mathbf{a},\mathbf{b}] = E\left[(\mathbf{a} - \mathbf{m_a})\,(\mathbf{b} - \mathbf{m_b})^T\right] = \sigma_a\sigma_b\sqrt{1 - \delta_{ab}^2}\,\mathbf{R}.$$

The joint covariance matrix for the difference-based model is therefore

$$\mathbf{K_w} = \begin{bmatrix} \sigma_a^2\mathbf{R} + \sigma_\mu^2\mathbf{I} & \sigma_a\sigma_b\sqrt{1 - \delta_{ab}^2}\,\mathbf{R} \\ \sigma_a\sigma_b\sqrt{1 - \delta_{ab}^2}\,\mathbf{R} & \sigma_b^2\mathbf{R} + \sigma_\nu^2\mathbf{I} \end{bmatrix}. \tag{6}$$

Notice that covariance matrices of the correlation-based and difference-based models are actually equivalent, with $\delta_{ab}^2 = 1 - \rho_{ab}^2$. It is interesting to consider the traditional correlation based and difference based similarity measures in this context. The correlation-based measures (cross-correlation, correlation coefficient) estimate ρ_{ab}, and the difference-based measures (sum of squared differences, sum of absolute differences) estimate δ_{ab}. Seen in the light of the abovementioned models these are essentially equivalent approaches, but differ with respect to the practical considerations associated with estimators: variance, bias, robustness and computational economy. The proposed model suggests this interpretation of the traditional measures, but was not the rationale for their development. For the most part, authors viewed their similarity measures as a deterministic feature of the image pair, and not as the stochastic model parameter that had to be estimated in order to establish whether a match or mismatch model was in force. Without any loss of generality the cross-correlation coefficient ρ_{ab} will be used as the match-determining parameter from this point onward.

The process in Figure 1 also represents a convenient way of synthesising image pairs. It can transform two independent random $N(0, \mathbf{R})$ images (generated using a procedure like the one outlined by Johnson [12]) into an image pair with specified inter-image correlation. Figure 2 shows examples of noise-free image pairs synthesised with different match correlation coefficients. The individual images are first order Markov random fields with a one-step spatial correlation coefficient of $\rho = 0.8$.

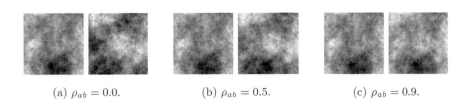

(a) $\rho_{ab} = 0.0$.   (b) $\rho_{ab} = 0.5$.   (c) $\rho_{ab} = 0.9$.

Fig. 2. Synthesised image pairs

4   The Optimal Test for Matching

The hypothesis test for matching can now be based on a match and a mismatch model, where these two models differ only in the value of parameter ρ_{ab}. Different hypothesis testing scenarios are distinguished by the *a priori* information available about ρ_{ab} under the match and mismatch hypotheses, and by the extent of *a priori* knowledge about the other model parameters, $\varphi = \left[\mathbf{R}, \sigma_a^2, \sigma_b^2, \sigma_\mu^2, \sigma_\nu^2\right]$. If ρ_{ab} and φ are well-known, then the hypotheses are simple and the optimal test is based on the likelihood ratio statistic [13]. Here the simple match and mismatch hypotheses share the pdf $p_\mathbf{w}(\mathbf{w}|\rho_{ab})$, where $H_1 \iff \rho_{ab} = \rho_1$

and $H_0 \iff \rho_{ab} = \rho_0$, and the likelihood ratio test (LRT) is

$$l\left(\mathbf{w}\right) = \frac{p_{\mathbf{w}}\left(\mathbf{w}|\rho_{ab} = \rho_1\right)}{p_{\mathbf{w}}\left(\mathbf{w}|\rho_{ab} = \rho_0\right)} \underset{H_0}{\overset{H_1}{\gtrless}} \lambda, \tag{7}$$

where λ is a scalar decision threshold that is determined by the optimality criterion (eg. minimum error rate, Neyman-Pearson, or minimax rule). In practise the test is often written in terms of a sufficient statistic $s\left(\mathbf{w}\right)$ and modified decision threshold $\acute{\lambda}$.

In order to derive the LRT statistic, the image-pair pdf under match and mismatch hypotheses must be substituted into equation (7) and expressed in a convenient form for implementation. This task can be simplified considerably if the images have independent, identically distributed (iid) pixels, which is the case in general if calculation of the LRT statistic is preceded by the Karhunen-Loève transform (KLT) on each image, yielding $\acute{\mathbf{u}} = \mathbf{V}^T\mathbf{u}$ and $\acute{\mathbf{v}} = \mathbf{V}^T\mathbf{v}$, where $\mathbf{V}$ is the KLT matrix. Expressed in terms of these iid images, the simplified LRT statistic can be written as [14]

$$s\left(\acute{\mathbf{u}}, \acute{\mathbf{v}}\right) = \sum_{i=1}^{n^2} \beta_i \left(\acute{u}_i - m_{\acute{u}_i}\right)\left(\acute{v}_i - m_{\acute{v}_i}\right) - \alpha_i \left(\left(\acute{u}_i - m_{\acute{u}_i}\right)^2 + \left(\acute{v}_i - m_{\acute{v}_i}\right)^2\right), \tag{8}$$

$$\text{where} \quad \alpha_i = \frac{k_i^2\left(\rho_1^2 - \rho_0^2\right)}{\left(1 - k_i^2\rho_0^2\right)\left(1 - k_i^2\rho_1^2\right)} \quad, \quad \beta_i = 2\frac{k_i\left(\rho_1 - \rho_0\right)\left(1 + k_i^2\rho_0\rho_1\right)}{\left(1 - k_i^2\rho_0^2\right)\left(1 - k_i^2\rho_1^2\right)}$$

$$\text{and} \quad k_i = \frac{\sigma_a\sigma_b\omega_i}{\sqrt{\left(\sigma_a^2\omega_i + \sigma_\mu^2\right)\left(\sigma_b^2\omega_i + \sigma_\nu^2\right)}}.$$

Note that $m_{\acute{u}_i}$ and $m_{\acute{v}_i}$ are pixels in the iid mean images of $\mathbf{u}$ and $\mathbf{v}$, respectively, and that ω_i is the eigenvalue of $\mathbf{R}$ associated with the eigenvector in the i-th column of $\mathbf{V}$ (the KLT matrix).

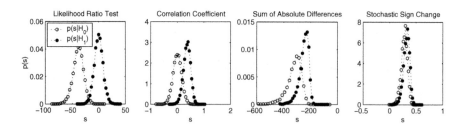

Fig. 3. Monte Carlo histograms under match and mismatch hypotheses

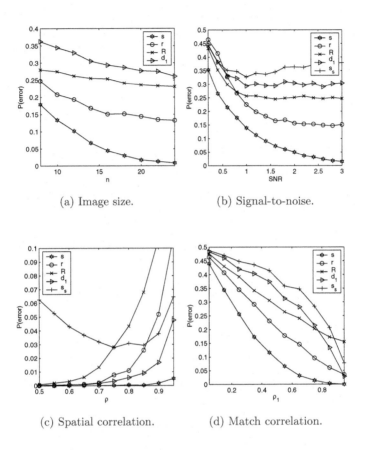

(a) Image size. (b) Signal-to-noise.

(c) Spatial correlation. (d) Match correlation.

Fig. 4. Error rate comparisons of the LRT statistic (s), correlation coefficient (r), cross correlation (R), sum of absolute valued differences (d_1) and stochastic sign change (s_s). Each experiment used 10^4 image pairs with zero mean, unit variance, $SNR = \frac{\sigma_a}{\sigma_\eta} = 2$ and spatial correlation $\rho = 0.95$ by default. Match correlation in (a), (b) and (d) $\rho_1 = 0.99$ and in (c) was $\rho_1 = 0.60$. Mismatch correlation was $\rho_0 = 0$ in all cases

5   Simulation Experiments

Monte Carlo simulation experiments can now explore the performance of the LRT and other common similarity measures. Figure 3 compares the match- and mismatch-conditional pdfs, and reveals that the minimum error rate (the overlap of the pdfs) is lowest for the LRT statistic. Figure 4 explores the minimum error rates over ranges of four parameters in the image pair model. In all cases the LRT statistic is superior. This is to be expected, since it is the model-optimal solution. The comparison is unfair on the SSC measure since the experiments do not test occlusion tolerance. What is interesting is the extent of the LRT's superiority under ideal conditions, which suggests that there is scope to improve on the common measures. Huber describes near-optimal performance under the classical model as one of the desirable characteristics of a robust procedure [15, p. 5]. So despite the fact that the LRT is optimal under a fairly unrealistic classical model, its performance under this model can be used as a benchmark for other methods.

6   Conclusion

We have presented a linear model for the image pair and have derived the simplest case of the optimal hypothesis test for matching based on this model. Monte Carlo experiments under ideal conditions show this test's superiority, the extent of which suggests that there is scope for improvement on the common measures. Ongoing work is exploring the performance of the LRT and other measures with a more comprehensive set of experiments, investigating the efficient implementation of the LRT statistic, and applying principles of optimal matching to specific problems such as block matching, face recognition and image retrieval.

References

1. M. Boninsegna and M. Rossi, "Similarity measures in computer vision," *Pattern Recognition Letters*, vol. 15, pp. 1255–1260, 1994. 61
2. R. Brunelli and S. Messelodi, "Robust estimation of correlation with applications to computer vision," *Pattern Recognition*, vol. 28, pp. 833–841, June 1995. 61
3. A. Venot, J. F. Lebruchec, and J. C. Roucayrol, "A new class of similarity measures for robust image registration," *Computer Vision, Graphics, and Image Processing*, vol. 28, pp. 176–184, 1984. 61
4. D. N. Bhat and S. K. Nayar, "Ordinal measures for image correspondence," *IEEE Transactions on Pattern Analysis and Machine Intelligence*, vol. 20, pp. 415–423, Apr. 1998. 61
5. P. Viola and W. M. Wells III, "Alignment by maximization of mutual information," in *International Conference on Computer Vision*, pp. 16–23, June 1995. 61
6. T. Buzug and J. Weese, "Similarity measures for subtraction methods in medical imaging," in *18th Annual International Conference of the IEEE EMBS*, p. 140, 1996. 61
7. G. P. Penny, J. Weese, J. A. Little, P. Desmedt, D. L. G. Hill, and D. J. Hawkes, "A comparison of similarity measures for use in 2D-3D medical image registration," in *First Conference on Medical Image Computing and Computer Assisted Intervention*, vol. 1496, (Cambridge, MA, USA), pp. 1153–1161, 1998. 62
8. E. H. W. Meijering, W. J. Niessen, and M. A. Vergiever, "Retrospective motion correction in digital subtraction angiography: A review," *IEEE Transactions on Medical Imaging*, vol. 18, pp. 2–21, Jan. 1999. 62
9. P. Aschwanden and W. Guggenbül, "Experimental results from a comparative study on correlation-type registration algorithms," in *International Workshop on Robust Computer Vision*, no. 2, pp. 268–289, Mar. 1992. 62

10. B. R. Hunt, "Nonstationary statistical image models (and their application to image data compression)," *Computer Graphics and Image Processing*, vol. 12, pp. 173–186, 1980. 62

11. P. B. Chapple and D. C. Bertilone, "Stochastic simulation of infrared non-Gaussian terrain imagery," *Optics Communications*, no. 150, pp. 71–76, 1998. 62

12. G. E. Johnson, "Constructions of particular random processes," *Proceedings of the IEEE*, vol. 82, pp. 270–285, Feb. 1994. 65

13. D. Kazakos and P. Papantoni-Kazakos, *Detection and Estimation*. Computer Science Press, 1990. 65

14. G. S. Cox, *"Designing Hypothesis Tests for Digital Image Matching"*. PhD thesis, University of Cape Town, December 2000. 66

15. P. J. Huber, *Robust Statistics*. John Wiley and Sons, 1981. 68

On the Coupled Forward and Backward Anisotropic Diffusion Scheme for Color Image Enhancement

Bogdan Smolka[1]* and Konstantinos N. Plataniotis[2]

[1] Department of Automatic Control, Silesian University of Technology
Akademicka 16 Str, 44-101 Gliwice, Poland
bsmolka@ia.polsl.gliwice.pl
[2] Edward S. Rogers Sr. Department of Electrical and Computer Engineering
University of Toronto, 10 King's College Road, Toronto, Canada
kostas@dsp.toronto.edu

Abstract. The use of low-level visual features to search and retrieve information in the multimedia databases has drawn much attention in the recent years [1,2]. Many of the existing techniques of image retrieval are based on image segmentation, which is a difficult task in many practical situations due to image noise and various compression artifacts. In this paper a novel approach to the problem of edge preserving smoothing, which allows to break an image into a set of homogeneous regions, is proposed and evaluated. The new algorithm is based on the combined forward and backward anisotropic diffusion with incorporated time dependent cooling process. This method is able to efficiently remove image noise while preserving and enhancing image edges. The proposed algorithm can be used as a first step of different techniques, which are based on color, shape and spatial location information, to search and retrieve information from multimedia databases.

1 Anisotropic Diffusion

Perona and Malik [3] formulate the anisotropic diffusion filter as a process that encourages intraregional smoothing, while inhibiting interregional denoising. The Perona-Malik (P-M) nonlinear diffusion equation is of the form [3-7] :

$$\frac{\partial}{\partial t} \mathbf{I}(x,y,t) = \nabla \left[c(x,y,t) \nabla \mathbf{I}(x,y,t) \right],\tag{1}$$

where $\mathbf{I}(x,y,t)$ denotes the color image pixel at position (x,y), t refers to time or iteration step in the discrete case and $c(x,y,t)$ is a monotonically decreasing conductivity function, which is dependent on the image gradient magnitude

* This work was partially supported by KBN grant 8 T11E 013 19 and NATO Collaborative Linkage Grant LST. CLG. 977845 7T11A01021

M. S. Lew, N. Sebe, and J. P. Eakins (Eds.): CIVR 2002, LNCS 2383, pp. 70–80, 2002.

$c(x, y, t) = f\left(||\nabla \mathbf{I}(x, y, t||\,)\right)$ such as:

$$c_1(x, y, t) = \exp\left\{-\left(\frac{||\nabla \mathbf{I}(x, y, t)||}{\beta}\right)^2\right\},$$

$$c_2(x, y, t) = \left\{1 + \left(\frac{||\nabla \mathbf{I}(x, y, t)||}{\beta}\right)^2\right\}^{-1} \tag{2}$$

which were introduced in the original paper of Perona and Malik [3]. The parameter β is a threshold parameter, which influences the anisotropic smoothing process. Using the notation $g = ||\nabla \mathbf{I}(x, y, t)||$, $s = g/\beta$, where $||\cdot||$ denotes the vector norm, we obtain following formulas for the conductivity functions: $c_1(x, y, t) = \exp\left(-s^2\right)$, $c_2(x, y, t) = 1/(1 + s^2)$. It is evident that the behavior of the anisotropic diffusion filter depends on the gradient threshold parameter β. To show the influence of β it is helpful to define a flux function: $\Phi(x, y, t) = c(x, y, t)||\nabla \mathbf{I}(x, y, t)||$. With the flux function defined above, Eq. 1 can be rewritten as $\frac{\partial}{\partial t}\mathbf{I}(x, y, t) = \nabla \Phi(x, y, t)$. The flux functions Φ_1 and Φ_2 corresponding to conduction coefficients c_1 and c_2 are shown in Fig. 1.

As it is easy to notice in Fig. 1b, the flow increases with the gradient strength to reach a maximum and then decreases slowly to zero. This behavior implies that the diffusion process maintains homogenous regions since little smoothing flow is generated for low image gradients. In the same way, edges are preserved because the flow is small in regions where the image gradient is high.

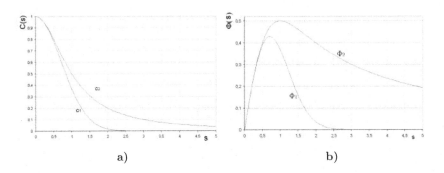

a)   b)

Fig. 1. Dependence of the conductivity functions c_1 and c_2 (**a**), and the respective flux functions Φ_1 and Φ_2 (**b**), on the value of the normalized image gradient s

2 Forward-and-Backward Diffusion

The conductance coefficients in the P-M process are chosen to be a decreasing function of the signal gradient. This operation selectively smoothes regions

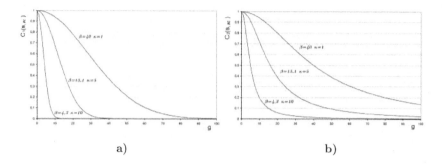

a)   b)

Fig. 2. Dependence of the conductivity functions on the iteration step and the image gradient g for the c_1 and c_2 conductivity functions, (forward diffusion, $\beta_1 = 40$, $\gamma = 0.8$)

that do not contain large gradients. In the Forward-and-Backward diffusion (FAB), a different approach is taken. Its goal is to emphasize the extrema, if they indeed represent singularities and do not come as a result of noise. As we want to emphasize large gradients, we would like to move "mass" from the lower part of a "slope" upwards. This process can be viewed as moving back in time along the scale space, or reversing the diffusion process. Mathematically, we can change the sign of the conductance coefficient to negative: $\frac{\partial}{\partial t} \mathbf{I}(x, y, t) = \nabla \left[-c(x, y, t) \nabla \mathbf{I}(x, y, t) \right]$, $c(x, y, t) > 0$. However, we cannot simply use an inverse linear diffusion process, because it is highly unstable. Three major problems associated with the linear backward diffusion process must be addressed: explosive instability, noise amplification and oscillations.

One way to avoid instability explosion is to diminish the value of the inverse diffusion coefficient at high gradients. In this way, when the singularity exceeds a certain gradient threshold it does not continue to affect the process any longer. The diffusion process can be also terminated after a limited number of iterations. In order not to amplify noise, which after some pre-smoothing, can be regarded as having mainly medium to low gradients, the inverse diffusion force at low gradients should also be eliminated. The oscillations should be suppressed the moment they are introduced. For this, a forward diffusion force that smoothes low gradient regions can be introduced to the diffusion scheme.

The result of this analysis is that two forces of diffusion working simultaneously on the signal are needed - one backward force (at medium gradients, where singularities are expected), and the other, forward one, used for stabilizing oscillations and reducing noise. These two forces can actually be combined to one coupled forward-and-backward diffusion force with a conductance coefficient possessing both positive and negative values. In [8-10] a conductivity function that controls the FAB diffusion process has been proposed

$$c_{FAB}(g) = \begin{cases} 1 - (g/k_f)^n & , \quad 0 \leq g \leq k_f \\ \alpha \left[((g - k_b)/w)^{2m} - 1 \right], & k_b - w \leq g \leq k_b + w \\ 0 & , \quad \text{otherwise} \end{cases} \quad (3)$$

where g is an edge indicator (gradient magnitude or the value of the gradient convolved with the Gaussian smoothing operator), k_f, k_b, w are design parameters and $\alpha = k_f/(2k_b)$, $(k_f \leq k_b)$ controls the ratio between the forward and backward diffusion. The dependence of such a defined conductance coefficient on the value of the gradient indicator is shown in Fig. 4b.

In the P-M equation, an "edge threshold" β is the sole parameter, the FAB process described in [8-10] is modeled by a parameter which regulates forward force k_f, two parameters for the backward force (defined by k_b and width w), and the relation between the strength of the backward and forward forces α. Essentially k_f is the limit of gradients to be smoothed out and is similar in nature to the role of β parameter of the P-M diffusion equation, whereas the k_b and w define the backward diffusion range.

In this study we propose two more natural conduction coefficients directly based on the P-M approach:

$$c_{1_{FAB}}(s) = 2\exp\left(-s_1^2\right) - \exp\left(-s_2^2\right), \quad c_{2_{FAB}}(s) = \frac{2}{1 + s_1^2} - \frac{1}{1 + s_2^2}. \quad (4)$$

The plots of the $c_{1_{FAB}}$ and $c_{2_{FAB}}$ diffusion coefficients are shown in Figs. 3, 4a. In the diffusion process smoothing is performed when the conductivity function is positive and sharpening takes place for negative conduction coefficient values.

3   Cooling Down of the Diffusion Process

Various modifications of the original diffusion scheme were attempted in order to overcome stability problems. Yet, most schemes still converge to a trivial solution (the average value of the image gray values) and therefore require the implementation of an appropriate stopping mechanism in practical image processing. In case of images contaminated by Gaussian noise, a common way of denoising is the usage of nonlinear cooling, which depends on the gradient, where large gradients cool faster and are preserved. In this study four simple time-dependent conduction coefficients were used:

$$c_1(g, t) = \exp\left[-\frac{g}{\beta(t)}\right]^2, \quad c_2(g, t) = \frac{1}{1 + \left(\frac{g}{\beta(t)}\right)^2}, \quad (5)$$

$$c_{1_{FAB}}(s, t) = 2\exp\left[-\frac{g}{\beta_1(t)}\right]^2 - \exp\left[-\frac{g}{\beta_2(t)}\right]^2, \quad (6)$$

$$c_{2_{FAB}}(s,t) = \frac{2}{1 + \left(\frac{g}{\beta_1(t)}\right)^2} - \frac{1}{1 + \left(\frac{g}{\beta_2(t)}\right)^2}. \tag{7}$$

where $g = ||\nabla \mathbf{I}(x, y, t)||$ is the L_1 or L_2 norm of the color image vector in the RGB space, $\beta_i(t+1) = \beta_i(t) \cdot \gamma, \gamma \in (0, 1]$, $\beta_i(1)$ is the starting parameter, $i = 1, 2$, $\beta_1(t) < \beta_2(t)$.

The scheme depends only on two (in case of forward or backward diffusion) or three (in case of FAB diffusion) parameters: initial values of starting β_i parameters and the cooling rate γ. Setting γ to 1 means, that there is no cooling in the system. As γ decreases, the cooling is faster, less noise is being filtered but edges are better preserved. Figures 3a and 3b illustrate the dependence of diffusion coefficients $c_1(s, t)$ and $c_2(s, t)$ on iteration step t. The behaviour of the diffusion coefficients $c_{1_{FAB}}(s, t)$ and $c_{1_{FAB}}(s, t)$ are compared in Fig. 4.

If the cooling coefficient γ is lower than 1, then the gradient threshold $\beta(t)$ decreases with time, allowing lower and lower gradients to take part in the smoothing process. As time advances, only smoother and smoother regions are being filtered, whereas large gradients can get enhanced due to local inverse diffusion. The scheme converges to a steady state for $\beta \to 0$, which means that no diffusion is taking place.

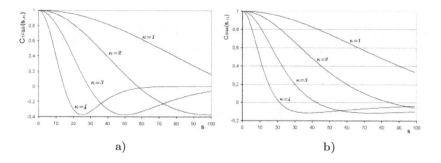

a)   b)

Fig. 3. Dependence of the conductivity functions on the iteration number κ and the normalized image gradient s for the c_{1FAB} and c_{2FAB} conductivity functions, (forward and backward diffusion), for $\beta_1(1) = 40$, $\beta_2(1) = 80$ and $\gamma = 0.5$. Note, that because of low $\gamma = 0.5$ already in the second iteration, the conductivity functions have negative values for large enough gradients

4   Experimentations and Results

In this paper a novel approach to the problem of edge preserving smoothing is proposed. The experiments revealed that better results of noise suppression

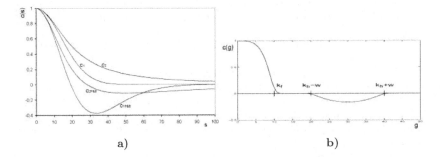

a)   b)

Fig. 4. a) Comparison of the shape of the proposed forward and backward diffusion conductivity functions, **b)** the forward and backward conductivity function proposed in [8-10]

using the FAB scheme were achieved using the conductivity function c_2 from the original P-M approach (7). This is due to the shape of the coupled forward and backward conductivity shown in Fig. 3b, which allows more efective image sharpening.

The efficiency of the proposed technique is presented in Fig. 5, where two images are enhanced using the purely backward and FAB anisotropic techniques. The ability of the new algorithm to filter out noise and sharpen the color images is shown in Fig. 6, where the original color images were contaminated with Gaussian noise ($\sigma = 30$) and restored with the FAB anisotropic diffusion scheme. Another examples of the new filter efficiency are shown in Figs. 8 and 9.

The results confirm excellent performance of the new method, which could be used for the enhancement of noisy images in different techniques which are based on color, shape and spatial location information to search and retrieve information from multimedia databases.

References

1. Special Issue on Content Based Image Retrieval, IEEE PAMI, Vol. 8, No. 8, August, 1996
2. Special Issue on content based image retrieval, IEEE Computer Vol. 28, No. 9, September, 1995
3. P. Perona, J. Malik, Scale space and edge detection using anisotropic diffusion, IEEE PAMI, Vol. 12, No. 7 pp. 629-639, July 1990
4. R. Whitaker, G. Gerig, Vector-Valued Diffusion, Ed. B. M. ter Haar Romeny, "Geometry-Driven Diffusion in Computer Vision", Kluwer Academic Press, pp. 93-134, 1994

5. M. J. Black, G. Sapiro, D. H. Marimont and D. Heeger, Robust anisotropic diffusion, IEEE Transactions on Image Processing, Vol. 7 (3), pp. 421-432, March 1998
6. J. Weickert, "Anisotropic Diffusion in Image Processing", Stuttgart-Germany: Teubner-Verlag, 1998
7. P. Blomgren, T. Chan, Color TV: Total variation methods for restoration of vector valued images, IEEE Transactions on Image Processing, March 1998, Special issue on Geometry Driven Diffusion and PDEs in Image Processing
8. G. Gilboa, Y. Zeevi, N. Sochen, Anisotropic selective inverse diffusion for signal enhancement in the presence of noise, Proc. IEEE ICASSP-2000, Istanbul, Turkey, vol. I, pp. 211-224, June 2000
9. G. Gilboa, Y. Zeevi, N. Sochen, Resolution enhancement by forward-and-backward nonlinear diffusion process, Nonlinear Signal and Image Processing, Baltimore, Maryland, June 2001
10. G. Gilboa, Y. Zeevi, N. Sochen, Signal and image enhancement by a generalized forward-and-backward adaptive diffusion process, EUSIPCO-2000, Tampere, Finland, September 2000

Fig. 5. Illustration of the new combined forward and backward anisotropic diffusion scheme. Top: color test images, below images enhanced with the pure backward diffusion and at the bottom images enhanced with the FAB diffusion scheme presented in the paper

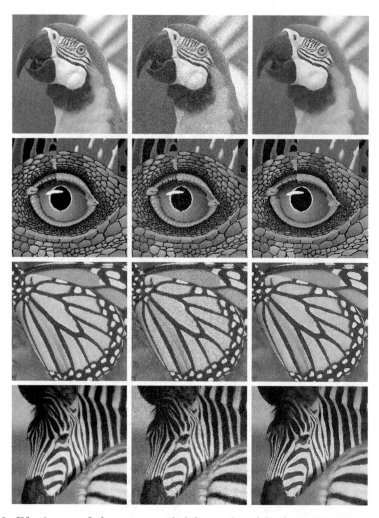

Fig. 6. Efectivness of the new coupled forward and backward anisotropic diffusion scheme. Lef column: color test images, in the center images contaminated with additive Gaussian noise ($\sigma = 30$ added independently on each RGB channel), to the right images enhanced with the new FAB anisotropic diffusion scheme (7)

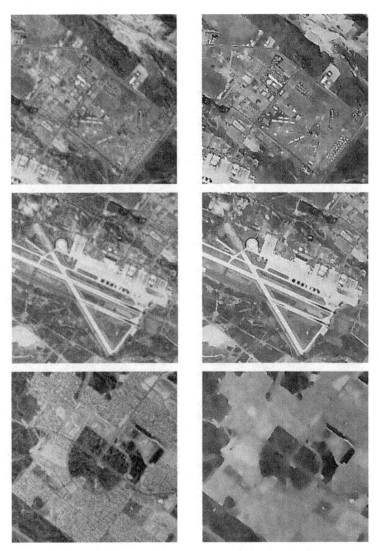

Fig. 7. Illustration of the efficiency of the new filtering scheme. Left column: multispectral satellite images, besides images processed with the new method

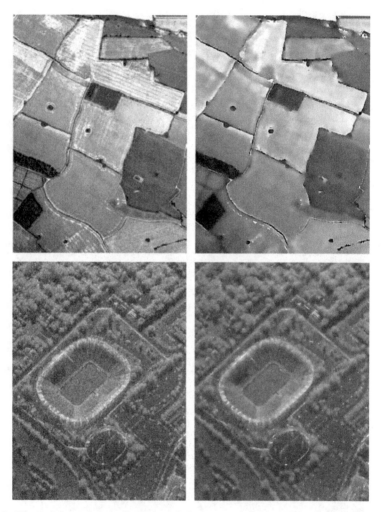

Fig. 8. Illustration of the efficiency of the new combined forward and backward anisotropic diffusion scheme. Left: color aerial image and below a test SAR image, to the right images enhanced with the new filtering scheme presented in the paper

Multiple Regions and Their Spatial Relationship-Based Image Retrieval

ByoungChul Ko and Hyeran Byun

Department of Computer Science, Yonsei University
Shinchon-Dong Seodaemun-Gu, Seoul, Korea
{soccer1,hrbyun}@cs.yonsei.ac.kr
http://vip.yonsei.ac.kr/Frip

Abstract. In this paper, we present a new multiple regions and their spatial relationship-based image retrieval method. In this method, a semantic object is integrated as a set of related regions based on their spatial relationships and visual features. In contrast to other ROI (Region-of-Interest) or multiple region-based algorithms, we use the Hausdorff Distance (HD) to estimate spatial relationships between regions. By our proposed HD, we can simplify matching process between complex spatial relationships and admit spatial variations of regions, such as translation, rotation, insertion, and deletion. Furthermore, to solve the weight adjust problem automatically and to reflect user's perceptual subjectivity to the system, we incorporate relevance feedback mechanism into our similarity measure process.

1   Introduction

Spatial relationships between regions often capture the most relevant and regular part of information in an image. However, determining similarity according to spatial relationships is generally complex and as difficult as semantic object-level image segmentation. Spatial relationships are usually vague, in the sense that they are not represented by a single crisp statement but rather by a set of contrasting conditions that are concurrently satisfied with different degrees [1]. Even though a few Region-based Image Retrieval (RBIR) systems [2,3,4] retrieve images based on region information, since humans are accustomed to utilizing object-level concepts (e.g., human) rather than region-level concepts (e.g., face, body and legs) [5], object-based content analysis is more reasonable work. Although much work has been done in decomposing images into semantic objects [2,3,5], meaningful and accurate object segmentation is still beyond current computer vision technique. Therefore, it is essential to integrate related regions into an object in order to provide object or multiple region-level query environments to the user.

There are a lot of works to develop similarity-matching algorithms that capture spatial and region information in an image as shown in Table 1.

We do not mention the detail description of related works in this paper due to the limitation of pages.

M. S. Lew, N. Sebe, and J. P. Eakins (Eds.): CIVR 2002, LNCS 2383, pp. 81–90, 2002.

Table 1. Major multiple regions and their spatial relationship-based image retrieval approaches

Authors	Year	Approach
Li, et al.	2000	-Integrate Region Matching (IRM) -Image segmentation -Spatial relationship between regions -Allows one region to be compared with several regions
Smith, et al.	1999	-Composite Region Template (CRT) -Image segmentation -Spatial relationship between regions -Decomposes image into regions and represents them as strings
Moghaddam, et al., Tian, et al.	1999 2000	-Region(s)-of-Interest (ROI) -Divides a image into nxn non-overlapping blocks -No image segmentation -Spatial relationship between ROI
Chang, et al.	1987	-2D string -Encoding a bi-directional arrangement of a set of objects into a sequential structure -Spatial relationship between symbols

In this paper, we propose a new multiple region-level image retrieval algorithm based on region-level image segmentation and their spatial relationships as a part of our FRIP [4] (Finding Region In the Pictures) system. We name this system Integrated FRIP (IFRIP). In this system, a semantic object is integrated as a set of related regions based on their spatial relationships and visual features, such as color, scale, shape, etc. Then, the user queries an object or multiple regions and according to our spatial matching algorithm, most similar images are returned. Furthermore, we incorporate relevance feedback in order to reflect the user's high-level query and subjectivity to the system and compensate for performance degradation due to imperfect image segmentation.

The rest of the paper is organized as follows. Section 2 provides a brief overview of original FRIP system. Our proposed similarity measurement for multiple regions and additional relevance feedback algorithms for performance improvement are discussed in Section 3. Section 4 then shows experimental results, exploring the performance of our technique using real-world data with brief conclusion.

2   A Brief Overview of FRIP System

This system includes our robust image segmentation and description scheme from each segmented region. For image segmentation, in this paper we update our previous image segmentation algorithm [4,14]. In our new image segmentation algorithm, the image segmentation process consists of two levels. At the first level, we segment an image using three types of adaptive circular filters based on the amount of image texture and CIE-L*a*b* color information the image carries. At the second (iterative) level, small patches of an image can be merged into adjacent similar regions by region merging and region labeling. From the segmented image, we need to extract features from each region. The contents of image should be extracted and stored as an index. Features (color, texture, normalized area (NArea), location and shape) of each region are used to describe the content of an image.

- Color: We extract average and standard deviation from each RGB colors space instead of color histogram in order to reduce the storage space.
- Texture: We choose the Biorthogonal Wavelet Frame (BWF) [19] as the texture feature. From the first-level high-pass filtered images, we calculate X-Y directional amplitude of each region.
- NArea: Feature NArea is defined the number of pixels of a region divided by the image size.
- Location: After region segmentation, the centroid is calculated from each region. Location of each region is defined by this centroid.
- Shape: For efficient shape matching, we use two properties. One is the eccentricity for global shape feature and the other is our Modified Radius-based Signature (MRS) for local shape feature.

After five features are extracted, they must be normalized over a common interval to place equal emphasis on every feature score since the single distance may be defined on widely varying intervals. Five feature vectors are normalized as soon as features are extracted from segmented regions using Gaussian normalization method.

3 Similarity Measurements for Multiple Regions and Their Spatial Relationship

In contrast to other algorithms related to multiple regions and their spatial relationships, our system uses Haudorff Distance in order to estimate spatial relationship between regions. By our updated Hausdorff Distance, we simplify similarity-matching process between complex spatial relationships and admit spatial variation of regions. Therefore, we can retrieve similar regions regardless of their spatial variation, and provide flexibility to the system if their relationships are changed. Furthermore, to solve the weight adjust problem automatically and to reflect user's perceptual subjectivity to the system, we incorporate relevance feedback mechanism into our similarity measure process.

3.1 Similarity Measure For Multiple Regions Using a Modified Hausdorff Distance (HD)

To measure spatial similarity between query and target regions, we apply the Hausdorff Distance (HD) measure to the FRIP system. The HD measure computes a distance value between two sets of edge points extracted from the object model and a test image. Because the HD measure is simple and insensitive to changes of image characteristics, it has been investigated in computer vision, object recognition and image analysis [12]. The classical HD measure between two point sets $Q = \{q_1,...,q_{Nq}\}$ and $T = \{t_1,...,t_{Nt}\}$ of finite sizes Nq and Nt, respectively, is defined as

$$H(Q,T) = \max(h(Q,T), h(T,Q)) \tag{1}$$

where $h(Q,T)$ and $h(T,Q)$ represent the directed distances between two sets Q and T. Here, $h(Q,T)$ is defined as $h(Q,T) = \max_{q \in Q} \min_{t \in T} \|q - t\|$ and $\|\|$ is some underlying norm on the points of Q and T. The function $h(Q,T)$ is called the directed Hausdorff distance from Q to T and $h(Q,T)$ in effect ranks each point of Q based on its distance to the nearest point of T. That is, each point of T is ranked by the distance to the nearest point of Q, and the largest ranked point determines the distance [15]. The Hausdorff distance $H(Q,T)$ is the maximum of $h(Q,T)$ and $h(T,Q)$. Thus, it measures the degree of mismatch between two sets by measuring the distance of the point Q that is farthest from any point of T and vice versa. However the original HD measure is sensitive to noise, Modified HD (MHD) [13] is proposed.

In our system, we regard two sets of points as regions. That is, a set $Q = \{qr_1,...,qr_{Nq}\}$ represents query regions in an image Q and a set $T = \{tr_1,...,tr_{Nt}\}$ represents target regions in an image T. Given query regions, find candidate regions from target image based on weighted sum of distances by Equation (2) with two different features, color ($d_{Q,T}^C$) and NArea ($d_{Q,T}^{NArea}$) distance. The $d_{Q,T}^C$ and $d_{Q,T}^{NArea}$ between corresponding query (qr) and target (tr) region are measured by Euclidean distance (see, reference [14]).

$$d(qr_i, tr_j) = w_1 \cdot d_{Q,T}^C + w_2 \cdot d_{Q,T}^{NArea} \tag{2}$$

Using Equation (2), one region in the query image has 1: N distances (for example, in Fig. 1-(a), $q1$ against $t1$, $t2$, and $t3$) and only one region that has the smallest distance less than a pre-defined threshold, is added to a candidate target set T_C (in Fig. 1 $T_C = \{t1, t2\}$). Therefore, even though some regions are inserted in target image, if their features are not almost the same with query regions, the number of candidate target set is the same or less than the number of query set. In addition, if one region is labeled as a candidate region, it is removed from target region set T to avoid selecting twice at one time. In Equation (2), we set w_1 and w_2 as 0.7 and 0.3 to emphasize on color property. After identifying the candidate target regions, we estimate the spatial distances between query region set Q and candidate region set T_C using our proposed Equation (3) instead of MHD. Since MHD still computes directed distance between two sets without considering their resemblance, we modify it like follows:

$$h(Q,T) = \frac{1}{Nq} \sum_{q \in Q} \min_{t \in T_C} (f \cdot d_t^p(q)), \; h(T,Q) = \frac{1}{Nt} \sum_{t \in T_C} \min_{q \in Q} (f \cdot d_q^p(t)) \tag{3}$$

In Equation (3), Nq and Nt are the total number of regions included in query region set Q and candidate region set T_C, respectively. $d_t^p(q)$ and $d_q^p(t)$ represents spatial distance between the centroid of query (C^{qr}) and candidate target region (C^{tr}) according to

$$d_t^p(q) = \sqrt{(C_x^{qr} - C_x^{tr})^2 + (C_y^{qr} - C_y^{tr})^2} \tag{4}$$

and f is a cost function for eliminating dissimilar regions and defined by

$$f = \begin{cases} 0.2 & d_{qr,tr}^C \leq \tau \\ 1 & d_{qr,tr}^C > \tau \end{cases} \tag{5}$$

Here $d_{qr,tr}^C$ represents the color distance between the corresponding query and target region, and τ is a threshold for color distance. Weights for cost function f are determined by experiments that give best results. Equation (5) means that the two regions bear little resemblance regardless of their spatial distance when $d_{qr,tr}^C > \tau$. For example, in Fig. 1 (a), even though region q2 of query image and region t1 of target image have a shorter spatial distance (d21) than the distance between q2 and t2 (d22), because color distance between q2 and t1 over the minimum threshold τ, their final spatial distance becomes longer (1.0×d21) than that (0.2×d22) of q2 and t2. In this case, the overall spatial distance of h (Q,T) is estimated by (0.2×d11+0.2×d22)/2.

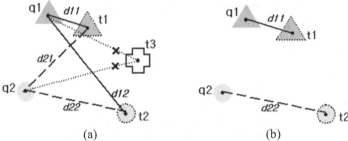

(a)   (b)

Fig. 1. An example of modified Hausdorff distance h (Q,T) in case of translation and insertion, (a) partial distances between the query region set Q and translated (t1, t2) and inserted (t3) region set T . t3 is not selected to the candidate target set T_C by Equation (2), (b) the final distance of h (Q,T) and h(T,Q) : (0.2×d11+0.2×d22)/2

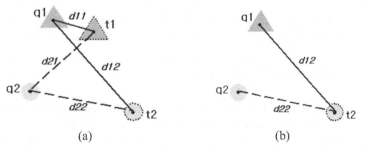

(a)   (b)

Fig. 2. An example of modified Hausdorff distance *h (Q,T)* in case of deletion (*t1*), (a) the same number of query and target regions (*Nq = Nt*) (b) the final distances of *h (Q,T)* is (1.0×d12+0.2×d22)/2, and h (T,Q) is (0.2×d22)/1

In case of deletion of regions, for example, if $t1$ is deleted in Fig. 2, $h(Q,T)$ is estimated by $(1.0×d12+0.2×d22)/2$. Therefore, $h(Q,T)$ is increased according as the deletion of regions. On the contrary, because $h(T,Q)$ has the smaller distance $((0.2×d22)/1)$ than $h(Q,T)$, the final distance $H(Q,T)$ is determined as not $h(T,Q)$ but $h(Q,T)$ by Equation (1). By using the proposed HD and cost function f, we can simplify the multiple region-level comparison problems and provide invariance to the system on translation and rotation as well as insertion and deletion of regions.

The overall distance between query region set (Q) and target region set (T) is estimated by weighted linear combination of color, texture, NArea, shape and $H(Q,T)$ distance using Equation (6). Here, each feature distance is only estimated among the best-matched regions between the set of query region and candidate target region.

In our system, we incorporate our Integrated FRIP with Modified Probabilistic Feature Relevance Feedback (MPFRL) [14] to interactively improve retrieval performance and allow progressive refinement of query results according to the user's specification. In MPFRL, we use previous relevance feedback as *history* to help to improve the performance and reduce user's time investment. Here, our history is consisted of two parts: One is the positive history and the other is the negative history. Here, *history* is generated by user's selection (action): at each iteration, if the user marks the k images as relevant or irrelevant, these images are recorded in the positive history H_P and negative history H_N respectively.

$$if \ \ y \in H_P\{h_i,.., h_n\}$$

$$D(x,y) = [\sum_{i=1}^{F} w_i \cdot \frac{d_i(f_i(x), f_i(y))}{Nr} + H(Q,T)] / \hat{K} \quad (6)$$

$$else \ \ if \ \ y \in H_N\{h_i,.., h_n\}$$

$$D(x,y) = [\sum_{i=1}^{F} w_i \cdot \frac{d_i(f_i(x), f_i(y))}{Nr} + H(Q,T)] \cdot \hat{K}$$

$$else \ \ D(x,y) = \sum_{i=1}^{F} w_i \cdot \frac{d_i(f_i(x), f_i(y))}{Nr} + H(Q,T)$$

In the above Equation, x is an image including the query region(s) and y is an image in the database. $H_P\{h_i,.., h_n\}$ represents the set of relevant images and $H_N\{h_i,.., h_n\}$ represents a set of irrelevant images. $d_i(f_i(x), f_i(y))$ is the sum of i-th feature distance on best-matched regions between query and candidate region set.

$$d_i(f_i(x), f_i(y)) = \sum_{k=1}^{Nq} \min_{t \in Tc} d_i(qr_k, tr_t) \quad (7)$$

Nr has the total number of the best-matched regions (=the number of regions included in target set). The overall distance (D) of feature vectors is estimated by the weighted sum of five normalized distances (d_i, i: 1~5- *color, texture, NArea, shape, location*), Nr and $\hat{K}$ ($\hat{K}>0$). $\hat{K}$ is a constant decision parameter for top k. In this paper, we set $\hat{K}$ to 10.

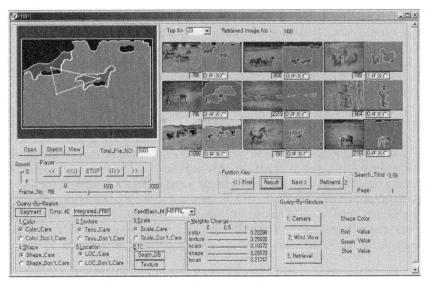

Fig. 3. An example of query processing (left: query regions and their spatial relations, right: retrieved images, O: relevant, X: irrelevant)

4 Experimental Results and Conclusion

We have performed experiment using a set of 3000 images from the Corel-photo CD, covering a wide variety of content ranging from natural images to graphic images without pre-selected categories. To do query, the user pushes segmentation button and selects one or some regions. After that, the system shows only selected regions and their spatial relationships. Then, the user pushes the 'Retrieval' button. From k images are retrieved, the user marks the images as relevant (o) or irrelevant (x) according to the user's subjectivity as shown in Fig. 3. By these feedback actions, the weights for feature distances are adjusted toward to the user's perception. In this experiment, we choose all user constraints (color-care/don't care, scale-care/don't care and shape-care/don't care) and top 20 nearest neighbors are returned that provide necessary relevance feedback. The test is performed on 11 different domains, which are selected from database images such as (1) tiger with green background, (2) eagle with blue sky, (3) sun with red sky, (4) two horses with green background, (5) airplane with blue sky, (6) car with roof, two wheels and body, (7) flower with green leaf, (8) human head with black hair, face and lip, (9) blue sky with a white cloud and sea, (10) traffic sign with a black arrow, (11) ship with sea. To validate the effectiveness of our approach, we compare the retrieval performance of our algorithm with 2-D string projection [10] and Flexible-Subblock algorithm [17] that is similar to ROI and histogram intersection method [18]. As shown in Fig. 4, our approach consistently outperforms the other algorithms at each query. Fig. 6 shows retrieval results of query number 9. As we can see from Fig. 3 and Fig. 6, our method permits variations of regions. If our method applies critical constraints to the system, many similar images are not retrieved in spite of their resemblance. We also carried out additional evaluation test between Integrated FRIP and FRIP using MPFRL. To test FRIP, we choose only one region from query regions such as, (1) tiger, (2) wing of eagle, (3) sun, (4) one horse, (5) airplane, (6) body of car, (7) flower, (8) face, (9) sky, (10) traffic sign, and (11) ship.

The average precision of FRIP is about 42% without relevance feedback. However, in case of Integrated FRIP, performance is superior as 50 %. Also, after 5 iterations, the average precision of FRIP is increased to 64%. However, as we can see from Fig. 5, the precision of Integrated FRIP is superior to original FRIP system as 76 % because it has additional information compared to FRIP except a query number 10. In case of query # 10, even though it has additional arrow information, many false positives are retrieved because other traffic signs included in our database have different color arrows or don't have an arrow. Regardless of these some errors, the overall retrieval performance of Integrated FRIP is superior to other methods. Furthermore, by the relevance feedback, we can allow progressive refinement of query results according to the user's specification and improve the retrieval performance.

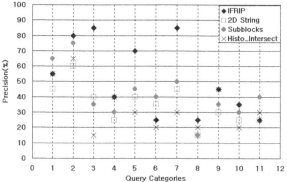

Fig. 4. The performance comparison of eleven queries by precision among IFRIP, 2D String, Subblock and Histogram intersection method (Precision)

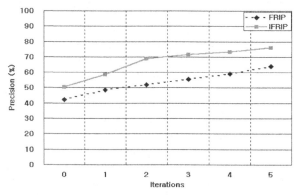

Fig. 5. Average performance of eleven queries between FRIP and Integrated FRIP (Precision)

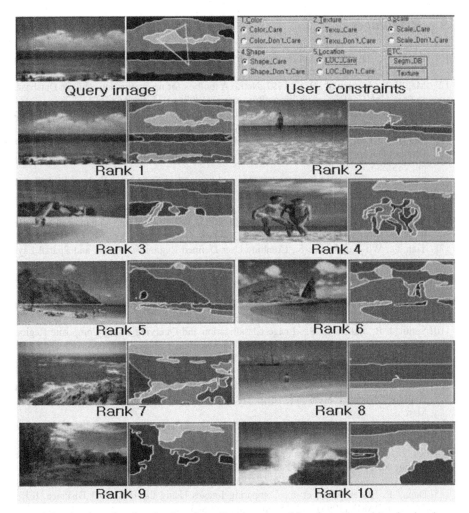

Fig. 6. Retrieval results after five iterations: Query region – blue sky with a white cloud and sea (left: original image, right: query regions and their spatial relationships). As shown in retrieval results, our method permits deletion or insertion of regions

Acknowledgements

This research was supported as a Brain Neuroinformatics Research Program sponsored by Korean Ministry of Science and Technology (M1-0107-00-0008).

5   References

[1] Bimbo, A. D., Visual Information Retrieval, Morgan Kaufmann Publishers, Inc. San Francisco, California, (2001).
[2] Ma, W. Y., Manjunath, B. S., 1997. Netra: A toolbox for navigating large image Database, In Proc., Int. Conference on Image Processing, (1997) vol. 1, pp. 568 -571.
[3] Carson, M. Thomas, S. et al., Blobworld: A system for region-based image indexing and retrieval. In Proc. Int. Conf. Visual Inf. Sys, (1999) pp. 509-516.
[4] Ko, B. C., Lee, H. S., Byun, H., Region-based Image Retrieval System Using Efficient Feature Description. In proceedings of the 15th ICPR., Sept., Spain, (2000) vol. 4, pp: 283-286.
[5] Xu. Y., Duygulu, P., Saber, E., Tekalp, A. M., and Yarman-Vural, F. T. Object Based Image Retrieval Based on Multi-Level Segmentation. IEEE Int. Conf. On Acoustics, Speech, and Signal Processing (ICASSP), (2000) vol. 4, pp. 2019-2022.
[6] Moghaddam, B., Biermann, H. and Margaritis, D., Defining Image Content with Multiple Regions-of-Interest. IEEE Workshop on CBAIVL, (1999) pp. 89-93.
[7] Tian, Q., Wu Y., Thomas S., Combine User Defined Region-of-Interest and Spatial Layout For Image Retrieval. In Proc. IEEE ICIP, (2000) vol. 3. pp. 746-749.
[8] Chang, S.-K., Shi, Q. Y. and Yan, C.Y., Iconic indexing by 2-D string. IEEE Trans. on PAMI.,(1987) 9(3), pp. 413-428.
[9] Bimbo, A. D. and Vicario, E., Using weighted spatial relationship in retrieval by visual content. IEEE Workshop on CBAIVL, (1998) June.
[10] Smith, J. R. and Li, C. S., Image Classification and Querying Using Composite Region Templates. Computer Vision and Image Understanding, (1999) vol. 75, pp. 165-174.
[11] Li, J., Wang, J. Z., Wiederhold G., IRM: Integrated Region Matching for Image Retrieval. ACM Multimedia, (2000) pp. 147-156.
[12] Sim, D. G., Kwon, O. K. and Park R. H., Object Matching Algorithms Using Robust Hausdorff Distance Measures. IEEE Trans. on Image Processing, (1999) vol. 8, No. 3, March.
[13] Dubuission, M. P. and Jain, A. K., A modified Hausdorff distance for object matching. In Proceeding of 12th ICPR, (1994) pp. 566-568, Oct.
[14] Ko, B. C, Peng, J. and Byun, H., Region-Based Image Retrieval Using Probabilistic Feature Relevance Feedback. Pattern Analysis and Application (PAA), (2001) vol. 4, 174-184.
[15] Daniel P. Huttenlocher, et al., Comparing Images Using the Hausdorff Distance. IEEE Trans. on PAMI, (1993) Vol. 15, no. 9, pp. 850-863.
[16] Smith, J. R. and Chang, S. F., Integrated Spatial and Feature Image Query. Multimedia systems journal, (1999) vol. 7, no.2, pp. 129-140, March.
[17] Ko, B.C, Lee, H. S and Byun H., Flexible subblock for Visual Information Retrieval, IEE Electronic letters, (2000) vol. 36, no1. pp. 24-25, January.
[18] Swain, M. J., Interactive indexing into image database. Proceeding of SPIE: Storage and Retrieval Image and Video Database, (1993) pp. 95-103, Feb.
[19] Sideney B., Ramesh A. G., Haitao G., Introduction Wavelets and Wavelet Transforms, A primer, Prentice-Hall, (1998).

Query by Fax for Content-Based Image Retrieval

Mohammad F. A. Fauzi and Paul H. Lewis

Intelligence, Agents and Multimedia Group
Department of Electronics and Computer Science
University of Southampton, UK
{mfaf00r,phl}@ecs.soton.ac.uk

Abstract. The problem of *query by fax* (QBF) and by other low quality images for art image retrieval is introduced. A reliable solution to the problem, the slow QBF algorithm, is presented but it is too slow for fast interactive retrieval situations. It is used as a yardstick for assessing the quality of a much faster technique, the fast QBF algorithm, based on the pyramid-structured wavelet transform (PWT). Both algorithms use the fact that the fax images are almost binary in nature. In the fast QBF method, the query (fax) image is used to identify what would be an appropriate threshold for the database image. This threshold value is then used to select an appropriate set of precomputed wavelet signatures to be used for matching the query and database image. Results of the algorithms are compared. The speed and accuracy of the new technique make it a valuable tool for the *query by fax* application.

1 Introduction

The motivation for this research comes from a requirement by some museums to respond to queries for pictorial information which are submitted to them in the form of fax images or other low quality monochrome images of works of art. The museums have databases of high resolution images of their artefact collections and the person submitting the query is asking typically whether the museum holds the art work shown or perhaps some similar work.

Typically the query image will have no associated metadata and was produced from a low resolution picture of the original art work. The resulting poor quality image, received by the museum, leads to very poor retrieval accuracy when the fax is used in standard *query by example* searches using, for example, colour or texture matching algorithms. Some examples of genuine fax images received by the museum together with the corresponding high resolution images they represent are shown in Fig. 3. Both good and poor quality faxes are shown for illustration, and will be used in our experiments. In this paper, we make the assumption that query images represent the full original and are not partial image queries.

To increase the speed of retrieval, most content-based retrieval systems use feature vectors as a medium for the similarity match. The feature vectors of all images in the database are computed and stored in advance in a unique feature

M. S. Lew, N. Sebe, and J. P. Eakins (Eds.): CIVR 2002, LNCS 2383, pp. 91–99, 2002.
© Springer-Verlag Berlin Heidelberg 2002

vector database. During the retrieval process, the feature vector of the query image is computed and is compared to all the feature vectors in the database. Images are retrieved based on the distance measure between query and database image feature vectors.

Recently, wavelets have been used in many areas in image analysis. They deliver a precise and unifying framework for the analysis and characterization of a signal at different scales, hence they provide an excellent tool for multiresolution analysis. In [1], Mallat proposed the pyramid-structured wavelet transform (PWT), a computationally very fast method used widely in image analysis and matching.

In this paper we show that poor retrieval performance in *query by fax* can be overcome by using a modified approach to wavelet-based retrieval. Since the quality of the fax image is so low that it differs substantially from its original, applying the PWT to the original image will not produce a feature vector close enough to the feature vector from the query. The fax image is first converted to a binary image before the PWT is applied. A similar conversion to binary is applied to each of the database images, choosing a threshold which makes them as close as possible to the binary fax image, before the PWT is applied.

At first, this method seems to be unsuitable for use in the feature vector database approach mentioned above, since it requires the feature vectors of the images in the database to be computed in advance. However we explain in section 3 how we can still use the same concept for this particular method, due to the compact nature of the wavelet signatures. Once we have created the binary images, the feature vectors are computed using the PWT, and the ones nearest to the query are selected.

The algorithm, which we call the fast QBF method, has the advantage of high retrieval accuracy while maintaining a low computational load. In section 4 we compare the performance of the algorithm with a slow QBF method that gives very high retrieval accuracy but which is computationally very intensive. The fast QBF algorithm almost matches the high accuracy of the slow QBF method but with a much lower computational load.

This paper is organized as follows. The next section will briefly review the wavelet transform algorithms to be used as part of our approach. Section 3 describes the stages of the fast QBF method and in section 4 the method is evaluated and compared with both a basic PWT algorithm and the slow but reliable QBF method. Finally, the conclusions are presented in section 5.

2   Review of Wavelet Transform

Wavelets were invented to overcome the shortcomings of the Fourier Transform. The fundamental idea is not only to analyse according to frequency but also according to scale, which is not covered by the Fourier transform. Most standard wavelets are based on the dilations and translations of one mother wavelet, ψ, which is a function with some special properties.

In image processing terms, the dilations and translations of the mother functions ψ are given by the wavelet basis function:

$$\psi_{(s,l)}(t) = 2^{-s/2}\psi(2^{-s}x - l).$$

The variables s and l are integers that dilate and orientate the mother function ψ to generate a family of wavelets, such as the Daubechies wavelet family [3,7]. The scale index s indicates the wavelet's width, and the location index l gives its position in 2-D.

Wavelets can be divided into orthogonal and non-orthogonal. They are orthogonal if their basis functions are mutually orthogonal. Because of the orthogonality, there exist a series of coefficients that can represent the wavelet decomposition for the whole family of a particular wavelet. The series of coefficients is named the Quadrature Mirror Filter (QMF). This concept, along with the theory of filter banks, is the basis for producing the famous fast algorithm PWT decomposition by Mallat [1].

Quadrature mirror filter is a filter that can be either low-pass or high-pass just by changing sign and rearranging its coefficient. In 2-D wavelet decomposition, the QMFs are used as both highpass and lowpass in both horizontal and vertical directions, followed by a 2 to 1 subsampling of each output image. This will generate four wavelet coefficient images, i.e. the *low-low*, *low-high*, *high-low* and *high-high* (*LL*, *LH*, *HL*, *HH* respectively) subbands. The process is repeated on the *LL* channel, and the 2-D PWT decomposition is constructed. Fig. 1(a) illustrates the PWT decomposition up to 3 levels, while Fig. 1(c) shows the result of PWT decomposition (also referred to as the wavelet coefficients) on the image in Fig. 1(b). For detailed information on wavelet theory, please refer to [2,4,5].

In this paper, the PWT wavelet coefficients are used to compute the feature vector of an image. The information obtained from this is of great importance as it shows the dominant frequency channels of a particular image and is particularly useful in similarity matching.

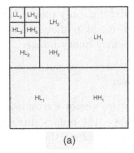

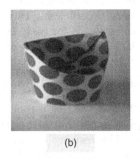

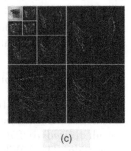

(a)   (b)   (c)

Fig. 1. (a)Wavelet decomposition using PWT. (b)An image. (c) Wavelet decomposition of image in (b)

3   Similarity Retrieval Using Fax Images

In this section we describe the various stages of the fast QBF algorithm. The binary image thresholding technique and the PWT feature vector computation is discussed, along with all the necessary requirements for performing an image retrieval task.

3.1   Binary Image Threshoding

As stated earlier, since the query images are almost binary, it is better to compute feature vectors in the binary domain. The query image can be converted to binary by thresholding in the middle of the grey scale range covered by the image. In order for the feature vector of a database image to be compared fairly with the feature vectors from the query, the database image must also be converted to binary. But the choice of threshold is not immediately obvious. For the original database image corresponding to the query, an appropriate threshold is probably one that produces a binary image with the same percentage of black (or white) pixels as the binary form of the query image. We could use this percentage for all the database images but it varies from query to query.

How do we match the percentage if we are to precompute the feature vectors from the binary versions of all the database images? Note that since the query image is the target and already effectively binary, it is the original image that must be made as close as possible to the binary query and not vice versa. One way to solve this problem is to convert each database image into a set of different binary images corresponding to different percentages of black pixels from 0 to 100%. If sufficient binaries are created, the binary query image will then be very similar to one of these binaries for the original image. We chose to create 99 binaries for each database image corresponding to percentages of black pixels from 1 to 99 in steps of 1%.

However, we do not need to store the binaries. Calculating the feature vectors for the database involves calculating the PWT for each of the binary images for each image in the database. This is implementable since the PWT is a fast algorithm and, more importantly, the feature vectors for each binary image have only a relatively small number of coefficients. During matching, only one of the sets of wavelet coefficients will be used for each database image, namely the set associated with the same black pixel percentage as the binary query image.

3.2   Feature Vector Computation and Comparison

As described earlier, the PWT is used as a tool for feature vector extraction. Since the PWT can be applied only on dyadic square image, the binary images are all resized to 256×256. The resizing can also be done before the binary conversion. The PWT algorithm is applied and the image is decomposed into 4 subimages or subbands (LL, LH, HL and HH). The LL band is decomposed further until the smallest subimages are of size 4×4, i.e. 6 levels of decomposition. This results in 19 different subimages or subbands.

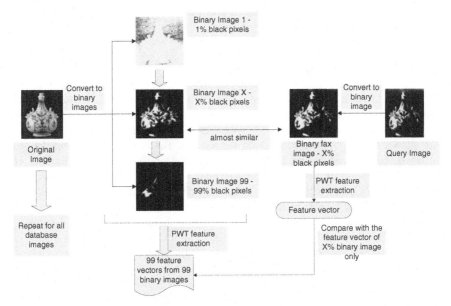

Fig. 2. Binary image matching between fax image and its original

Once the wavelet coefficients of a binary image are available, features are computed from each subband, resulting in 19 features for each binary image. The mean, μ is the energy measure used to compute the features. Let the image subband be $W_{mn}(x, y)$ and mn denotes the specific subband, m is the decomposition level and $n=1,2,3,4$ indicates the LL, LH, HL, HH bands respectively, then the μ_{mn} is calculated by:

$$\mu_{mn} = \frac{1}{N_{mn}^2} \int \int |W_{mn}(x, y)| \; dx \, dy.$$

where N is the length of a particular subband mn. The feature vector f for a particular binary image is therefore:

$$f = [\mu_{mn}], \; n \neq 1 \; except \; for \; the \; coarsest \; level, \; i.e. \; m=6,$$

$$= [\mu_{12}, \mu_{13}, \mu_{14}, \mu_{22}, ...\mu_{61}, \mu_{62}, \mu_{63}, \mu_{64}]$$

The feature vectors for the database images will have $99x19=1881$ coefficients, although only 19 will be used for comparison in each retrieval task. The distance classifier used is the Euclidean minimum distance. The distance between 2 features i and j is given by:

$$d(i, j) = \sqrt{\sum_m \sum_n \left[\mu_{mn}^{(i)} - \mu_{mn}^{(j)} \right]^2}$$

Once the distances are computed, the images will be retrieved in order of increasing distance from the query image.

4   Performance Evaluation

Experiments have been conducted using each of 20 genuine fax images as the query, and a database consisting of 1058 images of various types and sizes, including the original images of the 20 fax images. The fax images and their originals are shown in Fig. 3. The evaluation is based on the ability of the algorithm to retrieve the original image when the fax version of the original is used as the query. The results for the fast QBF algorithm in table 1 show the retrieval position of the original image among the 1058 database images, using *Daubechies* family wavelets with 8 vanishing moments as the wavelet bases.

The table also shows the results obtained by using a basic *query by example* retrieval with the same PWT features but calculated from the raw query and database images without the initial binarisation stage. It can be seen that the basic *query by example* algorithm is particularly poor for fax queries, but the retrieval results obtained using the fast QBF algorithm are very encouraging. All the original images are retrieved within the top 5. This is a good result considering the poor quality of fax images and the reasonably large image database used. The results suggest that the distance between the fax images and their originals are very close and should still produce good results for a larger image database. Different wavelet bases were also tested in this experiment, and it was found that the choice of wavelet base (*Haar* [7] and *Battle-Lemarie* [7] family) has little effect on retrieval result. However the *Daubechies* wavelet gives a slightly better result, probably because it is compactly supported in both the time and frequency domain [6].

To further compare the performance of our fast QBF algorithm, we implemented an alternative slow QBF method for *query by fax*. However, this algorithm is not appropriate for interactive retrievals since it does not use feature vectors as a medium for comparing images but uses pixel by pixel comparison resulting in a very high computational load. It is not possible to do any precomputing without using large amounts of storage for all database images so all calculations are performed during the retrieval process. Hence it is a very slow method. The method is used as a yardstick for comparison since, although computationally intensive, it achieves very high accuracy retrieval.

The slow QBF method is similar to the fast QBF algorithm except that, instead of using the PWT on the binary image, a pixel by pixel comparison is made between the binary fax and the binary of the database image with the appropriate percentage of black pixels. Both binaries are resampled to the same size (*64x64* pixels). The number of matching pixels, as a percentage of the total number of pixels, is used as the similarity measure and the database images are then retrieved in decreasing order of similarity.

Table 1 also shows the results for this slow QBF method. As expected, the slow QBF algorithm gives very good retrieval results. Almost all the originals

Fig. 3. The fax images used in this experiment and their originals

Table 1. Retrieval results using 20 fax images on database of 1058 images

Query Image No.	Rank of Original		
	Basic PWT Technique	Slow QBF Technique	Fast QBF Technique
1	104	1	1
2	369	1	1
3	15	1	1
4	21	1	3
5	272	1	1
6	130	1	1
7	258	1	1
8	2	1	3
9	502	1	1
10	302	20	2
11	603	1	1
12	299	1	1
13	60	1	1
14	495	1	4
15	500	1	2
16	339	1	1
17	15	1	2
18	264	1	4
19	1	1	1
20	1	1	1

are returned as the first match. Table 2 compares the average time taken for retrieving images from the database of 1058 images with the basic PWT algorithm and the slow and fast QBF algorithms. The times are for a 700 MHz Xeon processor. From table 1 and table 2 it can be seen that the fast QBF algorithm almost equals the slow QBF method in terms of retrieval performance, but involves a much smaller computational load. In other words, we can say that the fast QBF method integrates the high accuracy of the slow QBF method with the low computational load of the basic PWT method.

5   Conclusion

The main aim of this paper was to investigate the possibility of retrieving similar images from a large database using low quality monochrome images such as fax images as the query, with speed and high accuracy as main objectives. The wavelet transform, combined with the multiple binary thresholding and matching method provides an appropriate platform to tackle this problem, and the objectives have been achieved successfully. Also, the main purpose of the algorithm is to find either the exact parent image or similar images in the database. So in our application, the confusion level in the database is not a crucial factor.

Table 2. Comparison of speed between the three algorithms

Time taken to retrieve images (in seconds)	Basic PWT Technique	Slow QBF Technique	Fast QBF Technique
	1	130	1

None of the algorithms presented attempt to address partial (sub-image) queries and the issue of rotation and translation invariance. We are currently working on the possibility of partial query handling, as well as rotation and translation invariant versions to improve the system, both for *query by fax* and for general retrieval applications.

Acknowledgments

The first author is grateful to the Department of Electronics and Computer Science for financial support. We are also grateful to the EU for support under grant number IST_1999_11978 (The Artiste Project), and to one of our collaborators, the Victoria and Albert Museum, for use of their images.

References

1. Mallat, S.: A theory for multiresolution signal decomposition: The wavelet representation. In IEEE Transactions on Pattern Analysis and Machine Intelligence. **11(7)** (1989) 674-693   92, 93
2. Daubechies, I.: The wavelet transform, time-frequency localization and signal analysis. In IEEE Transactions on Information Theory **36(5)** (1990) 961-1005   93
3. Graps, A.: An introduction to wavelets. In IEEE Computational Science and Engineering **2(2)** (1995) 50-61   93
4. Ngunyen, T., Gunawan, D.: Wavelet and wavelet-design issues. In Proceedings of ICCS **1** (1994) 188-194   93
5. Mallat, S.: Wavelets for a vision. In Proceedings of IEEE **84(4)** (1996) 604-614   93
6. Nason, G.: A little introduction to wavelets. In IEE Colloquium on Applied Statistical Pattern Recognition (1999) 1/1-1/6   96
7. Wavelab Toolbox ver. 802 for Matlab. At http://www-stat.stanford.edu/ wavelab/ 93, 96

Spectrally Layered Color Indexing

Guoping Qiu[1,2] and Kin-Man Lam[2]

[1] School of Computer Science, The University of Nottingham
qiu@cs.nott.ac.uk
[2] Center for Multimedia Signal Processing, The Hong Kong Polytechnic University
enkmlam@polyu.edu.hk

Abstract. Image patches of different roughness are likely to have different perceptual significance. In this paper, we introduce a method, which separates an image into layers, each of which retains only pixels in areas with similar spectral distribution characteristics. Simple color indexing is used to index the layers individually. By indexing the layers separately, we are implicitly associating the indices with perceptual and physical meanings thus greatly enhancing the power of color indexing while keeping its simplicity and elegance. Experimental results are presented to demonstrate the effectiveness of the method.

1 Introduction

An effective, efficient, and suitable representation is the key starting point to building image processing and computer vision systems. In many ways, the success or failure of an algorithm depends greatly on an appropriately designed representation. In the computer vision community, it is a common practice to classify representation schemes as low-level, intermediate-level and high level. Low-level deals with pixel level features, high level deals with abstract concepts and intermediate level deals with something in between. Whilst low level vision is fairly well studied and we have a good understanding at this level, mid and high level concepts are very difficult to grasp, certainly extremely difficult to represent using computer bits. In the signal processing community, an image can be represented in the time/spatial domain and in the frequency/spectral domain. In contrast to many vision approaches, signal processing is more deeply rooted in mathematical analysis. Both time domain and frequency domain analysis technologies are very well developed, see for example many excellent textbooks in this area, e.g., [8]. A signal/image can be represented as time sequence or transform coefficients of various types, Fourier, Wavelet, Gabor, KLT, etc. These coefficients often provide a convenient way to interpret and exploit the physical properties of the original signal. Exploiting well-established signal analysis technology to represent and interpret vision concepts could be a fertile area for making progress.

Content-based image and video indexing and retrieval have been a popular research subject in many fields related to computer science for over a decade [1]. Of all the challenging issues associated with the indexing and retrieval tasks, "retrieval rele-

M. S. Lew, N. Sebe, and J. P. Eakins (Eds.): CIVR 2002, LNCS 2383, pp. 100-107, 2002.

vance" [7] is probably most difficult to achieve. The difficulties can be explained in a number of aspects. Firstly, relevance is a high level concept and is therefore difficult to describe numerically/using computer bits. Secondly, traditional indexing approaches mostly extract low-level features in a low-level fashion and it is therefore difficult to represent relevance using low-level features. Because low-level features can bear no correlation with high level concepts, the burden has to be on high-level retrieval strategies, which is again hard. One way to improve the situation is to develop numerical representations (low-level features) that not only have clear physical meanings but also can be related to high level perceptual concepts. Importantly as well, the representations have to be simple, easily to compute and efficient to store.

There is apparent evidence to suggest that the human vision system consists of frequency sensitive channels [6]. In other words, when we see the visual world, we perform some form of frequency analysis among many other complicated and not yet understood processing. Following the frequency analysis argument, it can be understood, that when a subject is presented an image in front of her, she will "decompose" the image into various frequency components and process each component with different processing channels (presumably in a parallel fashion). It is convenient to view such a process as decomposing the image into different layers, each layer consists of an image the same size as the original one, but only certain frequency components are retained in each layer, i.e., a band-pass filtered version of the original image. On each layer, only those grid positions where the pixels has certain "busyness" will have values while other grid positions will be empty. It is to be noted that the notion of layered representation was used in [9] as well, but [9] dealt with motion and high level concept and is not related to what we are proposing in this paper.

By decomposing an image into spectrally separated layers, we have applied the concept to the development of simple indexing features for content-based image indexing and retrieval (simplicity and effectiveness are important considerations in this paper). The organization in the rest of the paper is as follows. In section 2, we present the idea. Section 3 presents an algorithm. Section 4 presents experimental results and section 5 concludes our presentation.

2   The Idea

We are interested in developing efficient and effective image indexing features for content-based image indexing and retrieval. Ideally, the indexing features should be chosen in such a way that simple retrieval methods, such as computing simple metric distance measures in the feature space will produce good results. It is also well known that simple low-level features will bear little correlation with perceptual similarities if only simplistic distance measures are used. Our basic idea is to associate simple low-level features, such as color, with perceptual and physical meanings.

Color is an effective cue for indexing [3] which is well know. Because of its simplicity and effectiveness, it is an attractive feature. However, it is also well known that simple usage of color, i.e., color histogram, has significant drawbacks [1]. Researchers have tried various ideas of combining color with other features for indexing, see e.g., [1]. Our idea is to group colors according to their associated physical and perceptual properties.

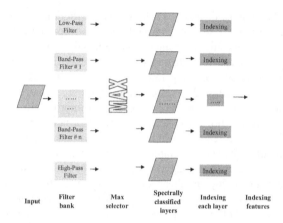

Fig. 1. Schematic of spectrally layered image indexing

The well-known opponent color theory [2] suggests that there are three visual pathways in the human color vision system and the spatial sharpness of a color image depends mainly on the sharpness of the light dark component of the images and very little on the structure of the opponent-color image components. In terms of perceptual significance of an image, the sharpness or roughness of an image region determines its perceptual importance. In other words, if two areas of an image contain the same color, then the difference/or similarity of the regions are separated/identified by their spatial busyness.

Digital signal processing researchers have developed a wealth of technologies to analyze physical phenomena such as sharpness/roughnessof a signal/image. The most effective way is frequency analysis, technologies ranging from FIR filter to filter banks are well studied [8]. A busy/sharp area is associated with higher frequency components, and a flat area has lower frequency distributions. A busy area may be associated with textured surfaces or object boundaries, a flat area may be associated with backgrounds or interior of an objects. Therefore a red color in a flat area may signify a red background or large red objects with flat surface, and a similar red color in a busy area may be indications of red colored textured surface or red object boundaries. Based on these observations and reasoning, we propose a spectrally layered approach to image indexing. A schematic is illustrated in Fig. 1.

Let x be the input image array, h_k be the impulse response of a band-pass filter (including the low-pass and high pass filters). Then the output of each band pass filter is

$$y_k(i,j) = x(i-l, j-m) * h_k(l,m) \qquad (1)$$

where * denotes convolution. For each pixel position, the MAX selector will identify the filter that produces the largest output, which is used to form the spectrally classified images. Let L_k be the kth layer image corresponding to the kth filter, then

$$L_k(i,j) = \begin{cases} x(i,j), & \text{if } y_k(i,j) = MAX\big(y_1(i,j), y_2(i,j), \cdots, y_n(i,j)\big) \\ Empty, & \text{Othewise} \end{cases} \qquad (2)$$

Indexing is then performed on each layer to obtain the indexing feature vector, I_f, for the image

$$I_f = \{I_f(L_1), I_f(L_2), \cdots, I_f(L_n)\} \qquad (3)$$

To summarize therefore, an image is first passed through a filter bank (each filter of the filter bank covers a specified spectral bandwidth). The output of the filter bank is used to classify the pixels. Pixels in an area with similar spatial frequencies are then retained on the same layer. Each layer, which contains only those pixels in areas with similar frequency distributions, is used to form its own index. The aggregation of the feature indices from all the layers then forms the overall index of the image. In this way, we have effectively classified the images according to the frequency contents of the image areas and indexed them separately. When such a strategy is used to match two images, the features from areas of similar spatial roughness are matched. That is, we will be matching flat area features in one image to the features in the flat areas of another image, and similarly, busy area features will be mapped to busy area features. When simple image features such as color are used for indexing this strategy should work very effectively. We introduce an implementation in the next section.

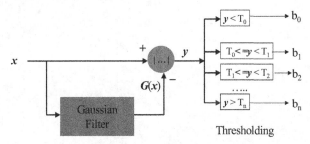

Fig. 2. A simple layer classification method

3   An Algorithm

One of the consideration factors in our current work is efficiency. Although it is possible to introduce various features to index each layer, we present a simple and yet effective algorithm for the implementation of the scheme of the last section. The task involves two aspects. The first is how to implement the filter bank scheme, the second is how to index each layer.

Filter bank is a well-studied area in image processing. However, we realize that the spectra classification can be implemented in a variety of ways. The essence is to classify pixels in areas with similar frequency characteristic into the same layer. We here present a simple non-filter bank based spectral classification method. This is illustrated in Fig. 2. Notice that only the Y component of the image is used in the classification process. This is because the sharpness of the image is mostly contained in this component. An image x is low-pass filtered first by a Gaussian kernel. This low-passed version is then subtracted from the original image. The difference is then rectified (i.e., taking the absolute values). Multiple thresholds are then applied. The binary images are obtained as (4) and the layers are formed as in (5)

$$b_k(i,j) = \begin{cases} 1, & \text{if } T_{k-1} <= |x(i,j) - G(x(i,j))| < T_k \\ 0, & \text{otherwise} \end{cases} \tag{4}$$

$$L_k(i,j) = \begin{cases} x(i,j), & \text{if } b_k(i,j)=1 \\ Empty, & \text{Othewise} \end{cases} \tag{5}$$

Clearly, **y** is a Laplacian image. The magnitude of the image indicates the sharpness of the image area. A busy area, which contains many sharp changes will result in large differences between the original image and the Gaussian smoothed image. A flat area will result in small difference between the original and the Gaussian image. Therefore the pixel magnitude in the Laplacian is an indication of the roughness of the image area surrounding the pixel. The scheme therefore effectively classifies pixels in areas with similar roughness into the same layer.

Fig. 3. Top left original image. Top right Laplacian. Bottom row, layers form with different thresholding. The black pixels are empty areas of the layer

Once we have classified pixels into different layers, we can use a simple color histogram [3] to index each layer. Let H_k be the histogram of the L_k, then the indexing feature of image x is expressed in (6) and we call (6) layered color indexing (*LCI*).

$$H(x) = \{H_k\}, k = 0,1,...n \tag{6}$$

An explanation of (6) is in order. Instead of building one histogram, we are building multiple color histograms for an image, each taking colors from pixels in areas with similar sharpness. It is therefore immediate that (6) will be more powerful than the original simple color histogram. Because we not only index the color, but also the colors are associated with their surface roughness. Fig. 3 shows an example image. It is seen here that the pixels of the Tiger's skin actually have a similar roughness. Therefore such representation will allow not only matching the entire contents of the image, but also allow different weightings being given to different layers depending on the user's requirements. In this particular example, if a user is more interested in retrieving tigers, then more weighting can be given to the features of the last layer.

Some reader may be concerned about the complexity of our method. It is in fact very simple. The filtering and classification (thresholding) can be done very fast. The dimension of the color histogram does not have to be high. We implemented our experiments with 4-layer and 64-color LCI (256-dimension histogram, which has similar complexity to other state of the art methods, e.g., color correlogram [3]) and we have observed very good performances. We report detailed experiments in the next section.

4   Experimental Results

We used a subset of 5000 images from the Corel color photo collection in our experiment. To build the database, we used 4 layers and a 64-color quantizer to build the layered color indexing. Each image was therefore represented by a 256-cell histogram. The Gaussian filter of Burt and Aldenson [5] was used (the coefficients were: 0.05, 0.25, 0.4, 0.25, 0.05). The threshold values were chosen empirically, however, we observed that the performance was not very sensitive to small variations of threshold values. We first compiled a rough statistics on the histograms of a few hundreds of Laplacian images. We found the values are mostly concentrated in $0 - 18$ interval. The thresholds used in the results below were $y < 6$, $6 <= y < 12$, $12 <= y < 18$, $y > 18$. To compare the similarity of two histograms, a L_1-norm relative distance metric [4] was used, which is defined as: Let $X = (x_1, x_2, ...x_n)$, $Y = (y_1, y_2, ...y_n)$, then the L_1-norm relative distance between X and Y is

$$D(X,Y) = \sum_i \frac{|x_i - y_i|}{1 + x_i + y_i} \tag{7}$$

We performed experiments to evaluate the retrieval precision and recall performance. As a comparison, we have also implemented color correlogram (CC) which has a $D = \{1, 3, 5, 7\}$. The same 64 colors were used in all experiments. Therefore the complexity of the two methods was the same and the colors used were exactly the same.

4.1   Retrieval Precision

In this experiment, we collected 328 query images, each had a unique answer. The results of our method and that of the color correlogram's are shown in table 1. Based on the same color and the same complexity, LCI clearly outperformed CC.

Table 1. Image retrieval precision results. The table should be interpreted as: For CC method, 112 queries (out of 328) found their unique answers in the first rank, 181 queries found their unique answers within the first 10 returns etc. The average rank of all 328 answers is 118. Similar interpretation applies to the LCI method

Methods	Ranks				
	1	<=10	<=30	<=50	Mean Rank
LCI	139	226	258	272	69
CC	112	181	216	233	118

4.2   Recall

In this experiment, we collected 99 query images, each had a set of hand labeled "correct" answers. The number of correct answers for different queries ranged from 3 to 30. Let Q_i be the ith query image and let $Q_i(1)$, $Q_i(2)$ ... $Q_i(N_i)$ be the N_i "correct" answers to the query Q_i. We define the following accumulated recall measures

$$A\operatorname{Re}call(l) = \sum_i \left(\frac{\left|\left\{Q_i(j) \mid rank(Q_i(j)) < l\right\}\right|}{N_i} \right) \qquad (8)$$

ARecall(l) is a weighted score of how many correct answers are returned in the first l positions, accumulated over all queries. If there are more correct answers, the weighting is lower (propotional to the inverse of the number of correct answers). It is therefore a measure of recall performance. A higher value of ARecall indicates a better performance. A similar measure was used in [4], but instead of measuring each individual query, we measure the accumulated performance for many queries, a fairer measure. Fig.4 shows the recall performance of LCI and CC. It is seen that LCI performed better. Some examples of retrieved images are shown in Fig. 5.

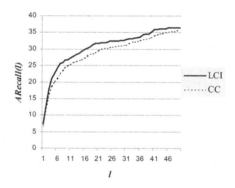

Fig. 4. Recall performance of the LCI and CC methods

5   Concluding Remarks

In this paper, we have introduced a new method for color indexing. It can be considered as an extension to the classic color indexing. The method significantly enhances the power of color indexing but at the same time retains its simplicity and elegance. The concept introduced in section 2 can be extended to other indexing features as well. We presented a simple implementation algorithm which has shown excellence performance and better than a state of the art technique with similar complexity.

(a)

(b)

Fig. 5. Examples of retrieved images. (a) results of CC method. (b) results of LCI method. The first image is the query

References

1. A. W. M. Smeulders et al, "Content-based image retrieval at the end of the early years", IEEE Trans PAMI, vol. 22, pp. 1349 - 1380, 2000
2. P. K. Kaiser and R. M. Boynton, Human Color Vision, Optical Society of America, Washington DC, 1996
3. M. Swain and D. Ballard, "Color Indexing", International Journal of Computer Vision, Vol. 7, pp. 11-32, 1991
4. J. Huang et al. "Spatial color indexing and applications", International Journal of Computer Vision, pp. 245 - 268, 1999
5. P. J. Burt and E. H. Adelson, "The Laplacian pyramid as a compact image code", IEEE Trans. Commun., vol. 31, pp. 532 – 540, 1983
6. Blakemore and F. W. Campbell, "On the existence of neurones in the human visual system selectively sensitive to the orientation and size of retinal images", Journal of Psychology, vol. 204, pp. 237 – 260, 1969
7. Y. Rui et al. "Relevance feedback: A power tool for interactive content-based image retrieval", IEEE Trans. CSVT, 1998, pp. 644 - 655
8. A. N. Akansu and R. A. Haddad, Multiresolution signal decomposition, Academic Press, 1992
9. J. Wang and E. H. Aldelson, "Representing moving images with layers" IEEE Trans on Image Processing, 1994, pp. 625 - 637

Using an Image Retrieval System
for Vision-Based Mobile Robot Localization

Jürgen Wolf[1], Wolfram Burgard[2], and Hans Burkhardt[2]

[1] Department of Computer Science, University of Hamburg
2527 Hamburg, Germany
jwolf@informatik.uni-hamburg.de
[2] Department of Computer Science, University of Freiburg
79110 Freiburg, Germany
{burgard,burkhardt}@informatik.uni-freiburg.de

Abstract. In this paper we present a vision-based approach to mobile robot localization, that integrates an image retrieval system with Monte-Carlo localization. The image retrieval process is based on features that are invariant with respect to image translations, rotations, and limited scale. Since it furthermore uses local features, the system is robust against distortion and occlusions which is especially important in populated environments. The sample-based Monte-Carlo localization technique allows our robot to efficiently integrate multiple measurements over time. Both techniques are combined by extracting for each image a set of possible view-points using a two-dimensional map of the environment. Our technique has been implemented and tested extensively using data obtained with a real robot. We present several experiments demonstrating the reliability and robustness of our approach.

1 Introduction

Localization is one of the fundamental problems of mobile robots. The knowledge about its position allows a mobile robot to efficiently fulfill different useful tasks like, for example, office delivery. In the past, a variety of approaches for mobile robot localization has been developed. They mainly differ in the techniques used to represent the belief of the robot about its current position and according to the type of sensor information that is used for localization. In this paper we consider the problem of vision-based mobile robot localization. Compared to proximity sensors, which are used by a variety of successful robot systems, cameras have several desirable properties. They are low-cost sensors that provide a huge amount of information and they are passive so that vision-based navigation systems do not suffer from the interferences often observed when using active sound- or light-based proximity sensors. Moreover, if robots are deployed in populated environments, it makes sense to base the perceptional skills used for localization on vision like humans do.

Over the past years, several vision-based localization systems have been developed. They mainly differ in the features they use to match images. For example, Basri and Rivlin [1] extract lines and edges from images and use this

M. S. Lew, N. Sebe, and J. P. Eakins (Eds.): CIVR 2002, LNCS 2383, pp. 108–119, 2002.
© Springer-Verlag Berlin Heidelberg 2002

information to assign a geometric model to every reference image. Then they determine a rough estimate of the robots position by applying geometric transformations to fit the data extracted from the most recent image to the models assigned to the reference images. Dudek, Zhang, and Sim [5,18] use a neural network to learn the position of the robot given a reference image. One advantage of this approach lies in the interpolation between the different positions from which the reference images were taken. Kortenkamp and Weymouth [10] extract vertical lines from camera images and combine this information with data obtained from ultrasound sensors to estimate the position of the robot. Paletta et al. as well as Winters et al. [14,22] consider trajectories in the Eigenspaces of features. A recent work presented by Se et al. [16] uses scale-invariant features to estimate the position of the robot within a small operational range. Furthermore, there are different approaches [11,12,20] that use techniques also applied for image-retrieval to identify the current position of the robot. Whereas all these approaches use sophisticated feature-matching techniques, they are not applying any filtering techniques to estimate the pose of the robot. The approach presented by Dellaert et al. [4] apply a probabilistic method for mobile robot pose estimation denoted as Monte-Carlo localization. Their system, however, requires an accurate ceiling mosaic of the robot's environment.

In this paper we present an approach that combines techniques from image retrieval with Monte-Carlo localization and thus leads to a robust vision-based mobile robot localization system. Our image retrieval system uses features that are invariant with respect to image translations, image rotations, and scale (up to a factor of two) in order to find the most similar matches. These features consist of histograms based on features of the local neighborhood of each pixel. This makes the localization system robust against occlusions and dynamics such as people walking by. To incorporate sequences of images and to deal with the motions of the robot our system applies Monte-Carlo localization which uses a sample-based representation of the robot's belief about its position. During the filtering process the weights of the samples are computed based on the similarity values generated by the retrieval system and according to the visibility area computed for each reference image using a given map of the environment. The advantage of our approach is that the system is able to globally estimate the position of the robot and to recover from possible localization failures. Our system has been implemented and tested on a real robot system in a dynamic office environment. In different experiments it has been shown to be able to globally estimate the position of the robot and to accurately keep track of it.

This paper is organized as follows. In the following section we present the techniques of the image-retrieval system used to compare the images grabbed with the robot's cameras with the reference images stored in the database. In Section 3 we briefly describe Monte-Carlo localization that is used by our system to represent the belief of the robot. In Section 4 we describe how we integrate the image retrieval system with the Monte-Carlo localization system. Finally, in Section 5 we present various experiments illustrating the reliability and robustness of the overall approach.

2   Image Retrieval Based on Invariant Features

In order to use an image database for mobile robot localization, one has to consider that the probability that the position of the robot at a certain point in time exactly matches the position of an image in the database is virtually zero. Accordingly, one cannot expect to find an image that exactly matches the search pattern. In our case, we therefore are interested in obtaining similar images together with a measure of similarity between retrieved images and the search pattern.

Our image retrieval system simultaneously fulfills both requirements. The key idea of this approach, which is described in more detail in [17,21], is to compute features that are invariant with respect to image rotations, translations, and limited scale (up to a factor of two). To compare a search pattern with the images in the database it uses a histogram of local features. Accordingly, if there are local variations, only the features of some points of the image are disturbed, so that there is only a small change in the histogram shape. An alternative approach might be to use color histograms. However, this approach suffers from the fact that all structural information of the image is lost, as each pixel is considered without paying attention to its neighborhood. Our database, in contrast, exploits the local neighborhood of each pixel and therefore provides better search results [17,21].

In the remainder of this section we give a short description of the retrieval process for the case of grey-value images. To apply this approach to color images, one simply considers the different channels independently. Let $\mathbf{M} = \{\mathbf{M}(x_0, x_1), 0 \leq x_0 < N_0, 0 \leq x_1 < N_1\}$ be a grey-value image, with $\mathbf{M}(i, j)$ representing the grey-value at the pixel-coordinate (i, j). Furthermore let G be a transformation group with elements $g \in G$ acting on the images. For an image $\mathbf{M}$ and an element $g \in G$ the transformed image is denoted by $g\mathbf{M}$. Throughout this paper we consider the group of Euclidean motions:

$$(g\mathbf{M})(i, j) = \mathbf{M}(k, l) \quad \text{with} \quad \begin{pmatrix} k \\ l \end{pmatrix} = \begin{pmatrix} \cos\varphi & -\sin\varphi \\ \sin\varphi & \cos\varphi \end{pmatrix} \begin{pmatrix} i \\ j \end{pmatrix} - \begin{pmatrix} t_0 \\ t_1 \end{pmatrix}, \quad (1)$$

where all indices are understood modulo N_0 resp. N_1.

In the context of mobile robot localization we are especially interested in features $F(\mathbf{M})$ that are invariant under image transformations, i.e., $F(g\mathbf{M}) = F(\mathbf{M}) \forall g \in G$. For a given grey-value image $\mathbf{M}$ and a complex valued function $f(\mathbf{M})$ we can construct such a feature by integrating over the transformation group G [15]. In particular, the features are constructed by generating a weighted histogram from a matrix $\mathbf{T}$ which is of the same size as $\mathbf{M}$ and is computed according to

$$(\mathbf{T}[f](\mathbf{M}))(x_0, x_1) = \frac{1}{P} \sum_{p=0}^{P-1} f\left(g(t_0 = x_0, t_1 = x_1, \varphi = p\frac{2\pi}{P})\mathbf{M}\right). \quad (2)$$

Since we want to exploit the local neighborhood of each pixel, we are interested in functions f that have a local support, i.e., that only use image values

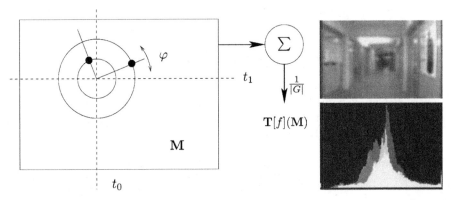

Fig. 1. Calculation of $\mathbf{T}[f](\mathbf{M})$ for $f = M(0,3)\cdot M(4,0))$, feature matrix (upper right image) and histogram (lower right image)

from a local neighborhood. Our system uses a set of different functions $\mathcal{F}$ with $f(\mathbf{M}) = \mathbf{M}(0,0)\mathbf{M}(0,1)$ as one member.

The kernel function defines the impact of surrounding pixels on the local feature of each coordinate. Obviously, for $f = M(0,0)$ the matrix of local features is given by $\mathbf{T}[f](\mathbf{M}) = \mathbf{M}$ and hence the global feature $F(\mathbf{M})$ simply is a grey-level histogram.

Figure 1 illustrates the calculation of $\mathbf{T}[f](\mathbf{M})$ for the kernel function $f = M(0,3)\cdot M(4,0)$. This function considers for each pixel (t_0, t_1) all neighboring pixels with distance 3 and 4 and with a phase shift of $\pi/2$ in polar representation. The corresponding grey-levels are multiplied and $(\mathbf{T}[f](\mathbf{M}))(t_0, t_1)$ is the average over all angles φ.

For each monomial $f \in \mathcal{F}$, we generate a weighted histogram over $\mathbf{T}[f](\mathbf{M})$. These histograms are invariant with respect to image translations and rotations and robust against distortion and overlapping and therefore well-suited for mobile robot localization based on images stored in a database. The upper resp. lower right image of Figure 1 shows the feature matrix resp. the resulting histogram for $f = M(0,3)\cdot M(4,0)$ extracted from the upper left image of Figure 2. In this figure each color channel has been calculated separately so that the chart contains an overlay of three histograms.

The global feature $F(\mathbf{M})$ of an image $\mathbf{M}$ consists of a multi-dimensional histogram constructed out of all histograms computed for the individual features $\mathbf{T}[f](\mathbf{M})$ for all functions in $\mathcal{F}$. Please note that the invariant features may be extracted with sub-linear complexity (based on a Monte-Carlo integration over the Euclidean motion) without the need of any feature extraction or segmentation.

The similarity between the global feature $F(\mathbf{Q})$ of a query image $\mathbf{Q}$ and the global feature $F(\mathbf{D})$ of a database image $\mathbf{D}$ is then computed using the intersection-operator normalized by the sum over all m histogram bins of $F(\mathbf{Q})$:

$$\bigcap_{\mathbf{norm}} (F(\mathbf{Q}), F(\mathbf{D})) = \frac{\sum\limits_{k \in \{0,1,\ldots,m-1\}} \min(F(\mathbf{Q})_k, F(\mathbf{D})_k)}{\sum\limits_{k \in \{0,1,\ldots,m-1\}} F(\mathbf{Q})_k}, \qquad (3)$$

where $F(\mathbf{M})_k$ denotes the value of the k-th (linear ordered) bin of the multi-dimensional histogram. Compared to other operators, the normalized intersection has the major advantage that it also allows to match partial views of a scene with an image covering a larger fraction. Figure 2 shows an example of a database query (upper left image) and the corresponding answer.

Fig. 2. Query image (upper left image) and the seven images with the highest similarity to it. The similarities from left to write, top-down are 81.67%, 80.18%, 77.49%, 77.44%, 77.43%, 77.19%, and 77.13%

3   Monte-Carlo Localization

To estimate the pose $l \in L$ of the robot in its environment, we apply a Bayesian filtering technique also denoted as *Markov localization* [2] which has successfully been applied in a variety of successful robot systems. The key idea of Markov localization is to maintain the probability density of the robot's own location $p(l)$. It uses a combination of the recursive Bayesian update formula to integrate measurements o and of the well-known formula coming from the domain of Markov chains to update the belief $p(l)$ whenever the robot performs a movement action a:

$$p(l \mid o, a) = \alpha \cdot p(o \mid l) \cdot \sum p(l \mid a, l') \cdot p(l'). \qquad (4)$$

Here α is a normalization constant ensuring that the $p(l \mid o, a)$ sum up to one over all l. The term $p(l \mid a, l')$ describes the probability that the robot is at position l given it executed the movement a at position l'. Furthermore, the quantity $p(o \mid l)$ denotes the probability of making observation o given the robot's current location is l. It highly depends on the information the robot possesses about the environment and the sensors used. Different kinds of realizations can be found in [13,8,19,2,9]. In this paper, $p(o \mid l)$ is computed using the image retrieval system described in Section 2.

To represent the belief of the robot about its current position we apply a variant of Markov localization denoted as Monte-Carlo localization [4,6]. In Monte-Carlo localization, which is a variant of the well-known Condensation algorithm [7], the update of the belief generally is realized by the following two alternating steps:

1. In the **prediction step**, we draw for each sample a new sample according to the weight of the sample and according to the model $p(l \mid a, l')$ of the robot's dynamics given the action a executed since the previous update.
2. In the **correction step**, the new observation o is integrated into the sample set. This is done by bootstrap resampling, where each sample is weighted according to the likelihood $p(o \mid l)$ of making observation o given sample l is the current state of the system.

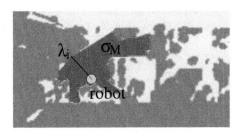

Fig. 3. Visibility area σ_M extracted for a reference image. The circle corresponds to the position of the robot when the image was grabbed in the environment depicted in Figure 4 (lower left portion). The position of the closest obstacle in the direction of the optical axis is indicated by λ_i

4   Using Retrieval Results for Robot Localization

The image retrieval system described above yields images that are most similar to a given sample. In order to integrate this system with a Monte-Carlo localization approach, we need a technique to weight the samples according to the results of the image retrieval process. The key idea of our approach is to extract a visibility region σ_M for each image M in the image database. In our current system, the visibility area of an image M corresponds to all positions in a given metric map of the environment from which the closest object in M in the direction of the optical axis is visible.

We represent each σ_M by a discrete grid of poses and proceed in two steps: First we apply ray-tracing to compute the position λ_i of the closest object on the optical axis according to the position of the robot when this image was grabbed. Then we use a constrained region growing technique to compute the visibility area $\sigma_{\mathbf{M}}$ for $\mathbf{M}$. Throughout this process only those points are expanded, from which λ_i is visible. Figure 3 shows a typical example of the visibility area for one of the images stored in our database.

In Monte-Carlo localization one of the crucial aspects is the computation of the weight ω_i of each sample. In many systems this weight is chosen as the likelihood $p(o \mid l_i)$ [4,6] where l_i is the position represented by the sample and o is the measurement obtained by the robot. In the context of vision-based localization, however, $p(o \mid l_i)$ generally is hard to assess because of the high dimensionality of the image space. In our system, we use the similarity measure ξ_i of each image $\mathbf{M}_i$ to weight the samples in the corresponding visibility area $\sigma_{\mathbf{M}_i}$. Before we assign a similarity measure ξ_i to a sample, we need to check, whether the sample lies in the visibility area σ_i of image $\mathbf{M}_i$. At this point it is important to note, that each sample represents a possible pose of the robot, i.e., a three-dimensional state consisting of the $\langle x, y \rangle$-position and orientation ϕ. Thus, in order to appropriately weight the samples we also have to consider the orientation of that sample. For example, if the heading direction of pose represented by a sample is too far off, the image stored in the database cannot be visible for the robot.

In our system we compute the weight ω of a sample according to

$$\omega = \sum_{i=1}^{n} I(\langle x, y \rangle, \sigma_i) \cdot d(\psi) \cdot \xi_i, \tag{5}$$

where $\psi \in [-180; 180)$ is the deviation of the heading ϕ of the sample from the direction to λ_i. Furthermore, d is a function which computes a weight according to the angular distance ψ. Finally, $I(\langle x, y \rangle, \sigma_i)$ is an indicator function which is 1 if $\langle x, y \rangle$ lies in σ_i and 0, otherwise.

In our current implementation we use a step function so that only such areas are chosen, for which the angular distance $|\psi|$ does not exceed 5 degrees. Please note that Equation (5) rests on the assumption that the images in the database cover different aspects of the environment. For example, if the database contains two images taken from the same or a similar pose, then the weights of the samples lying in the intersection of both visibility areas would be weighted too high compared to other samples for which there is only one image. Although this independence assumption is not always justified, we did not observe any evidence in our experiments, that this made the robot overly confident in being at a certain position.

5 Experiments

The system described above has been implemented on our mobile robot Albert, an RWI B21 robot equipped with a stereo camera system, and tested intensively in real robot experiments. The image database used throughout the experiments contained 936 images. They were obtained by steering the robot

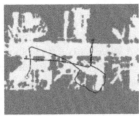

Fig. 4. Map of the office environment used to carry out the experiments and trajectory of the robot (ground truth) (left image). Trajectory of the robot according to the odometry data (center image). Positions of the robot estimated by our system during global localization (right image)

through the environment and grabbing sets of images from different positions in the environment. Our system is highly efficient since it only stores the histograms representing the global features. The overall space required for all 936 images therefore does not exceed 4MB. Furthermore, the retrieval process for one image usually takes less than .6 secs on an 800MHz Pentium III. Our current implementation (described in detail in [23]) updates the belief in each iteration in time $O(k^2 + n \cdot k)$, where k is the number of samples contained in the sample set and n is the number of reference images stored in the database.

Fig. 5. Typical images captured by Albert during the global localization experiment

5.1 Global Localization

The experiment described in this section is designed to demonstrate that our system allows the robot to reliably estimate the global position of a mobile robot within its environment and to reliably keep track of it afterwards. During the experiment the robot was moving with speeds up to 30cm/sec through our office environment. The left image of Figure 4 shows a part of the map of the environment and the trajectory of the robot during this experiment. Also shown in green/grey is an outline of the environment. The significant error in the odometry obtained from the robot's wheel encoders is shown by the center image of the same figure. In this experiment we used a sample set consisting of 5000 samples that were initialized using a uniform distribution. Some of the images perceived by the robot during this experiment are depicted in Figure 5. The right image of Figure 4 shows the trajectory estimated by our system. Obviously, the system is able to quickly determine the position of the robot and to reliably keep track of it afterwards despite of the dynamic aspects. Please note that since we use the sample mean to estimate the robot's pose, in the beginning the estimated position is always in the center of the map, which is not shown entirely in this figure. One side-effect of using the sample mean is that the trajectories estimated by our system during global localization generally contain a line going from the center from the map to the true position of the robot. This corresponds to the situation in which the system has discovered the true position of the robot and happens after the integration of the fourth image in this particular example.

5.2   Effect of the Image Retrieval System

The visibility areas extracted for the reference images (see Section 4) introduce constraints on the possible locations of the robot while it is moving through the environment. In principle, the visibility areas characterize the free space in the environment. Therefore, just by knowing the odometry information one can often infer the position of the system. Since a robot cannot move through obstacles, the trajectory that minimizes the tradeoff between the deviation from the odometry data and the number of times the robot moves through obstacles corresponds to the most likely path of the robot and thus indicates the most likely position of the vehicle. In fact, it has already been demonstrated that these constraints can be sufficient to globally localize a robot [3].

The goal of the experiment described in this section therefore is to demonstrate that the localization capabilities significantly depend on the exploitation of the image retrieval results. Again, we evaluated the global localization capabilites but without utilizing the results from image retrieval process. More specifically, all samples obtained the same weight as long as the were located within an arbitrary visibility region. Accordingly, the outcome resulted only from the odometry data and from the constraints introduced by the visibility areas. The left image of Figure 6 shows a typical plot of the localization error if only the constraints imposed by the visibility regions are used (left image). As can be seen from the figure, the system is unable to localize the robot solely based on this information. However, if the image retrieval results are used, the localization accuracy is quite high and the robot is quickly able to determine its absolute position in the environment (see right image of Figure 6).

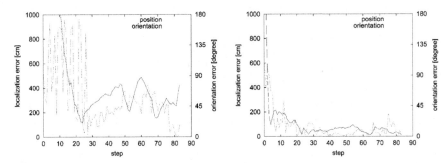

Fig. 6. Typical localization error during the *global localization* experiment affected by implied constraints (left picture) and additional using of retrieval results

6   Conclusions

In this paper we presented a new approach to vision-based localization of mobile robots. Our method uses an image retrieval system based on invariant features. These features are invariant with respect to translation, rotation, and scale (up to a factor of two) so that the system is able to retrieve similar images even if only a small part of the corresponding scene is seen in the current image. This approach is particularly useful in the context of mobile robots, since a robot often observes the same scene from different view-points. Furthermore, the system uses local features and therefore is robust to changes in the scene. To represent the belief of the robot about its pose, our system uses a probabilistic approach denoted as Monte-Carlo localization. The combination of both techniques yields a robust vision-based localization system with several desirable properties previous approaches are lacking. It is able to globally estimate the position of the robot and to reliably keep track of it.

Acknowledgments

This work has partly been supported by the EC under contract number IST-2000-29456.

References

1. R. Basri and E. Rivlin. Localization and homing using combinations of model views. *Artificial Intelligence*, 78(1-2), 1995. 108
2. W. Burgard, A. B. Cremers, D. Fox, D. Hähnel, G. Lakemeyer, D. Schulz, W. Steiner, and S. Thrun. Experiences with an interactive museum tour-guide robot. *Artificial Intelligence*, 114(1-2), 2000. 112, 113

3. W. Burgard, D. Fox, D. Hennig, and T. Schmidt. Estimating the absolute position of a mobile robot using position probability grids. In *Proc. of the National Conference on Artificial Intelligence (AAAI)*, 1996. 116

4. F. Dellaert, W. Burgard, D. Fox, and S. Thrun. Using the condensation algorithm for robust, vision-based mobile robot localization. *Proc. of the International Conference on Computer Vision and Pattern Recognition (CVPR)*, 1999. 109, 113, 114

5. G. Dudek and C. Zhang. Vision-based robot localization without explicit object models. In *Proc. of the International Conference on Robotics & Automation (ICRA)*, 1996. 109

6. D. Fox, W. Burgard, F. Dellaert, and S. Thrun. Monte Carlo localization: Efficient position estimation for mobile robots. In *Proc. of the National Conference on Artificial Intelligence (AAAI)*, 1999. 113, 114

7. M. Isard and A. Blake. Contour tracking by stochastic propagation of conditional density. In *Proc. of European Conference on Computer Vision*, pages 343–356, 1996. 113

8. L.P. Kaelbling, A. R. Cassandra, and J. A. Kurien. Acting under uncertainty: Discrete Bayesian models for mobile-robot navigation. In *Proc. of the International Conference on Intelligent Robots and Systems (IROS)*, 1996. 113

9. K. Konolige. Markov localization using correlation. In *Proc. of the International Joint Conference on Artificial Intelligence (IJCAI)*, 1999. 113

10. D. Kortenkamp and T. Weymouth. Topological mapping for mobile robots using a combination of sonar and vision sensing. In *Proc. of the National Conference on Artificial Intelligence (AAAI)*, 1994. 109

11. B. Kröse and R. Bunschoten. Probabilistic localization by appearance models and active vision. In *Proc. of the International Conference on Robotics & Automation (ICRA)*, 1999. 109

12. Yoshio Matsumoto, K. Ikeda, M. Inaba, and H. Inoue. Visual navigation using omnidirectional view sequence. In *Proc. of the International Conference on Intelligent Robots and Systems (IROS)*, 1999. 109

13. I. Nourbakhsh, R. Powers, and S. Birchfield. DERVISH an office-navigating robot. *AI Magazine*, 16(2), 1995. 113

14. L. Paletta, S. Frintrop, and J. Hertzberg. Robust localization using context in omnidirectional imaging. In *Proc. of the International Conference on Robotics & Automation (ICRA)*, 2001. 109

15. H. Schulz-Mirbach. Invariant features for gray scale images. In G. Sagerer, S. Posch, and F. Kummert, editors, *17. DAGM - Symposium "Mustererkennung"*. Springer, 1995. 110

16. S. Se, D. Lowe, and J. Little. Vision-based mobile robot localization and mapping using scale-invariant features. In *Proc. of the International Conference on Robotics & Automation (ICRA)*, 2001. 109

17. S. Siggelkow and H. Burkhardt. Image retrieval based on local invariant features. In *Proceeding of the IASTED International Conference on Signal and Image Processing*, 1998. 110

18. R. Sim and G. Dudek. Learning visual landmarks for pose estimation. In *Proc. of the International Conference on Robotics & Automation (ICRA)*, 1999. 109

19. R. Simmons, R. Goodwin, K. Haigh, S. Koenig, and J. O'Sullivan. A layered architecture for office delivery robots. In *Proc. of the First International Conference on Autonomous Agents*, Marina del Rey, CA, 1997. 113

20. I. Ulrich and I. Nourbakhsh. Appearance-based place recognition for topological localization. In *Proc. of the International Conference on Robotics & Automation (ICRA)*, 2000. 109

21. R. Veltkamp, H. Burkhardt, and H.-P. Kriegel, editors. *State-of-the-Art in Content-Based Image and Video Retrieval*. Kluwer Academic Publishers, 2001. 110

22. N. Winters, J. Gaspar, G. Lacey, and J. Santos-Victor. Omni-directional vision for robot navigation. In *Proc. IEEE Workshop on Omnidirectional Vision, South Carolina*, 2000. 109

23. J. Wolf. Bildbasierte Lokalisierung für mobile Roboter. Master's thesis, Department of Computer Science, University of Freiburg, Germany, 2001. In German. 115

JPEG Image Retrieval Based on Features
from DCT Domain

Guocan Feng[1,2] and Jianmin Jiang[1]

[1] Department of Electronic Imaging & Media Communications
University of Bradford, Bradford, BD7 1DP, UK
{g.c.feng,j.jiang1}@bradford.ac.uk
[2] Department of Mathematics, Zhongshan University
Guangzhou, China 510275

Abstract. To improve efficiency of the compressed image retrieval, the techniques of the direct feature extraction in compressed domain are extensively emphasized. An algorithm directly computing moments from DCT domain is presented. Employing the algorithm, An image retrieval system based on the JPEG image is developed. The system are robust for the translation, rotation and scale transform with minor disturbance. The theoretical analysis and experimental results also demonstrate the system has good performance whether in retrieval efficiency or effectiveness.

1   Introduction

At present, almost all visual information are stored in compressed formats, among which the format defined by Joint Picture Expert Group (JPEG) is one of the most typically formats which widely used on Internet or image database [2]. For the feature detection of the compressed image, the conventional approaches are to decode the image to the pixel domain first, then to perform the existed feature extraction techniques based on the pixel domain [6]. It is not only time consuming, but also increase the computation complexity. However, it becomes more and more important to improve the retrieval efficiency of the image retrieval/indexing in large compressed image database. Therefore, direct feature extraction algorithms are extensively emphasized [6][13]. As we know, the inverse DCT (IDCT) is the last step of the JPEG decoder [2], while DCT itself is a one of the best filters for the feature extraction. Thus, the direct feature extraction from DCT domain in JPEG image can disregard the any necessity of decomposing the image and then exploring its features in pixel domain. It will greatly alleviate the computational cost caused. Along with the idea, An algorithm directly computing moments from DCT domain is proposed in this paper. Employing the algorithm, a fast JPEG image retrieval system is developed and evaluated. The experimental results demonstrate the presented system has good performance in retrieval efficiency and efficiency.

M. S. Lew, N. Sebe, and J. P. Eakins (Eds.): CIVR 2002, LNCS 2383, pp. 120-128, 2002.
© Springer-Verlag Berlin Heidelberg 2002

2   Direct Computation of Statistical Parameters in DCT Domain

Some statistical parameters are commonly employed to describe the image characteristics. Among those statistical parameters, the moments or their functions are kind of
the most fundamental description of texture and shape[1][4][12]. Hu [1] first derived
results illustrating the algebraic invariant of 2-dimensional moments. Thereafter, it is
widely applied in feature detection and representation in pattern analysis because of its
invariant to translation, scale and orientation [5]. The lower order moments, such as
mean and variance, are one of the most essential and concise, which can provide an
indication of how uniform a region is. Therefore these parameters extensively used as
statistical descriptor to depict texture feature in region segmentation[14].

Since segmentation techniques based statistical characteristic usually are considered on a proper window. According to the gotten statistical parameters on the neighbor window, the further operation is to decide whether merge window or split the
window. Whereas the image or video compression standard based on DCT such as
JPEG, MPEG are usually to implement compression technique in block with size 8×8.
This conduces to more suitable for discussing the image segmentation in the block
based on statistical descriptors whatever region merging or region split if the statistical
parameters can directly obtained on the block. In order to achieve the goal, this section develops a direct computation of the two lower order moments: the mean and
variance, on DCT domains. And then, the parameters on merged or split windows will
be discussed.

2.1   Calculation of the Mean μ and Variance σ^2 in DCT Domain

Texture description techniques can be grouped into three large classes: statistical,
spectral and structural. Statistical descriptors are based on region's statistical characteristics, such as histogram, their extensions and their moments; they measure contract,
granularity and coarseness. The fundamental variables, mean μ and variance σ^2(or
called 1st moment and 2nd central moment), provide an indication of how uniform or
regular a region is. Thus they are regarded as the ones of the most useful statistical
parameters in the texture analysis of an image processing. In fact, in consideration of
merging regions, μ and variance σ^2 are employed to measure the whether growing
regions keep the homogenous[7][8].

Assume $X=x(i,j)$ $(i,j =0,1, N-1)$ are intensity values on N×N block of an image (The
8x8 block adopted in JPEG format). The kth moment m_k, and central moment μ_k of X
on this block is defined to be the expectation of X^k as blow.

$$m_k = E(X^k) = \frac{1}{N^2}\sum_{i=0}^{N-1}\sum_{j=0}^{N-1}x^k(i,j) \tag{1}$$

$$\mu_k = E[(X-E(X))^k] = \frac{1}{N^2}\sum_{i=0}^{N-1}\sum_{j=0}^{N-1}[x(i,j)-E(X)]^k \tag{2}$$

1st order moment m_1 and 2nd order central moment μ_2 are so-called mean (μ) and
variance (σ^2) on this block respectively. m_1 represents the average intensity on the

block, while μ_2 represents the difference of intensity among the pixel on this block. In fact, in terms of the DCT definition in NxN block, we can derive the following equation,

$$m_1 = \frac{1}{N}C(0,0) \tag{3}$$

$$\mu_2 = \frac{1}{N^2}\sum_{u,v=0(u\neq 0 \text{ and } v\neq 0)}^{N-1}C^2(u,v) \tag{4}$$

where $C(u,v)$ is the DCT coefficient at (u, v) in $N\times N$ block The concise expressions (3) and (4) show that the mean m_1 (μ) is directly derived from DC coefficient, while variance μ_2 (σ^2) is just the average of summation of all squared AC coefficients. The computational cost of the proposed algorithm is less than that computing from the pixel domain. In fact, the operations of multiplication and addition in calculation of μ_2 is greatly reduced since $C(u,v)$ on down-right of the block are approximately zero and they are usually omitted. In our system, only first 32 coefficients of each block along zigzag are taken into account. Furthermore, the proposed algorithm is without necessity of decoding. The comparison of computational cost between traditional algorithm and the proposed is listed as shown in Table 1.

Table 1. Comparison of computation cost between the conventional algorithm and the proposed method

Moment	Compute from pixel domain			Proposed algorithm		
	+	×	Decoding	+	×	Decoding
m_1	63	1	Yes	0	1	No
μ_2	64	66		≤ 62	≤ 64	

2.2   Mean and Variance on Different Size of the Local Windows

In texture analysis, a new problem could occur from that which size of the local window is proper. This causes that we need often to know the statistical parameters on the double size window or half size window from the given window.

i)   The computation of the parameters in the merged window: If the 4 neighbored $N\times N$ windows' DCT coefficients, their means m'_1 and variance σ'^2 can be given as follow,

$$m'_1 = \frac{1}{4N}\sum_{k=1}^{4}C_k(0,0) \tag{5}$$

$$\mu'_2 = \frac{1}{4N^2}\sum_{k=1}^{4}\sum_{u,v=0(u\neq 0 \text{ and } v\neq 0)}^{N-1}C_k^2(u,v) \tag{6}$$

where $C_k(u,v)$ is the DCT coefficient at (u, v) of kth $N\times N$ block.

ii)   The computation of the parameters in the split window: Assume NxN (N is even) block is divided into 4 $[N/2]\times [N/2]$ subblocks. As described in [9], the DCT coefficients between maternal block and its subblocks satisfies

$$\begin{pmatrix} C_s^{0,0} & C_s^{0,1} \\ C_s^{1,0} & C_s^{1,1} \end{pmatrix} = 2DC_bD^T \tag{7}$$

where C_b and $C_s^{i,j}$ $(i,j=0,1)$ are the matrices of the DCT coefficients in $N \times N$ block and its subblocks respectively; D, an orthogonal matrix of the parameters with dimensions $N \times N$, can be given in advance. Because D is sparse matrix and the value of C_B along zigzag gradually reduce, the only up-left $[N/2] \times [N \ N/2]$ needs considering to save computation cost. As a result, the parameters in 4 sub-blocks can be determined by the DCT coefficients in the matrices $C_s^{i,j}$ according to equation (3) and (4).

The above conclusion illustrates that the statistical parameters of 16x16 can be gotten from its 4 8x8 subblock. Meantime, the corresponding parameters in 4x4 sub-block can be proximately gotten from itself maternal block. This algorithm of the statistical parameters of the decomposed or merged windows can be applied in region growing and region splitting directly from DCT domain instead of via IDCT first. As a result, the computation cost will reduce greatly.

3   The Implement of The Image Retrieval System

According to the JPEG compression standard, its decoding steps are (a) decode via Huffman entropy table, (b) dequantize the gotten data, and (c) perform IDCT trans-formation on the obtained block of the DCT coefficients. And then, the conventional context based image retrieval (CBIR) system can conduct the procedure of key ex-traction based on pixel domain [6]. To improve the retrieval efficiency, we develop a method of the key construction based on the above parameters m_1 and μ_2 so that IDCT step is disregarded. The image retrieval system is designed as shown in Fig. 1.

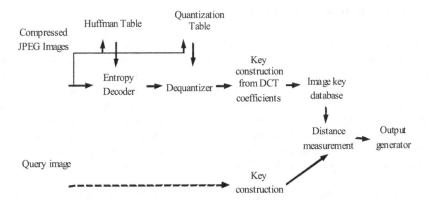

Fig.1. The diagram of the image retrieval system

Key Construction

Mean m_1 represents average intensity on the 8x8 block, which contains color information; while variance μ_2 or standard deviation σ ($= \sqrt{\mu_2}$) is regarded a character being abundant in texture information which is usually exploited in image segmentation [4, 7]. m_1 and σ establish a 2-D m_1-σ space, which satisfies $0 \leq m_1 \leq 255$ and $0 \leq \sigma < 128$. In the proposed system, the key of an image is constructed by using two parameters as following steps.

Divide m_1-σ space into 28 unequal partitions (subspaces), among which, intensity mean m_1 equally divided 4 non-overlapping entries, denote them as 0, 1, 2,3 respectively, and σ is divided unequally into 7 non-overlapping entries and denote them as 0,1, ..., 6. There are totally 28 entries by combining two divisions and concatenate as a vector with 28 components. For a given JPEG image, decode it into 8x8 blocks of DCT coefficients as described in Fig. 1. Calculate m_1 and σ of all blocks according to equation (6) and (8) and count the numbers f_r (r=0,1,...,27) within the 28 subspaces (entries). Normalize the vector $\{f_0, f_1, ..., f_{27}\}$ by dividing the number of blocks $\sum_{r=0}^{27} f_r$. The gotten vector with 28 components is regarded as the key of the given compressed image.

Similarity Measurement

Since the size of used image database image database is usually very large, the Weight Distance Transform(WDT) technique is employed in order to save cost of computing distance between keys[10]. Because the WDT leaves out all operations of multiplication used by Euclidean Distance, has only operations of additions, it will speed up the retrieval.

4   Performance Evaluation

Along with the above design, a content based image retrieval system (CBIR) is implemented by using C++. Usually there are two factors to evaluate the performance of a CBIR system [3]. One is the retrieval efficiency which focuses on the speed of retrieval. The other is the retrieval effectiveness which emphasizes the accuracy of the retrieval. As above discussion, because the proposed system omits the step of IDCT in JPEG decoder, and directly generates a query key from DCT domain, it saves a large of computational cost than the conventional system. The generating a key database of 500 JPEG compressed images takes only 23 seconds on Pentium II computer with 350 MHz CPU by the proposed algorithm, while it takes 51 seconds after a completely decompressed.

To investigate the robustness of the proposed system, we first design two experiments to assess the influence on the key construction by the geometric transform and by the different windows size.

Effect on Key Construction by Rotation, Translation, Scale Transform

From the definition, the both of parameters m_1 and σ are not invariant to rotation, translation and scale transform. If only considering rotation locally, the squared block is almost invariant to the rotation. However the rotation of the image usually is global instead of the block rotation. Thus those transforms will cause retrieval error. To examine the realistic effect on key construction caused by those transforms in the proposed system, we create a set of images by rotation, translation and scale from an original image and 100 references as test. Using original image as query image to retrieval, the results show the best is the case by rotated in 90° because it is really same. The followings are 4 cases by translated in 1-4 pixel, then in turn cases by rotated in 60°, 45° and 30°. The last are case of scale transform, and the first 10 ranks fits the correct items. The results illustrate the effect on the key construction by those transformation in minor disturbance are insignificant. The system can keep to be proximately invariant to those transform.

Effect on the Retrieval Result by Size of the Local Window

In order to ensure optimized performances and to extract local or global feature from DCT domain, new problem is that what size of local window is more proper. In this experiment, we choose respectively the sizes of 4x4, 8x8 and 16x16 as the test window, compute the corresponding parameters along with section 2 and then construct key according to section 3. Fig. 2 is the retrieval sample for 3 cases. The experiments show the performance of three cases are very similar, however the performance in 8x8 window is the best in efficiency and effectiveness.

(a) Query image   (b) Retrieved results by 4x4 block

(c) Retrieved results by 8x8 block   (d) Retrieved results by 16x16 block

Fig.2. The retrieval results on different window sizes

On the other hand, we perform two another experiments to demonstrate the system effectiveness.

i)   Faloutsos [3] proposed a variation of normalized precision to assess the performance of the retrieval system. Here, we employ parameters AVRR, IAVRR, and the ration of AVRR to measure our system. Ratio of *AVRR* is defined as the ratio of *AVRR* to *IAVRR*, which give a measure of the effectiveness of the system retrievals. When the ratio is equal to 1, the result of retrieval is the most ideal.

In our experiment, we selected 5 query images from a database with 100 images as test. The experimental results are listed as shown in Table 2. Notice the AVRR =5.68, and Ratio of AVRR=1.84, which is better than that in [3][11]

ii)   To investigate in which rank the first relevant image is retrieved, we randomly select 32 images from a database with 524 images as query images, and record the total number of the retrieval images in each rank from 0 to 7 (exclusive the query image). The two below parameters are computed as depicting the system performance. One is the ratio of retrieval (RR (r)) which is equal to the percentage that number of retrieved relevant images in r^{th} rank takes in. The other is the accumulation ratio at rank r ($AR(r)$) which is defined as

$$AR(r) = \frac{Number \ of \ found \ relevant \ image \ before \ rank \ (r)}{Total \ number \ of \ query \ image} \times 100\% \qquad (8)$$

The experimental results are plotted as shown in Fig. 3. The results show (a) the relevant images amass the front of ranks; (b) the probability of the relevant image ranged at 0th rank over 80%; and (c) the probability of at lest one relevant image ranged first 3 top rank is close to 97%. This means the system can retrieve acquire at lest one relevant image within the first 3 ranks with a very high probability. Fig. 4. displays a typical example that visually illustrate the system has good ability of image retrieval.

Table 2. The measurement of the retrieval effectiveness

Query image	1	2	3	4	5	Average
Number of the relevant images	4	8	5	4	16	
AVRR	9/4	39/8	13/5	19/4	223/16	5.68
Ratio of *AVRR*	1.50	1.39	1.30	3.17	1.86	1.84

5   Conclusions

In this paper, we have derived an algorithm directly and effectively computing 1st and 2nd moments from DCT domain, and given the relation of the moments between a various size of windows. This proposed algorithm can save a lot of computation cost because no IDCT involved. Employing the proposed algorithm, we develop an image retrieval system for JPEG image. The experiments show the system is almost insensitive to translation, rotation and scale transform with minor disturbance. The theoreti-

cal analysis and experimental results also demonstrate the system has good performance in retrieval efficiency and in effectiveness.

The authors wish to acknowledge the financial support from The Council for Museums, Archives & Libraries in the UK under the research grant LIC/RE/082, and partial
support from NSFC under the grant No. 60175031.

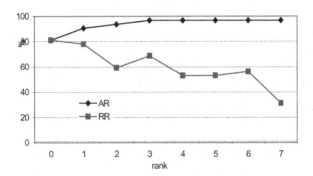

Fig. 3. The retrieval ratio and accumulation retrieval ratio

Fig. 4. The example retrieved by the proposed system

References

1. M. K. Hu, Visaul pattern recognition by moment invariants, *IRE Trans. in Information Theory*, Vol.8, 179-187, 1962
2. G. K. Wallace, The JPEG still picture compression standard, *Communication of the ACM*, Vol.34, No.4, pp31-45,1991
3. C. Faloutsos, R. Barber and M. Flickner et al, Efficient and effective querying by image content, *Journal of Intelligent Information System,* Vol.3, No.3, pp.231-262, 1994
4. R. Chantal and J. Michel, New minimum variance region growing algorithm for image segmentation, *Pattern Recognition Letters*, Vol.18, No. 3, pp.249-258, 1997
5. B. M. Methtre, M. S. Kankanhalli, and W. F. Lee, Shape measures for content based image retrieval: A comparison, *Information Processing & Management*, Vol.33, No.3, pp319-337,1997
6. M. K. Mandal, F. Idris and S. Panchanatha, A critical evaluation of image and video indexing techniques in the compressed domain, *Image and Vision Computing*, Vol.17, pp.513-529, 1999
7. A. P.Mendonca and E. A. B. da Silva, Segmentation approach using local image statistics, *Electronics Letters,* Vol. 36, No. 14, pp. 1199-1201, 2000
8. Pitas, Digital image processing algorithms, Prentice Hall, New York, 1993
9. G. Feng and J. Jiang, Image spatial transformation in DCT domain, *ICIP'2001*, pp.836-839, Greece, Oct. 2001
10. D. Borgeors, Another comment on a "note on distance transformations in digital images", *Image Understanding*, Vol. 54, No.2. pp.201-306, 1991
11. D. H. Lee and J. J. Kin, A fast contest-based indexing and retrieval technique by the shape information in large database, Journal of systems and software, Vol.56, pp.165-182,2001
12. R. Fazel-Rezai and W. Kinsner, Texture analysis and segmentation of image using fractals, *Canadian Conference on Electrical and Computer Engineering,* Vol.2 pp.786-791, 1999
13. B. Shen and Ishwar K. Sethi, "Direct feature extraction from compressed images", *SPIE: Vol.2670 Storage & Retrieval for Image and Video Databases* IV , 1996
14. S. Y. Lai and W. K. Leow, Invariant texture matching for content-based image retrieval, *Proc. of Int. Conf. on Multimedia Modelling,* 1997

Image Retrieval Methods
for a Database of Funeral Monuments

A. Jonathan Howell and David S. Young

School of Cognitive and Computing Sciences, University of Sussex
Falmer, Brighton BN1 9QH, UK
{jonh,davidy}@cogs.susx.ac.uk

Abstract. This paper investigates the use of Gabor features in matching
2- and 3-D objects in photographic images, using a database of images
of funeral monuments from English churches. The technique, which can
be applied to large databases, allows an arbitrary patch of a reference
image to be matched to patches of each database image at a range of
scales and positions. We investigate the use of nonlinear preprocessing
to reduce the influence of lighting and surface reflectance on the match
results.

1 Introduction

In this paper, we address the task of matching sections of photographic images in
a large database. The database, which is currently under development, contains
images of English funeral monuments from the 16th–19th centuries; it is likely
to be a valuable resource for art historians, social historians and genealogists.
Effective ways of searching the database for particular kinds of structure will add
to its effectiveness, but there are significant difficulties associated with attempts
to use the image content for this purpose. Query and matching images can range
in complexity of shape from small, flat 2-D panels to large, 3-D assemblies of
statues and decoration; in addition, photography inside churches is constrained,
and it is not often possible to make lighting and camera position consistent.
Monuments which are similar in terms of structure and detail may vary in surface
colour and texture, see Figs. 1(a) and 2(a).

Current commercial solutions to such a task rely on text notation of such
images, using experts to devise a standard notation or typology of features.
Although this can result in rapid retrieval of queries, it does not allow for search
outside the standard notation, nor allow for wide variation of names and styles
of cultural artifacts.

Using image features rather than text notation is commonly termed Content-
Based Image Retrieval (CBIR) [2,8]. Although several useful CBIR research
prototypes have been developed, few allow the user to define query and matching
images with arbitrary size and position.

Related work on retrieval by appearance includes that of Ravela and Manmatha. Two versions have been developed, one for whole-image matching and

M. S. Lew, N. Sebe, and J. P. Eakins (Eds.): CIVR 2002, LNCS 2383, pp. 129–137, 2002.
© Springer-Verlag Berlin Heidelberg 2002

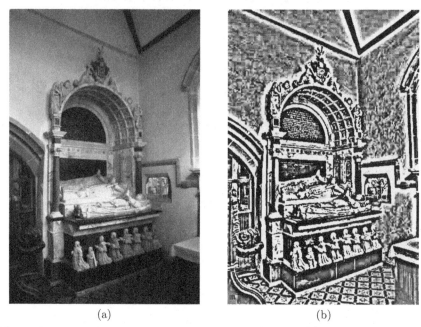

(a)   (b)

Fig. 1. Examples from the funeral monument database: (a) a context view, typically perspective, (b) 'flattened' version

one for matching selected parts of an image. The part-image technique involves filtering the image with Gaussian derivatives at multiple scales [7], and then computing differential invariants; the whole-image technique uses distributions of local curvature and phase [6]. Particularly useful is the ability to describe an image at varying levels of detail (useful in natural scenes where the objects of interest may appear in a variety of guises), and avoid the need to segment the image into regions of interest before shape descriptors can be computed.

Manjunath and Ma [5] used Gabor texture features for browsing aerial photographs, showing the Gabor method to be the most accurate of a range of other texture features.

In this work we are taking an appearance-based approach, with each patch of image represented as a vector of Gabor filter coefficients. This has been shown to usefully characterise the 3-D appearance of human faces [4] and temporal hand gesture trajectories [3] in a computationally effective way (approaching frame rate in online applications). This capability is particularly important for our chosen application, as the matching algorithm will be required to match 3-D objects over a range of views compared to the query image. Such an ability would necessarily be supported by the choice and range of image views in 3-D for each object in the image database: sufficient example views will then 'pave' the expected 'view sphere' of query image positions.

(a)   (b)

Fig. 2. Examples from the funeral monument database: (a) a detail view, typically orthogonal, (b) 'flattened' version

In Section 2, we describe the image database and the techniques used for
matching. We present preliminary results showing how the techniques compared
to human performance in Section 3. Finally, in Section 4, we draw some conclusions about our approach and further work.

2   Method

We computed Gabor coefficients starting both from grey-level images and from
images preprocessed to emphasise the local structure. We refer to these processed
images as 'flattened', since the grey-level range is compressed, see Figs. 1(b)
and 2(b).

Flattened images were generated as follows. The image was convolved with
a Gaussian mask to produce a smoothed version, and the smoothed image subtracted from the original image. The result was then compressed using the logistic function on each pixel value, with appropriate scaling so that extreme
positive and negative values are clipped to 0 and 1 [9]. Local contrasts are thus
emphasised, whilst the overall grey-level of a region is suppressed. The output is
very sensitive to local detail; thus Fig. 2(b) reveals artefacts in the plain region
surrounding the plaque resulting from the image's being stored in compressed
JPEG format.

For each image, a set of overlapping regions, each of size 200×200 or 300×300
pixels, was extracted. The degree of overlap was 5region size. Each region was
resized by averaging to 100×100 pixels and a set of Gabor coefficients found for
it. The coefficients were stored for later matching.

For each region, we applied Gabor masks at three scales and three orientations, equally spaced. At the largest scale the mask was was the size of image
region; for the intermediate scale the mask scale was halved and the mask applied to the four quadrants of the patch; for the smallest scale the mask size has
halved and the mask applied at 16 non-overlapping positions. We tried using

both the Gabor coefficients thus generated and the amplitudes produced combining the cosine and sine terms. Thus for each patch, four sets of coefficients were available for matching:

- **Grey-level** - a sparse grid of Gabor filters is applied to 'raw' grey-level pixels - 126 ((4×4 + 2×2 + 1) * 2 * 3) coefficients per image sample patch.
- **Flattened** - a sparse grid of Gabor filters is applied to 'flattened' grey-level pixels - 126 coefficients per image sample patch.
- **Amplitude Grey-level** - same as 'Grey-level', except that the root of squared sine/cosine values is computed - 63 ((4×4 + 2×2 + 1) * 3) coefficients per image sample patch.
- **Amplitude Flattened** - same as 'Flattened', except that the root of squared sine/cosine values is computed - 63 coefficients per image sample patch.

The matching algorithm computes the Euclidean distance between the query and matching images' coefficient vector, and the match value is expressed as the exponent of the negative of this distance (giving a value between 0 and 1, with 1 being a perfect match).

Although the computation of the Gabor coefficients for all image patches at all scales and offset positions can be a lengthy process, it only ever needs to be performed once for each image (assuming the choice of scales and offsets remain constant). This done, each query can be performed over the entire database (227 images) in around five seconds. Matching of coefficient vectors is a task that can be performed by highly optimised standard software, and we expect that, with an appropriate choice of such software, the matching could well be carried in very much less time.

For this work we are not considering colour information in the images, as it has not been possible to standardise illumination and colour over all database images. This is due to the images having been collected with a variety of methods: conventional 35mm colour film (digitised with PhotoCD) and digital camera, both with combinations of available light and flash. Any available light images can have arbitrary colour casting, due to light from stained-glass windows. In addition, colour information collected from the monuments is variable due to wear and damage over several centuries and the effects of dust and dirt.

3   Results

For the tests, a query image region is matched with the regions, at all scales and positions, from 227 database images, and the match values sorted by similarity. Only the best matching region from any given image is shown here; usually several closely-position regions returned matches with similar values, as would be expected given the degree of overlap.

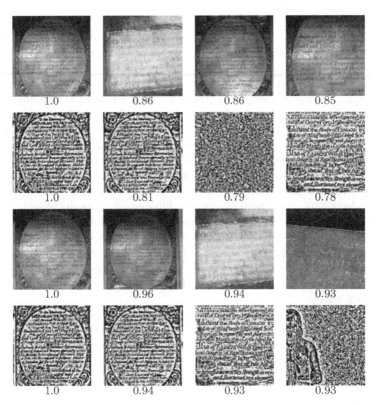

Fig. 3. Results for Test 1 (Oval Panel). Left column: search query (best match). 2–4th columns: further matches from the database, in order of decreasing match value. Rows: grey-level, flattened, grey-level amplitudes, flattened amplitudes. Numbers below images show match strength

3.1   Test 1: Oval Panel

This test uses an oval panel containing an inscription as the target. There are many panels with inscriptions in the database, and we would like to detect those with similar shapes and textures.

All the methods find the target image as the best match, and a different image of the same monument as the second-best or third-best match. In the other close matches, the grey-level methods have found patches where the grey-level distribution matches that of the target; the 'flattened' methods have found patches where the texture matches that of the target.

It is not clear, of course, which aspect of the monument will be most important to a given user, the shape of the plaque, the existence of text, or some other feature. It will be necessary to tailor the matching process to the requirements of users.

3.2   Test 2: Kneeling Figure

The second test uses 3-D objects. An important capability for our application is to be able to match such objects from a range of views, not just the specific one seen in the query image.

The grey-level methods both find two further images of the same object, from slightly different viewpoints (there are only 3 images of this object in the database). It is likely that the distribution of grey levels is the dominant aspect of the matching process; this is indicated by the final match displayed, where the match image does not appear similar to the target, but has generated a similar set of Gabor coefficients. The 'flattened' methods, however, tend to find regions with similar texture to the target, but with less regard to the shape or absolute range of grey levels.

Matching with amplitudes produces a different set of results to matching with individual cosine/sine coefficients, particularly on the flattened images. Combining the coefficients loses phase information, giving a degree of position independence at the largest filter scale, and reducing position sensitivity at the smaller scales. In the case of this image, where the shape is significant but there is not a characteristic small-scale texture, this compression of the coefficients has a significant effect.

One point which arises from this test is that the allowable range of scale and location can restrict the choice of query image more than the user would like: how much of the figure needs to be included to represent it sufficiently? The user would need to choose between a small area around the head or a larger area containing more of the body.

3.3   Test 3: Text Area

In the final test reported here, the target is a region of text from the centre of a plain slab with a long inscription. The target is thus dominated by a single characteristic texture.

In this case the methods all perform well, effectively discriminating between text and non-text regions. All the matches shown are from inscriptions. As before, only the closest matching region is shown from any given image for each method, although usually neighbouring regions will also match well. The same inscription appears more than once because several images of it were present in the database.

Compressing the coefficients to amplitudes has relatively little effect here. The text image is characterised by an overall texture, and the Gabor amplitudes represent it adequately when the phase information is lacking. Flattening the image may produce results that are more sharply tuned to the texture of the target, but this requires further investigation.

4   Conclusions and Further Research

Gabor coefficients provide a concise representation of image patches that may be used for effective texture and shape matching. The results of the present

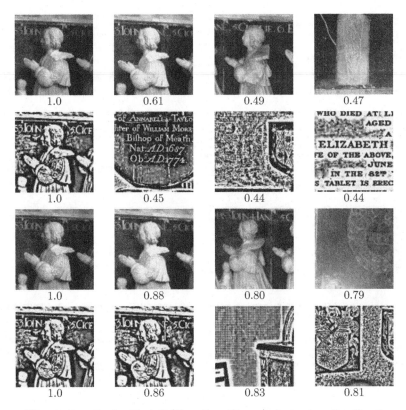

Fig. 4. Results for Test 2 (Kneeling Figure). Layout as for Fig. 3

study suggest that these techniques are worth investigating in the context of specialised databases such as the funeral monuments database. This approach may have practical benefits for the users of the database; it may also allow the development of methods that have more generality than those for, say, face recognition, without tackling the full complexity of general-purpose image databases or content-based image retrieval from the web.

This has been a pilot study, and to take it further, we need to carry out a more systematic survey of the effects of varying the parameters of the matching process. In particular, the number of coefficients calculated is somewhat arbitrary, and will be important to explore the trade-off between the number of orientations and scales and the accuracy and flexibility of matching. Although the 'flattening' method gives some protection against lighting and surface colouring effects, we need to look in more detail at how to obtain matches which are robust against orientation changes in 2-D and 3-D.

The most significant step, however, is to investigate more closely how techniques such as this can be tailored to the needs of users. In particular, methods such as relevance feedback [1] are required to allow the matching process to

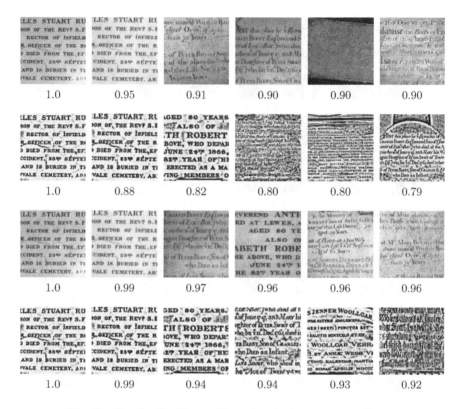

Fig. 5. Results for Test 3 (Text Area). Layout as for Fig. 3

adapt to the requirements of expert users. Our matching metric, based on the Euclidean distance between the coefficient vectors, is very straightforward but clearly leaves much scope for generalisation to a more adaptive approach.

Acknowledgements

This work was funded by the Arts and Humanities Research Board of the UK as part of the project *Towards a National Census of Funeral Monuments: A Pilot Study of East and West Sussex*. The images were collected by David Hickman, the Photography and Design Unit of the University of Sussex, and the authors.

References

1. I. J. Cox, M. L. Miller, T. P. Minka, T. Papathomas, and P. N. Yianilos. The Bayesian image retrieval system, PicHunter: Theory, implementation and psychophysical experiments. *IEEE Transactions on Image Processing*, 9:20–37, 2000. 135

2. J. P. Eakins and M. E. Graham. Content-based image retrieval. Report to the JISC Technology Applications Programme http://www.unn.ac.uk/iidr/research/cbir/report.html, 1999. 129

3. A. J. Howell and H. Buxton. Learning gestures for visually mediated interaction. In P. H. Lewis and M. S. Nixon, editors, *Proceedings of British Machine Vision Conference*, pages 508–517, Southampton, UK, 1998. BMVA Press. 130

4. A. J. Howell and H. Buxton. Learning identity with radial basis function networks. *Neurocomputing*, 20:15–34, 1998. 130

5. B. S. Manjunath and W. Y. Ma. Texture features for browsing and retrieval of image data. *IEEE Transactions on Pattern Analysis & Machine Intelligence*, 18:837–842, 1996. 130

6. S. Ravela and R. Manmatha. On computing global similarity in images. In *Proceedings of IEEE Workshop on Applications of Computer Vision (WACV98)*, pages 82–87, 1998. 130

7. S. Ravela and R. Manmatha. Retrieving images by appearance. In *Proc. of the IEEE International Conf. on Computer Vision (ICCV'98)*, pages 608–613, 1998. 130

8. C. C. Venters and M. D. Cooper. A review of content-based image retrieval systems. Report to the JISC Technology Applications Programme http://www.jtap.ac.uk/reports/htm/jtap-054.html, 2000. 129

9. D. Young. Straight lines and circles in the log-polar image. In *Proceedings of the 11th British Machine Vision Conference*, pages 426–435, 2000. 131

AtomsNet: Multimedia Peer2Peer File Sharing

Willem de Bruijn and Michael S. Lew

LIACS Media Lab, Leiden University
Niels Bohrweg 1, 2300 CA Leiden, The Netherlands
{wdebruijn,mlew}@liacs.nl

Abstract. Current Peer2Peer (P2P) systems such as Napster or Kazaa do not perform analysis on the content of the media but instead depend on manual text annotation. In the AtomsNet project we are investigating multi-modal content based browsing and searching methods for P2P retrieval systems. This is the first P2P system which performs analysis on the video content for browsing multimedia collections over large, distributed P2P networks.

1   Introduction

Peer2Peer (P2P) networks have had major attention recently and typically have millions of peers within their sub-network. The widely popular applications designed specifically for sharing information amongst peers have only emerged a few years ago. The best known example of P2P networks is the music sharing program called Napster (www.napster.com).

The strongpoint of Napster is very clear. Many more songs are available for free using this network than anywhere else. Live recordings, bootlegs and "rare takes" were all made available by music enthusiasts. Unfortunately Napster suffered from several weaknesses: legal problems and the difficulty in searching for files - the user had to search through an unordered bag of files.

Napster's shortcomings can be dealt with in several ways. We started this project in 2000, when no alternative to the Napster network had fully emerged. Our intention was to explore the possible solutions to the more general problem of multimedia retrieval. Since then many different P2P applications have been created, such as Kazaa, Morpheus and Gnutella. Taking into account the current state of the art, we will try to address which methods can be used to improve the aspects of multimedia retrieval over large scale P2P networks.

2   Background

The number of people using the internet has grown rapidly over the last couple of years. With this increase in public involvement a shift occurred in networking methodology. Traditionally, there existed a large gap between the suppliers and the consumers of information. In the popular client/server paradigm suppliers would be

M. S. Lew, N. Sebe, and J. P. Eakins (Eds.): CIVR 2002, LNCS 2383, pp. 138-146, 2002.
© Springer-Verlag Berlin Heidelberg 2002

active all of the time, servicing requests from clients at a fixed location. In an environment where many participants are both consumers and producers of information and no server can be trusted to be active, however, another way of communicating is necessary.

With the recent growth of available information on the internet and the lack of internal structure linking this data, a way of indexing information based not on hierarchy but on equality has to be found. The current situation renewed interest, both from the scientific community and from the general public, in distributed networks. Distributed networks, also referred to as peer-to-peer or P2P networks, index information in a decentralized manner, thus acknowledging the fact that no internal hierarchy exists.

There are other interesting WWW image search engines which have been described in the research literature. The WebSeek[17] system from Columbia University finds similar images and performs automatic text based category classification. The Webseer[3] system from the University of Chicago lets users search by the number of faces and by text queries. Taycher, Cascia, and Sclaroff[19] designed the ImageRover system to primarily use relevance feedback for the search process, and in the PicToSeek system, Gevers and Smeulders[4] search through the images using similar images and image features. Lew [10] describes one of the first web systems which finds images and video based on understanding of semantic categories.

In visual information retrieval there has been relevant previous research [1-22]. Picard[5] reported promising results in classifying blocks in an image into "at a glance" categories. What this means is that she investigated multiple model methods to classify an NxN block into categories which humans could classify without logically analyzing the content. Forsyth, et al. [2] found objects from feature blobs. More recently, Vailaya, Jain, and Zhang [21] have reported success in classifying images as city vs. landscape. They found that the edge direction features have sufficient discriminatory power for accurate classification of their test set. The commonality between these methods was using multiple features for object/concept detection. Regarding object detection, the recent surge of research toward face recognition has motivated robust methods for face detection in complex scenery. Representative results have been reported by Sung and Poggio[18], Rowley and Kanade[16], Lew and Huijsmans [11], and Lew and Huang[12].

When searching for an item one always has to describe certain characteristics that distinguish the desired item from others. These characteristics will then be cross-checked with the characteristics of the possible options to select the best answer. It doesn't matter if a person is searching for a house by asking directions or if he is trying to find a digital document on the internet, characteristic negotiation is the basic element of searching.

When dealing specifically with data, these characteristics can be seen as data describing the data. For this special type of information the term {\it metadata} is generally used, where meta stands for 'above'. In everyday situations this information is completely distinct from the object it details. When dealing with digital information, however, it all boils down to numbers. This means that metadata can be seen both as a description and an object itself.

Traditionally metadata has been separated from objects. The most widely used form of metadata is the filesystem structure used to index digital data on a harddrive. This data is identical for every file and therefore extremely brief. Such an implementation might be sufficient for a local filesystem where a small group of people handle the data, but it is an impractical indexing method for use in large heterogeneous environments. The large scale on which information is made available on the internet increase the demand for automation of resource identification to a level beyond that of basic filesystems.

Currently a few projects are underway to create a more flexible framework for describing data. Since standardization is essential for any system to become widespread the initiatives backed by international organizations have the highest chance of success. The Resource Description Format backed by the influential World Wide Web consortium, is a semantic layer placed on top of the meaningless XML syntax. In RDF, meaning is given to data by coupling RDF formatted content to specialized dictionaries. It is already being used to identify and rebroadcast news messages on the internet and by the Open Directory Project webdirectory. RDF has been designed specifically with the automation of internet resource discovery in mind. The developers expect RDF to become the framework underlining the so called Semantic Web, a global information network where computers can autonomously index information.

Despite the effort of various interest groups, metadata is for the larger part currently written in proprietary formats, eliminating interoperability between software systems. Future developers should try to use standardized frameworks as the RDF mentioned above as much as possible if the internet is to remain a global information domain.

Here we summarize the de facto standards and innovative implementations:

- **Napster** and **Gnutella** - mostly music sharing; parsing mp3 headers for text annotation.
- **KaZaa** and **Morpheus** - filesystem info + manually entered info that is saved between transfers
- **AudioGalaxy** - searching using knowledge about user behaviour (link multiple downloads from the same user as potentially connected, thus creating subcategories based on user).

Note that none of these systems uses content based analysis in the browse/search process. In fact, manual text annotation is a critical factor in all of their search engines. This has also led to many users download viruses because some malicious users used false text annotation for their viruses.

3   AtomsNet: Network Design and Content Based Analysis

AtomsNet was designed to be scalable to the entire Internet, which currently means approximately 700,000 networked computers. We describe the design of the interconnection network and the content based analysis in this section. The AtomsNet system allows the following query types:

- text keywords
- category based searching

It provides functionality to

- automatically categorize all of the files in a directory
    - you can drag and drop local directories for automatic analysis
- allow plug-ins for custom media analysis and browsing:
    - extract text annotation from mp3 files
    - smart text searches: "find videos starring Harrison Ford"
    - extract representative keyframes from MPEG video

3.1 How Do You Find Your Peers?

Since the Web is not an organized database, it is challenging if not impossible to comprehensively find all of the hyperlinks. The general procedure used by typical search engines is to begin with an initial set of hyperlinks, follow each hyperlink to a page, parse each page for new hyperlinks, and follow the new hyperlinks, repeat ad infinitum. This typically results in a breadth first search of the WWW as shown in Figure 1. It also means that the comprehensiveness of any search engine is limited to the sites which are reachable from the initial set of seed links. For example, if there are no hyperlinks connecting networks in the U.S. and China, then a search engine which begins from exclusively U.S. links will never find the network in China.

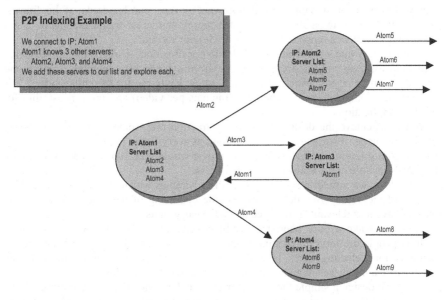

Fig. 1. P2P indexing example

In our scheme, AtomsNet is composed of atoms which are the peer computers. The typical atom is expected to have a slow connection of roughly 56Kbit/s to 1 Mbit/s. Atoms are connected in the virtual network by SuperAtoms which typically have 1 Mbit/s or greater network bandwidth.

In designing AtomsNet distributed network, the critical factor was to minimize the communications overhead between the computers. Our approach was to incorporate multiple tier hierarchical routing into the P2P search system. At the lowest tier, each atom was given sole responsibility for a list of hosts. Each atom sends its list of atoms to the next higher tier.

When an atom connects to the AtomsNet, it contacts a superatom for the list of active atoms and superatoms. The list of atoms and superatoms are sorted by topological network distance. When the atom posts a query, it is only searching the local list of atoms and superatoms. The superatom relays this information periodically to all of the child atoms. Hierarchical routing has the significant advantage that it is extendable to arbitrarily large networks as long as the bandwidths are sufficient.

3.2 Content Based Analysis

When processing a video, the client machine should have a minimal workload in P2P situations. Most video abstraction and summarization methods entail significant computational complexity which would result in the client machine being unusable for minutes or even hours.

Our goal is to extract a single frame from the entire video to represent the content. The reasoning for this is that each result will also return a pictorial keyframe so low bandwidth users presumably would not want dozens our hundreds of keyframes. A secondary goal is to be able to extract the representative keyframe at a rate of 3 full length MPEG movies (approx. 1GB per movie) per second.

There is only one work which is similar to our direction. Dufaux[15] implemented a method for extracting a single keyframe from a video by integrating motion and spatial activity analysis, which results in a single keyframe for the entire video. His process was computationally expensive (minutes per video) and not a possibility for our P2P application.

In order to achieve the desired keyframe extraction rate, some compromises clearly need to be made. First, we can not even look at every frame in the movies because it would require too much time - a gigabyte movie would take 100 seconds to read on a hard disk capable of a sustained 10 MB/s access.

Therefore, we turned to sampling N MPEG I-frames which are in the second/third quarter of the video. We do not process the beginning of the video because many movies have less relevant "setup" scenes in the early shots.

Here we make a pure heuristic. Subjectively, we assert that keyframes which contain people are more interesting than ones that do not. There have been several methods for detecting faces (see Rowley, et al. [16]) in images with complex backgrounds, but these are also typically computationally complex.

Thus, we design a simple face detector using color. Methods in the literature have used the well known color spaces for skin detection. Our novel contribution is to design a new color space specifically for skin detection.

In principle, we do not care about representing skin, but only classifying a pixel as skin color based on the pixel and a small region around it. The optimal linear classifier can be expressed as

$$\text{maximize} \quad \frac{k^t \Sigma_x k}{k^t \Sigma_n k} \tag{1}$$

where k is the vector containing the pixel and its neighbors. We do not go into detail about equation (1) because it is found in most pattern recognition books. Instead we explain how we use it. The process is that we create training sets which comprise small skin regions and small non-skin regions. We find the orthogonal basis which gives us the optimal linear classifier which satisfies equation (1). This basis is the new color space for optimally classifying color pixel regions as skin or non-skin.

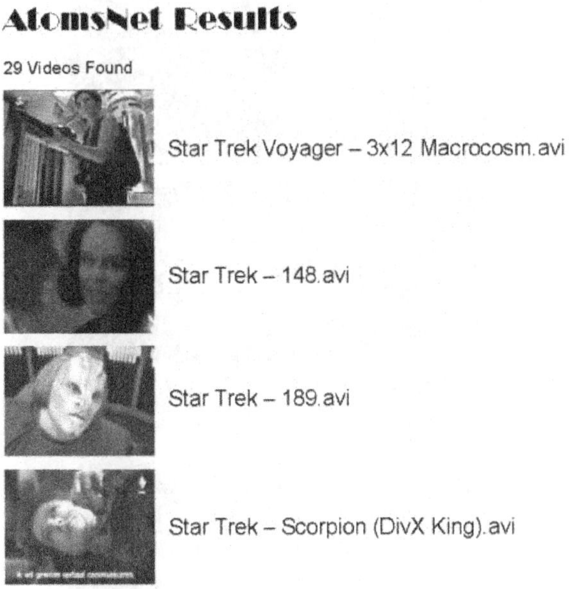

AtomsNet Results

29 Videos Found

Star Trek Voyager – 3x12 Macrocosm.avi

Star Trek – 148.avi

Star Trek – 189.avi

Star Trek – Scorpion (DivX King).avi

Fig. 3. An example of results from AtomsNet

A minimum distance classifier is used for deciding between skin/non-skin. After we classify each pixel as skin or non-skin, we select the keyframe which has the largest percentage of skin in it as shown in Figure 2.

For a PIII/800Mhz with a Seagate Cheetah Ultra160 73 GB harddrive, our system can analyze 9 keyframes per video and reach the goal of processing 3 full length videos per second. An example of the results page from AtomsNet for a video search on "star trek" is displayed in Figure 3.

4   Conclusions

In the AtomsNet multimedia P2P system, we needed to satisfy network considerations and allow browsing based on the content of the video. The main contribution of this paper toward video retrieval is the heuristic algorithm for finding a single keyframe to represent an entire video. It was designed intentionally to minimize computational complexity. The novel aspect of the keyframe selection algorithm was to create a new color space based on the theory of optimal linear classifiers. This new color space is optimized for classifying small pixel regions into the categories of skin or non-skin.

Fig. 2. Sampled frames ordered by most skin to least skin

There is ample room for improvements in P2P networks. The networking aspect of finding peers remains as a difficult problem. Browsing and searching for different media types are also major challenges. One advantage of the AtomsNet framework is that plugins can be created for each type of media. In the near future, our intention is to focus on the video analysis aspect toward extracting semantic descriptive information.

Acknowledgments

This project was assisted and supported by Altavista (Cambridge Research Lab), Magicbot, and the LIACS Media Lab. The authors would like to thank Michael Swain for beneficial and interesting discussions on searching the internet.

References

[1]   Del Bimbo, A., and P. Pala, "Visual Image Retrieval by Elastic Matching of User Sketches,"   IEEE Trans. Pattern Analysis and Machine Intelligence, February, pp. 121-132, 1997.

[2]   Forsyth, D., J. Malik, M. Fleck, T. Leung, C. Bregler, C. Carson, and H. Greenspan, "Finding Pictures of Objects in Large Collections of Images," Proceedings, International Workshop on Object Recognition, Cambridge, April 1996.

[3]   Frankel, C., M. Swain and V. Athitsos, "WebSeer: An Image Search Engine for the World Wide Web," Technical Report 96-14, University of Chicago, August 1996.

[4]   Gevers, T. and A. Smeulders, "PicToSeek: A Content-Based Image Search System for the World Wide Web," VISUAL'97, San Diego, December, pp. 93-100.

[5]   Picard, R.. "A Society of Models for Video and Image Libraries." IBM Systems Journal. 1996.

[6]   Hu, M., "Visual Pattern Recognition by Moment Invariants", IRA Trans. on Information Theory, vol. 17-8, no. 2, pp. 179-187, Feb. 1962.

[7]   Huijsmans, D. P., M. Lew, and D. Denteneer, "Quality Measures for Interactive Image Retrieval with a Performance Evaluation of Two 3x3 Texel-based Methods,"   International Conference on Image Analysis and Processing, Florence, Italy, September, 1997.

[8]   Kittler, J., M. Hatef, R. Duin, and J. Matas, "On Combining Classifiers," IEEE Trans. Patt. Anal. and Mach. Intel., vol. 20, no. 3, March 1998.

[9]   Kullback, S. "Information Theory and Statistics," Wiley, New York, 1959.

[10] Lew, M., "Next Generation Web Searches for Visual Content,"   IEEE Computer, November, pp. 46-53, 2000.

[11] Lew, M. and N. Huijsmans, "Information Theory and Face Detection," Proceedings of the International Conference on Pattern Recognition, Vienna, Austria, August 25-30, 1996, pp.601-605.

[12] Lew, M. and T. Huang, "Optimal Supports for Image Matching," Proc. of the IEEE Digital Signal Processing Workshop, Loen, Norway, Sept. 1-4, 1996, pp. 251-254.

[13] Ojala, T., M. Pietikainen and D. Harwood, "A Comparative Study of Texture Measures with Classification Based on Feature Distributions," vol. 29, no. 1, pp. 51-59, 1996.

[14] Petkovic, D., "Challenges and Opportunities for Pattern Recognition and Computer Vision Research in Year 2000 and Beyond, "Proc. of the Int. Conf. on Image Analysis and Processing, September, Florence, vol. 2, pp. 1-5, 1997.

[15] Dufaux, F. "Key Frame Selection to Represent a Video." ICIP, 2000.

[16] Rowley, H., and T. Kanade, Neural Network Based Face Detection, IEEE Trans. Patt. Anal. and Mach. Intell., vol. 20, no. 1, pp. 23-38, 1998.

[17] Smith, J. R. and S. F. Chang, "Visually Searching the Web for Content," IEEE Multimedia, 1997, pg. 12-20.

[18] Sung, K. K., and T. Poggio, Example-Based Learning for View-Based Human Face Detection, IEEE Trans. on Patt. Anal. and Mach. Intell, vol. 20, no. 1, pp. 39-51, 1998.

[19] Taycher, L., M. Cascia, and S. Sclaroff, "Image Digestion and Relevance Feedback in the ImageRover WWW Search Engine," VISUAL'97, December, San Diego, pp. 85-91.

[20] Tekalp, A. M., Digital Video Processing, Prentice Hall, New Jersey, 1995.

[21] Vailaya, A., A. Jain and H. Zhang, "On Image Classification: City vs. Landscape," IEEE Workshop on Content-Based Access of Image and Video Libraries, Santa Barbara, June 21, 1998.

[22] Wang, L. and D. C. He, "Texture Classification Using Texture Spectrum," Pattern Recognition 23, pp. 905-910, 1990.

Visual Clustering of Trademarks
Using the Self-Organizing Map

Mustaq Hussain[1], John Eakins[1], and Graham Sexton[2]

[1] School of Computing and Mathematics
University of Northumbria at Newcastle, NE1 8ST, United Kingdom
[2] School of Engineering
University of Northumbria at Newcastle, NE1 8ST, United Kingdom
{mustaq.hussain,john.eakins,graham.sexton}@unn.ac.uk

Abstract. This paper describes the experiments used to investigate ways in which digitised trademark images can be visually clustered on a 2-D surface, using the topological properties of the self-organizing map. Experiments were carried out on a set of original and edge detected binary trademark images, as well as their moment invariants, angular radial transformations and wavelet feature vectors. A radial based precision-recall measure was also used to evaluate the results objectively. Initial results are encouraging.

1 Introduction

The number and size of digital image collections has been rapidly increasing; a typical collection may contain hundreds of thousands of images, which represents an enormous quantity of information [1]. In an attempt to address some of the issues posed by such collections content-based image retrieval (CBIR) has become an active area of research. Images are usually stored and retrieved based on some automatically extracted feature that describes different visual cues such as colour, texture and shape. However, no single technique has been developed that can accurately describe a general image, instead researchers have developed a community of different models [2]. For example, shape analysis techniques include: global boundary based descriptors [3], [4], moment invariants [5], angular radial transformations (ART) [6], Zernike moments [7], Fourier descriptors [8] and wavelets [9].

Retrieval of trademark images has become an area of active research, in part due to the large number of trademark images that are geometric and/or abstract, deploying shape analysis techniques (colour is largely ignored). They also provide a test-bed for general image retrieval algorithms [10]; and generally they do not have to address the problem of which of the many feature models - such as colour, shape or texture - to use and how to integrate them [1].

CBIR trademark systems reported in the literature (see [1] for a recent review) are presented with a query image that produces a 1-D ranked list of similar trademark images. However, these 1-D lists can make it difficult to see how similar non-adjacent

M. S. Lew, N. Sebe, and J. P. Eakins (Eds.): CIVR 2002, LNCS 2383, pp. 147-156, 2002.
© Springer-Verlag Berlin Heidelberg 2002

images are related. One way to address this is to place images on a 2-D surface. This 2-D positioning can reflect the mutual distances between images in some feature space. Visualisation can provide visual clues as to why particular trademarks cluster around the query, and why others have been placed further away, in an effective and intuitive way for a user, by highlighting structures and patterns in the image (a technique described by Santini [11]).

Numerous clustering techniques are available [12]. The Self-Organising Map (SOM) was selected as a visualisation tool over others because it can achieve local and global ordering, it is a non-linear projector, new data points can be added without having to re-train the whole map and it is an unsupervised technique. Principal component analysis, on the other hand, is poor at non-linear projection, Sammon mapping - a type of multidimensional scaling - is computationally expensive even for moderate sized sets and can emphasise small pair-wise similarities, while k-means clustering only updates one data point at a time.

The aim of this study is to evaluate the effectiveness of the SOM as a visualisation tool based on different feature vectors. The remainder of this paper describes the SOM algorithm and the experiments carried out using the different feature extraction techniques. This is followed by a set of 2-D maps with trademarks images placed on them after SOM training, and how we tried to evaluate objectively the results using precision-recall measures. Finally, we discuss the results and how we are developing this work for larger maps and how we propose to extend our work to a hierarchical system.

2   Self-Organizing Map (SOM)

The Self-Organising Map [13] is an unsupervised topologically ordering neural network that can be used to visualise data. The SOM projects high dimensional data $\mathbf{x}$ from n-D input space onto a low dimensional lattice (usually 2-D for visualisation) of N neighbourhood connected (usually rectangular or hexagonal) nodes or neurons $\{\mathbf{r}_i \mid i=1,...,N\}$. Each neuron of the map, $i \in \mathbf{r}$, is connected to the n-D input $\mathbf{x}$ data by an n-D weight vector $\mathbf{m}_i = [m_{i1}, m_{i2}, ..., m_{in}]^{\mathrm{T}}$. During training, surrounding weight vectors are adapted bringing them closer to the input vector $\mathbf{x}$, ordering them topologically.

The algorithm used to train the SOM is summarised as:

1. **Initiation**. Randomly initialise all weight vectors $\mathbf{m}(0)$.
2. **Sample**. Stochastically draw a sample $\mathbf{x}(t)$ from the input distribution.
3. **Best Matching Unit** (BMU) $\mathbf{m}_c$ found $\mathbf{x}(t) - \mathbf{m}_c(t) = \min_i \| \mathbf{x}(t) - \mathbf{m}_i(t) \|$ over whole map $\forall i = 1,...,N$.
4. **Update Weights** around the BMU $\mathbf{m}_i(t+1) = \mathbf{m}_i(t) + \alpha(t)h_{ci}(t)[\mathbf{x}(t) - \mathbf{m}_i(t)]$ where $h_{ci}(t)$ is a Gaussian neighbourhood function with maximum adaptation at $\mathbf{m}_c$, $\alpha(t)$ determines the amount of adaptation and monotonically decreasing.
5. **Continue**. Repeat from stage 2, until the map has settled down.

The update process has the effect that neurons topologically close to $\mathbf{m}_c$ are activated by similar inputs. The SOM is therefore characterised by the formation of a topological map. Global ordered is achieved by the neighbourhood function initially covering a large area, gradually shrinking with time and fine tuning the local ordering.

3   Present Experiments

The UK's Trade Mark Registry provided a set of 10,000 abstract geometric trademark images for the evaluation of ARTISAN[14]. Within this set, 268 randomly selected images had been used in the initial evaluation of ARTISAN, containing four sets of test images {S1, S2, S3, S4}, which had been identified as being similar by Trade Mark Registry staff. This formed a collection of 31 test-bed trademarks that was used for our initial experiments.

3.1 Feature Extractions

Two versions of this 268-image test set were used: $\mathbf{D}_o$, original (*original_set*) images normalized to 128x128 pixels, and $\mathbf{D}_e$, the corresponding Laplacian edge detected (*edge_set*) images. The 31 test-bed images formed part of the collections.

As well as the original and edge detected images, moment invariants (*moments*) were investigated because of their ability to analysis whole images and regions. Angular radial transformations (*art*) were investigated as another region based descriptor which uses a complex angular transform on a unit disk to achieve invariance due to transformations, as it is also making up part of the MPEG-7 standard set of shape descriptors.

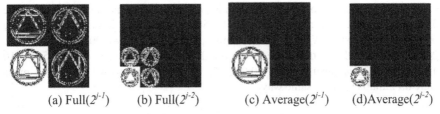

(a) Full(2^{j-1})   (b) Full(2^{j-2})   (c) Average(2^{j-1})   (d)Average(2^{j-2})

Fig. 1. Illustration of MRA Daubechies decomposition, at different levels

Multiresolution analysis (MRA) features using Daubechies and Haar (*daub_j/haar_j*) wavelets [9], [15] to *j*-levels of decomposition were extracted because of their ability to analyse images at different scales and resolutions, unlike the Fourier transform that cannot localise image features. The full (*full*) and average (*averg*) sub-image decompositions, and their mean and standard deviation (*meanSD*) were extracted as illustrated in Fig 1.

1. Full (*full*). This set consists of the average (low-level) sub-image being fully decomposed into its four sub-images. In Fig 1(b), the feature vector of the fully decomposed image to 2-levels of decomposition has 4 sub-images {P_2, D^2_2, D^1_2, D^3_2}, *daub_2full_orig*. This consists of the average details {P_2}, and the fine horizontal, vertical and diagonally sensitive details {D^2_2, D^1_2, D^3_2}, derived from the previous average details {P_1}, Fig 1(a)'s lower-left hand sub-image.
2. Average (*averg*). This feature vector, the average sub-image {P_2}, is a quarter of the length of the fully decomposed image, at the same level of decomposition.

---

* Automatic Retrieval of Trademark Images by Shape Analysis

In total 28 different training sets were used to train both the 31 and 268 test image sets.

The size of the resultant feature vector n determines the length of time required to train a SOM, as finding the BMU is of the order $O(n^2)$. Therefore, if a small feature vector can produce similar result to a large vector then it is advantageous in terms of searching and training time. For example the training time for the average MRA Daubechies image to 3 levels, *daub_3averg_orig*, is $1/16^{th}$ of the time of the original set, *original_set*. This can be significant for a 10,000 plus image set.

3.2 SOM Training

The parameters used to train the SOM were selected from empirical observations of the maps produced. The migration of the images was observed over different configurations. The initial leaning rate α_0 and neighbourhood size N_0 of $h_{ci}(0)$ effected the overall shape and development of the maps. With an initially small neighbourhood size, the BMU's region of influence h_{ci} is short, resulting in images forming tight clusters but not being able to push many images away. When $N_0=1$ (closest neighbours to BMU), initially isolated images remain isolated. If however the initial neighbourhood size is large, a larger region around the BMU is adapted resulting in many more images - lying outside this region - being pushed further to the edges. These observation were used to train the SOM maps to initially obtain a rough or global order then a fine ordering to obtain local ordering. For the final evaluation, the 31 test-bed images were trained on 12x15 SOM maps, with initial rough ordering parameters of $\alpha_0=0.1$, $N_0=5$ and fine tuning parameters of $\alpha_0=0.002$, $N_0=2$. For the 268 sets a SOM of 25x20 was used, with rough ordering parameters of $\alpha_0=0.5$, $N_0=10$ and fine tuning parameters of $\alpha_0=0.002$, $N_0=5$. All the hexagonal connected maps were all initialised in the same way.

3.3 Visual Analysis

Following training, the SOM's 2-D maps were displayed with the trademark images around their BMU. Fig 2 shows the 31 test-bed edge detected set, *daub_3full_edge*, created using Daubechies wavelets extracting all features at 3-levels of decomposition. The sets {S1, S2, S3, S4} and tentative cluster lines have been added to highlight the clusters. Many of the images can be found in or near their four prescribed groups. For example, the six triangular S1 images in the top-right hand corner and the nine S2 images in the lower-left hand quadrant. However, not all the images can be found in a single cluster, for example S3 is split between the four images running across the centre of the map and the two at the upper-left hand corner, and similarly for S4. A few of the images, such as S2 at the centre-right hand edge, are isolated and located some distance from their main cluster (the lower-left quadrant).

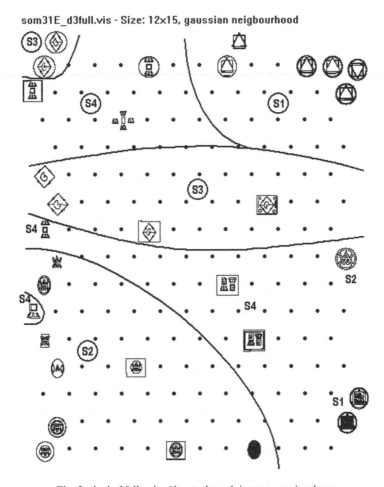

Fig. 2. *daub_3full_edge* 1k – trademark images on visual map

The original binary image map, *original_set*, also created similarly shaped clusters. However, its feature vector is 16k long and training took 45 minutes while *daub_3full_edge* was 1k long and took 2 minutes. The maps generated using moment invariants, *moments_orig* were not as visually as good as those from *daub_3full_orig* or *original_set*. However, training only took 3 seconds because their vector size is only 7 components long. The original ART maps *art_orig* (36 components) compared to moment invariants demonstrated better visual clusters, however the edge detected ART map *art_edge* were visually the worst of all the maps. This supports the observation that ART is suited to region based analysis [6].

Both the original D_o and edge detached D_e image sets produced clusters. However, because of the SOM's BMU rule (the magnitude of the Euclidean distance) it had a tendency to pull dark images together and push lighter images away. The edge detected SOMs were less affected by this effect as in Fig 3.

som268E_d3full.vis - Size: 25×20, gaussian neigbourhood

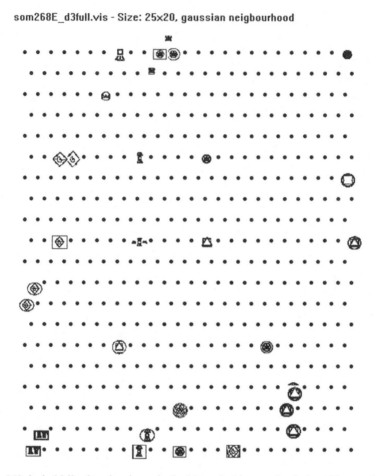

Fig. 3. *268 daub_3full_edge*, showing only the 31 test-bed images for clarity with more isolated clusters than the *daub_3full_edge* map

For the 268 test set, because of the increased number of trademark images involved clusters were not clearly visible. To allow the 31 test-bed images to be identified, they are the only images shown as in Fig 3. Here the 4 groups are distributed over a wider area. This is due in part to the larger image set, and increased number of similar feature vectors competing on the map. Images can find themselves initially distributed far from their cluster group. Therefore, during adaptation it can be difficult for an image to move closer to its main cluster group, because of the increased number of intermediate nodes in between. If these intermediate nodes themselves became BMUs they would push these outliner images further away resulting in isolated - or smaller - clusters. The larger map does however suggest that it is useful to show the general distribution, but a localised or a growing map approach may be better at training large and growing image sets.

3.4 Quantitative Evaluation

Previously precision-recall [16] measures have been used to objectively evaluate trademark image retrieval systems [14], [17]. Precision is the proportion of relevant items retrieved to the total retrieved, and recall is the proportion of relevant items retrieved to the query set.

Precision-recall values are traditionally calculated by selecting a maximum cut-off number of items N_{co} to be retrieved. With the SOM precision $P(r)$ recall $R(r)$ values can be calculated for different neighbourhood radii cut-off $r=N_{rco}$ [18] instead.

The average, over all N images, precision $P_{avg}(r) = \sum P(r)/N$ and recall $R_{avg}(r) = \sum R(r)/N$ values for increasing SOM neighbourhood radii for the edge detected and MRA decomposition of the 31 test-bed trademark image set $\mathbf{D_e}$ is plotted in Fig 4, alongside their vector size.

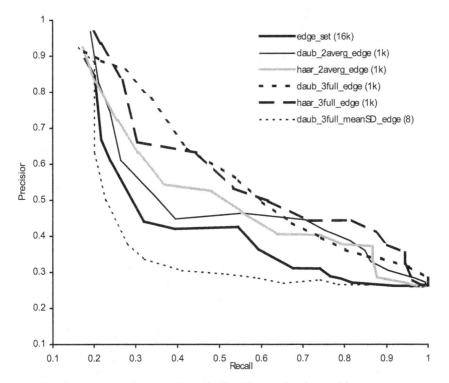

Fig. 4. Average precision-recall graphs for different edge detected feature vectors

The graph and computed results of the average MRA decomposition to 2-levels has average precision-recall values of (*daub_2averg_orig* 0.46Precision, 0.63Recal) and (*daub_2averg_edge* 0.46P, 0.65R) with vector sizes of 1k. However, better profiles and average values overall were obtained for the full MRA decomposition to 3-levels, (*daub_3full_orig* 0.49P, 0.65R) and (*daub_3full_edge* 0.50P, 0.68R). These also have a vector size of 1k. This suggests that MRA decomposition can produce better discrimination between images. Overall, the graph and computed results suggests that MRA appears to have better discriminating powers than the raw images (*original_set* 0.45P, 0.62R) and (*edge_set* 0.42P, 0.61R), and as the feature vectors is $1/16^{th}$ the size, it is $O(16)$ quicker to train. The edge detected D_e MRA training sets also yielded better average results, compared to the original image sets D_o.

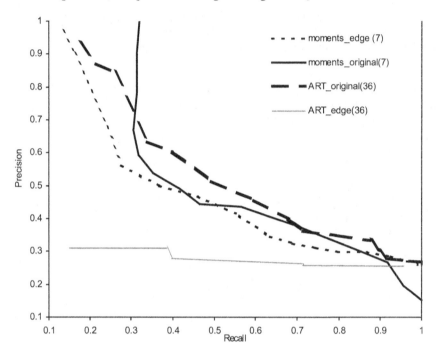

Fig. 5. Average precision-recall graphs for moment invariants and ART feature vectors

The original set's moment invariants, (*moments_orig* 0.47P, 0.64R), profile and average values are better than their edge detected moment invariants, (*moments_edge* 0.44P, 0.64R). If this is compared against ART, the other region based feature vector, the original profile and average values (*art_orig* 0.47P, 0.67R) is marginally better than moments invariants. However its edge detected set (*art_edge* 0.28P 0.60R) is very poor, which is almost horizontal as can be seen in Fig 5. In this work whole image ART is marginally better than moment invariants, although its map is visually better than moment invariants. This also agrees with observations that these two techniques are better suited to grey level and whole scene analysis rather than edge detected objects [6].

An anomaly that has occurred is that the graph and average values for edge detected moment invariants (*moments_edge* 0.44P, 0.64R) suggest that they are better than the raw edge detected (*edge_set* 0.42P, 0.62R) maps. However, by visually inspecting the SOM maps, *edge_set* appear to have better clusters than *moments_edge*. This suggests that, visual maps do not always agree with computational evaluation, further investigation is therefore needed.

Precision-recall profiles for the 268 sets were, as expected from the visual maps, not as good as the 31 set because of the distribution of the images over a wider region, and the capturing of more images as the radii is increased.

4   Conclusions and Further Work

These results demonstrate that for small images sets, small SOMs can be successfully used to visually cluster trademark images on a 2-D map. Results have shown that raw image clustering can be improved, if techniques such as MRA are used. However further work needs to be carried out to investigate which MRA wavelet technique – such as Haar (which favours horizontal and vertical descriptors) or Daubechies – is best suited for clustering trademarks.

If the image sets are large, larger maps are needed to allow more clusters to be distributed throughout the map. However, this may be impractical if the feature vectors are large too, as the training times would be very long. For the larger trademark sets cluster groups were being spread across a wider area, because of the increase in the number of components competing in the feature space. Therefore, different training schemes and SOM structures need to be investigated for larger trademark image sets.

To improve clustering of sets we are currently investigating a hierarchical [18] approach where small maps point to progressively larger maps further along the hierarchy. This should allow a larger set of trademarks to be distributed across larger maps, and achieve better visual clusters, as each low level map could be trained to specialise on particular patterns. As the cost of searching large map is larger, $O(n^2)$, we are also investigating how to limit the map sizes and therefore search space to small regions of activation through a growing hierarchical SOM approach [19].

We are also looking into adding more feature vectors, such as Zernike moments [7]. We also aim to compare our current whole-image feature approach with a component-based approach based on that used in ARTISAN [14]. This will entail building up feature maps of image components and then clustering images based on individual component positions on the feature map. In this way, we hope to be able to demonstrate that clustering using SOMs can prove a viable addition to more conventional retrieval techniques.

References

1.   Jain A. K., Vailaya A., Shape-Based Retrieval: A Case Study With Trademark Image Databases, Pattern Recognition, Vol. 31, No. 9, p1369-1390 (1998).
2.   Pentland A., Picard R. W., Sclaroff S., Photobook: Content-based manipulation of image databases, International Journal of Computer Vision, Vol. 18, No 3, p233-254 (1996).
3.   Gonzales R. C., Woods R. E., Digital Image Processing, Addison-Wesley (1993).

4.   Mehtre B. M., Kankanhalli M. S., Lee W. F., Shape Measures for Content Based Image Retrieval: a Comparison, Information Processing & Management, Vol. 33, No. 3, p319-337 (1997).
5.   Hu M. K., Visual Pattern Recognition by Moment Invariants, IRE Transactions Information Theory, Vol. IT-8, p179-187 (1962).
6.   Sikora T., The MPEG-7 Visual Standard for Content Description – An Overview, IEEE Transactions on Circuits and System for Video Technology, Vol. 11, No. 6, p696-702 (2001).
7.   Teh C. H.,Chin R. T., Image Analysis by Methods of Moment, IEEE Transactions of PAMI, Vol. 10, No 4, p496-513 (1988).
8.   Rui Y., She A. C., Huang T. S., Modified Fourier descriptors for shape representation – a practical approach, Proceedings of the First International Workshop on Image Databases and Multi Media Search (1996).
9.   Mallet S. G., A theory for multiresolution signal decomposition: the wavelet representation, IEEE Transactions of PAMI, Vol. 11, No. 7, p674-693 (1989).
10.   Ravel S., Manmatha R., Multi-modal retrieval of trademark images using global similarity, Internal Report, University of Massachusetts, Amherst (1998).
11.   Santini S., Jain R., The El Niño Database System, Proceedings of the IEEE International Conference on Multimedia Systems and Computing (June 1999).
12.   Duda R. O., Hart P. E., Stork D. G., Pattern Classification, John Wiley & Sons (2001 2nd Edition).
13.   Kohonen T., Self-Organizing Maps, Springer Verlag (2001 3rd Ed).
14.   Eakins J. P., Broadman J. M., Graham M. E., Similarity Retrieval of Trademark Images, IEEE MultiMedia, April-June, p53-63 (1998).
15.   Wang J. Z., Wiederhold G., Firschein, O., Wei S. X., Content-based image indexing and searching using Daubechies' wavelets, International Journal on Digital Libraries, Vol. 1, p311-328 (1997).
16.   Salton G., McGill M. J., Introduction to Modern Information Retrieval, McGraw-Hill. (1983).
17.   Alwis S., Austin J., An Integrated Framework for Trademark Image Retrieval using Gestalt Features and CMM Neural Network, IEE Image Processing and its Application, Conference Publication No 465, p290-295 (1999).
18.   Koskela M., Laaksonen J., Laakso S., Oja E., The PicSOM Retrieval System: Description and Evaluation, Proceedings of The Challenge of Image Retrieval, Third UK Conference on Image Retrieval, Brighton UK (2000).
19.   Hodge V. J., Austin J., Hierarchical Growing Cell Structures: TreeGCS, IEEE Knowledge and Data Engineering, Vol. 13, No 2 March/April (2001).

FACERET: An Interactive Face Retrieval System Based on Self-Organizing Maps

Javier Ruiz-del-Solar and Pablo Navarrete

Dept. of Elec. Eng., Center for Web Research – Dept. of Computer Science
Universidad de Chile, Santiago, CHILE
{jruizd, pnavarre}@cec.uchile.cl

Abstract. The basic problem in content-based image retrieval is the gap between the high-level descriptions used by humans to describe image contents and the low-level features, such as color, texture and shape, used to automatically index images in databases. This problem is even harder when there are non-trivial high-level human descriptions of the images, as in the case of face images. In these cases the employment of user interaction in the retrieval process is a good possibility to solve this task. In this context, FACERET, a content-based face retrieval system that uses self-organizing maps and user feedback is here introduced. Some simulations of the FACERET operation are also shown.

1   Introduction

Nowadays the massive access for information and the growing availability of digital image databases demand the existence of retrieval systems that can understand human high-level requests. For this reason content-based image retrieval is today an expanding and lively discipline. The aim of this article is to tackle this problem in the specific case of face images. The content-based retrieval of faces has multiple applications that exploit existing face databases. In this context, one of the most important tasks is the problem of searching a face without having an explicit image of it, but only its remembrance.

There are many alternatives in order to develop an efficient system for content-based face retrieval. In this work we have used the so-called *relevance feedback* approach to implement the here-proposed *FACERET* system. Under that approach previous human-computer interactions are employed to refine subsequent queries, which iteratively approximate the wishes of the user [1]. This idea is implemented using self-organizing maps. In particular, for a fast operation (training and retrieval) of our retrieval system (FACERET), a tree-structured self-organizing map (TS-SOM) [2] is employed for auto-organizing the face images in the database. Similar face images are located in neighbor positions of the TS-SOM. In order to be auto-organized, face images must be represented by feature vectors in the map. Based on one of the most successful approaches used in face recognition we choose to use PCA-projections for this task. Principal Component Analysis (PCA) projects face images onto a dimensional reduced space where the faces are well represented in a holistic sense.

M. S. Lew, N. Sebe, and J. P. Eakins (Eds.): CIVR 2002, LNCS 2383, pp. 157-164, 2002.
© Springer-Verlag Berlin Heidelberg 2002

In order to know in which part of the map the requested face is located, FACERET asks the user to select face images, which she considers are similar to the requested one, from a given set of face images. Then, FACERET shows the user new face images, which have neighbor positions in the map, respect to the ones selected by the user. The user and FACERET iterate until the interaction process converges, i.e. the requested face image is founded. A block diagram of our face retrieval system and its interaction with the user is shown in figure 1. In figures 5 and 6 are shown two real examples of the interactive face-retrieval process.

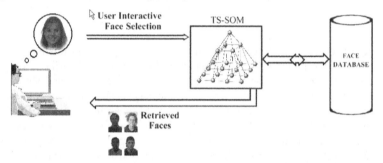

Fig. 1. Block diagram of the FACERET system

As can be noted, our approach for the interactive content-based face retrieval is based on the following assumptions:

1.   the face space forms a cluster in the whole image space,
2.   PCA-projections give a suitable representation of the image space cluster formed by the face images, and
3.   the PCA-representation together with the similarity measure used in the off-line TS-SOM training (usually Euclidean metric) is consistent with the criteria utilized by humans for the selection of similar faces.

The article is structured as follows. The FACERET system is described in section 2. In section 3 the properties of this system are studied based on simulation results. Finally, in section 4 some conclusions of this work are given.

2   FACERET

2.1   Tree Structured SOM – TS-SOM

The TS-SOM [2] is a tree structured vector quantization algorithm that uses self-organizing maps at each of its levels. In our system, as well as in PicSOM [1], all TS-SOM maps are two-dimensional. Figure 2 shows an example of a TS-SOM structure.

The TS-SOM training procedure begins with the highest map and it continues with the lower ones. As in any Self-Organizing Map, each node (or neuron) has a weight-vector that represents a certain area of the high-dimensional input space. In addition, the TS-SOM structure allows connections between each node (the parent) and four nodes in the next level (the children), as shown in Figure 2. The TS-SOM is trained using face images that are previously projected into a lower dimensional space using the PCA algorithm. In order to determine the weight-vectors of all nodes at all TS-SOM levels, using the training samples (projected face images), the TS-SOM algorithm carries out the following iteration procedure:

In the first level (that has one node) the weight-vector is the mean of all the training samples.

In the other levels, weight-vectors of the previous level are copied into the children of the current level (initialization).

The centroid associated to each node is determined as the mean of the closest training samples. The set of closest samples is found based on limited search, by comparing the distance of each sample to the node, with the distance of each sample to a selected set of nodes. This set is composed by the children nodes of the parent (best matching unit) of the previous level and the children of the parent's neighbors.

The new weight-vectors are calculated as the mean of the neighbor centroids, weighted by the number of closest samples (of each node) and a kernel function that gives more importance to the closest neighbors (usually a Gaussian function).

If the weight-vectors do not change more than a given value, then the procedure continues with the next TS-SOM level in step 1). If not, then it goes into step 2).

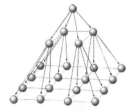

Fig. 2. Example of a two-dimensional TS-SOM structure of 3 levels

In order to use a self-organizing map for a content-based image retrieval we must worry about the complexity involved in the training process. A standard SOM map needs a long time to train huge maps using large databases. TS-SOM was originally intended as a fast implementation of the SOM. In fact, the TS-SOM algorithm works much faster because of the limited updating in each training level. Also the search space on the underlying SOM levels is restricted to a predefined portion of the map, just below the best-matching unit on the above SOM. Then, the complexity of searches in a TS-SOM structure with n units is $O(\log_p n)$, with p the number of children per node. This property explains the reasons for using this structure in our face retrieval system.

In figure 3 is shown an example of face auto-organization in a TS-SOM. In this figure a zoom into the lowest TS-SOM level is carried out.

2.2   Principal Components Analysis - PCA

PCA is a general method to identify the principal differences between signals and after that to make a dimensional reduction of them. In order to obtain the face vectors in the reduced space, which are used to auto-organize the faces in the TS-SOM, we need to obtain the projection axes in which exists the largest variance of the projected face images. Then, we repeat this procedure in the orthogonal space that is still uncovered, until we realize that there is no more variance to take into account. The theoretical solution of this problem is well known and is obtained by solving the eigensystem of the correlation matrix $\mathbf{R} \in R^{N \times N}$:

$$\mathbf{R} = E\left\{ (\mathbf{x} - \overline{\mathbf{x}})(\mathbf{x} - \overline{\mathbf{x}})^{\mathrm{T}} \right\} \tag{1}$$

where $\mathbf{x} \in R^N$ represent the normalized image vectors, $\overline{\mathbf{x}} \in R^N$ is the mean face image, and N is the original vector image dimension. The eigenvectors of this system represent the projection axes or eigenfaces, and the eigenvalues represent the projection variance of the correspondent eigenface. Then by sorting the eigenfaces in descendent order of eigenvalues we have the successive projection axes that solve our problem.

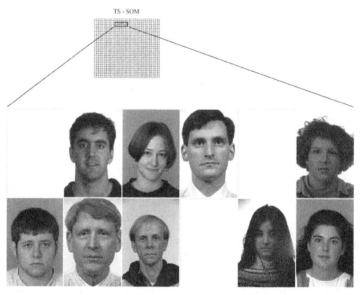

Fig. 3. Example of face auto-organization. Zoom at the lowest level of the TS-SOM., similar faces are located together (face images from FERET)

The main problem is that $\mathbf{R} \in R^{N \times N}$ is too big for a reasonable practical implementation. We have a database of NT face images (the training set), and then we need to estimate the correlation matrix just by taking the corresponding averages in the training set. Let $\mathbf{X} = \left[(\mathbf{x}^1 - \overline{\mathbf{x}})(\mathbf{x}^2 - \overline{\mathbf{x}}) \cdots (\mathbf{x}^{NT} - \overline{\mathbf{x}}) \right]$ be the matrix of the normalized training vectors. Then, the $\mathbf{R}$ estimator will be given by $\mathbf{R} = \mathbf{X}\mathbf{X}^{\mathrm{T}}$. We could say that the number of eigenfaces must be less than or equal to NT, because with NT training images all the variance must be projected into the hyperplane subtended by the training images. In other words the rank of $\mathbf{R}$ is less than or equal to NT. Thereafter they could have more null or negligible eigenvalues depending on the linear dependence of the vectors in the training set. In addition, the eigensystem of $\mathbf{X}^T\mathbf{X} \in R^{NT \times NT}$ has the same non-zero eigenvalues of $\mathbf{R}$, because $\mathbf{X}\mathbf{X}^T\mathbf{X}\,\mathbf{v}^k = \lambda_k\,\mathbf{X}\,\mathbf{v}^k$ represent both systems at the same time. Now we can solve the reduced eigensystem of $\mathbf{X}^T\mathbf{X} \in R^{NT \times NT}$. The correspondent eigenvalues are just the eigenvalues of the original system, and the eigenfaces are represented by $\mathbf{w}^k = \mathbf{X}\,\mathbf{v}^k$, and to be normalized they must be divided by $\sqrt{\lambda_k}$.

2.3   Content-Based Retrieval Procedure

FACERET is based in the PicSOM implementation [1]. In order to know in which part of the map the requested face is located (the map is auto-organized), FACERET asks the user to select face images that she considers are similar to the requested one, from a given set of face images. Then, FACERET shows the user new face images, which have neighbor positions in the map, respect to the ones selected for the user. The user and the FACERET iterate until the interaction process converges.

The main strategy in order to search a given face image on the TS-SOM maps is to use so-called selection matrices. In each level of the TS-SOM structure a selection matrix has the same dimension of the corresponding TS-SOM map. When the user selects the retrieved face images most similar to the face she is looking for, FACERET assigns positive values to the user's selected faces and negative values otherwise. The positive and negative values are then normalized so that their sum equals zero. Then, FACERET summed the obtained values to the nodes on the TS-SOM net in which the face images selected/ignored are respectively located. A low-pass filter is then applied on the selection matrix in order to spread the positive/negative response.

The interaction procedure between FACERET and a user is outlined in figure 4. At first FACERET shows a random set of face images from the database. The user selects the retrieved face images most similar to the face she is looking for. With this information FACERET updates all the selection matrices as previously described. Finally, the new set of retrieval face images is selected by choosing images whose corresponding nodes have the highest values in all the selection matrices. Several examples of selection matrices are shown in figures 5 and 6.

1. A random set of faces is presented to the user.
2. User interactive selection of faces.
3. System content-based face retrieval.
4. User analysis of retrieved faces.
4.a. Requested face was found   → Exit
4.b. Similar faces were found   → Go to 2
4.c. No similar faces were found
4.c.1. User tired   → Exit
4.c.2. User no tired (re initialization) → Go to 1

Fig. 4. Interaction procedure between FACERET and a user

The hypothesis assumed under the described approach is that the PCA-representation together with the similarity measure used in the off-line TS-SOM training (usually Euclidean metric) is consistent with the user-criteria used for the selection of similar faces. If this hypothesis is true, then the highest values in the selection matrix should move the searching process towards the zone in the TS-SOM map in which the searched face is located.

3   Simulations

In practical implementations FACERET needs a large face image database in which several similar face images are available. In other case the user will have no reference faces in order to find the person she is searching for.

For testing our system we choose to use 1196 face images of the gallery set from FERET database [3]. We employed a TS-SOM with 7 levels and 64x64 nodes in its lowest level. In Figure 5 and 6 are shown two examples of the real searching process for average cases. In these simulations six face images are shown in each iteration to the user. After the selection process, the selection matrices are computed and new images are chosen for the next iteration. These images are chosen by searching for face images not shown before, and closer to those points in which the selection matrices reach the highest values in all levels. Selected face images determine a set of zero-mean values that are summed to those points on selection matrices in which the selected face images are located. Then a low-pass filter is applied on each selection matrix in order to spread the selection values over each level.

4   Conclusions

In this work a content-based face retrieval system that uses self-organizing maps and user feedback was presented. To build this system we assumed that face images could be founded in large databases by searching for similar faces in Self-Organizing Maps. Also we assumed that the PCA projection methods work as a suitable representation of the image space cluster formed by face images, as well as a representation consistent with human criteria of similar faces. In our simulations we have realized that this kind of systems do work with large databases. As a future work we want to perform a detailed study of the properties of the proposed system.

Acknowledgements

This research was funded by Millenium Nucleous Center for Web Research, Grant P01-029-F, Mideplan, Chile. Portions of the research in this paper use the FERET database of facial images collected under the FERET program.

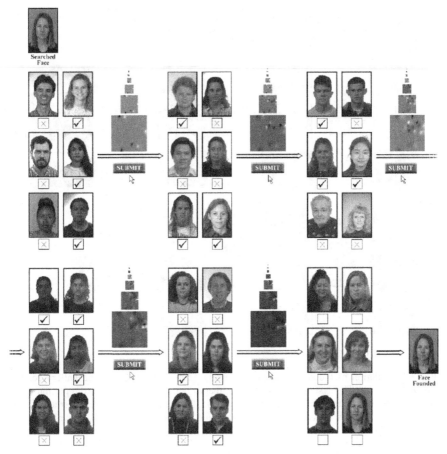

Fig. 5 Simulation of the searching process using 1196 face images of gallery set from FERET database. The system shows six face images per iteration. In this example the system shows the requested face image in the 6th iteration. The seven maps between iterations, show the selection matrices for each level that determine the face images to be shown in the next iteration

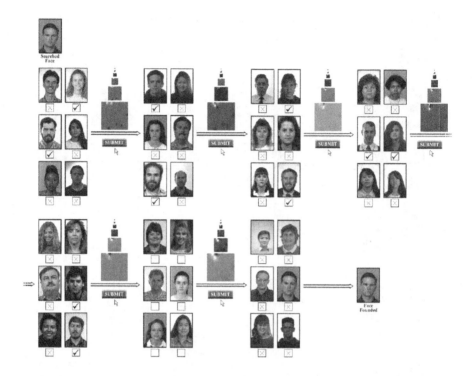

Fig. 6. Simulation of the searching process using 1196 face images of the gallery set from FERET database. The system shows six face images per iteration. In this example the system shows the requested face image in the 7th iteration. The seven maps between iterations, show the selection matrices for each level that determine the face images to be shown in the next iteration

References

[1] J. Laaksonen, M. Koskela, S. Laakso and E. Oja, "PicSOM – content-based image retrieval with self-organizing maps", *Pattern Recognition Letters*, vol. 21, 1199-1207, 2000.

[2] P. Koikkalainen, Oja E., "Self-organizing hierarchical feature maps", *Proc. Of 1990 Int. Joint Conf. on Neural Networks*, vol. II. IEEE, INNS, San Diego, CA, 1990.

[3] P. J. Phillips, H. Wechsler, J. Huang, and P. Rauss, "The FERET database and evaluation procedure for face recognition algorithms," *Image and Vision Computing J.*, Vol. 16, no. 5, 295-306, 1998.

Object-Based Image Retrieval
Using Hierarchical Shape Descriptor

Man-Wai Leung and Kwok-Leung Chan

Department of Computer Engineering and Information Technology
City University of Hong Kong, 83 Tat Chee Avenue, Kowloon, Hong Kong
m.w.leung@student.cityu.edu.hk
itklchan@cityu.edu.hk

Abstract. Shape is the most basic and convenient feature to describe objects. Retrieval by shape similarity is implemented in this project. Object shapes are segmented into tokens according to their local feature of minimum turn angle. User sketch is the query input and the retrieval algorithm matches the sketch with the nearest object in the database by using features distance. Scaling, rotation and missing sketch of objects are also considered in this paper. Together with the M-tree indexing, the system performance can be strengthened. However, many objects have similar outer shape boundary but different inner shapes. The retrieval accuracy will be affected by this situation. Hierarchical Shape Descriptor is proposed to solve the problem. It can distinguish similar outer boundaries but with different inner shapes objects. A completely new image retrieval system is implemented in order to accommodate the new image content descriptor. Our results show that the proposed system is fairly accurate and the Hierarchical Shape Descriptor is a better image content descriptor than the existing method using only the outer boundary.

1 Introduction

With the advances in multimedia technology, the size of on-line libraries of digital images is increasing. The user with a query picture in hand may want to retrieve all the similar images from the library. Effective access to the image library can be made based on the visual content of the image. Content-based image retrieval (CBIR) becomes a popular research topic. For an extensive review, see [1]. An image retrieval system usually focuses on some or all of the following problems [2]: description of image content; similarity measure; indexing and archiving. CBIR systems commonly use colour, shape, texture, or their combination as image content descriptors. Colour provides a global low-level statistical information of the image. Colour descriptor is compact which is good for similarity measure and indexing [3]. The limitations are that the retrieved results do not necessarily correspond to high-level semantics, and the descriptor does not handle spatial relationship between image

M. S. Lew, N. Sebe, and J. P. Eakins (Eds.): CIVR 2002, LNCS 2383, pp. 165-174, 2002.

regions. Shape, a common descriptor of objects, contains semantic information because human can recognize objects solely from their shapes. Shape description techniques can be coarsely classified into boundary-based and region-based [4]. The first approach is adopted in this project. Shape retrieval has been shown to be robust in the presence of partial occlusion, scaling and rotation [5]. Local protrusions are modelled by tokens representing the maximum curvature and orientation. Another shape descriptor, the curvature scale space (CSS), is developed and enhanced to represent 2D objects [6]. The use of curvature has two limitations. Firstly, it only provides local information of convexity/concavity of the object. Secondly, the calculation of curvature involves derivatives of boundary coordinates and is noise sensitive. Texture is often described in terms of structure and randomness. The use of texture in image modelling and retrieval has been performed in [7]. However, a purely texture-based retrieval system has a very limited application. Instead, practical retrieval systems accommodate various image descriptors, e.g. colour and texture [8], colour and shape [9], colour, shape, texture and sketches [10]. For effective organization of the image library, similarity measures mostly are of metric distance [11]. However, triangular inequality does not hold in human similarity perception. A good similarity measure should satisfy the contrasting requirements. Indexing and archiving become important problems when the digital libraries grow in size.

This project adopts the method proposed by Berretti et al. [5]. Modification is made to use a new image content descriptor called Hierarchical Shape Descriptor. The segmentation of shape boundary into tokens is described in the following section. The similarity measure is explained in Section 3. M-tree indexing is presented in Section 4. In Section 5, the idea of Hierarchical Shape Descriptor is introduced and some experimental results are shown in Section 6. Finally, conclusion is given in the last section.

2   Shape Segmentation (Shape Token)

The shape boundary is partitioned into segments, called token (τ_k) [5]. The partitions occur at the points where the turn angle [12] is locally minimum or at the change of +/- sign. Two token features, maximum turn angle and orientation, represent each token of the shape boundary.

2.1   Turn Angle

For each point in the shape boundary, the value of turn angle (θ) represents the angle turned of the line and the direction changed at the vertex. The definition of turn angle is illustrated in Fig.1. The range of θ is [-180, 180). The sign of θ shows the direction of the turn. Positive(negative) θ means the line turning clockwise(anti-clockwise). A minimum θ set, $T = \{t_i\}_{i=1}^{N}$, can be found for each shape boundary and we can assume that there is always a maximum θ, called m_k, between two consecutive minima t_k, t_{k+1}. For each segment, m_k represents the width of the token. As the result, wider(narrower) tokens have lower(higher) m_k values. The arrangement of tokens can affect shape perception. The orientation can reduce the number of arrangements of the tokens.

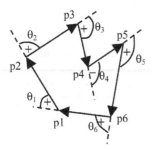

Fig. 1. Turn angle of the points in the shape boundary

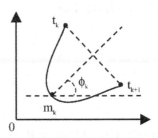

Fig. 2. The angle ϕ_k is the value of orientation for this token

2.2 Orientation

The orientation (ϕ) of each token, in polar coordinates, is the vector linking the median point of the two consecutive minima, t_k and t_{k+1}, to the point which has the value of t_{mk} with respect to the horizontal axis. The definition of token orientation is illustrated in Fig.2. The range of ϕ is [0, 360).

3   Token Distance and Shape Distance

The metric distance between τ_i and τ_j is used to measure their similarity.

$$d_{\text{turn angle}} = \left| \text{turn angle}(\tau_i) - \text{turn angle}(\tau_j) \right| \tag{1}$$

$$d_{\text{orientation}} = \left| \text{orientation}(\tau_i) - \text{orientation}(\tau_j) \right| \tag{2}$$

$$d_{ij} = \alpha d_{\text{turn angle}} + (1 - \alpha) d_{\text{orientation}} \tag{3}$$

The parameter α, in the range [0,1], determines the relative weight of turn angle and orientation distances. Assuming that shape A and shape B have n and m tokens respectively, $n<m$ (if $m<n$, then A and B are exchanged), and δ_s is the token distance threshold, the shape distance of A and B is calculated by

```
for each token a_i in A
    for each token b_j in B
            d_ij = d( a_i , b_j )
    for each token a_i in A
        match a_i with the nearest b_j that is not yet matched
        with any tokens in A
    if(∀i∈1,...,n   d_i = d_ij≤δ_s)



    then d(A,B)= Σ^n_{i=1} d_i
                ─────────
                    n
```

4   M-tree Indexing

The token set of each shape is stored in the database using M-tree indexing [13]. M-tree, which is built according to token distance, can reduce the searching time of a token as compared to a linear search. The routing object and the covering radius can help the query token to search for the nearest token at the leaf level quickly.

5   Hierarchical Shape Retrieval

There are plenty of objects in the world, most of them have similar outer shape boundaries. Therefore, the accuracy in retrieving objects with similar outer shapes but with different inner contents would be lowered. A more accurate image content descriptor has been developed in this project, called "Hierarchical Shape Descriptor". Objects with the same outer shape boundary but with different inner shapes can be distinguished by this approach. Hierarchical Shape Descriptor allows users to sketch more shapes other than the outer shape boundary. The number of inner shapes is unlimited. The retrieval time of Hierarchical Shape Descriptor would be longer than that of the retrieval using only one shape. This is due to more comparison for each object in the database. The performance can be improved by assuming that the inner shape(s) is not checked if the outermost shape is not similar (see Fig.3). Consider two objects A and B with n and m inner shapes respectively. The steps for hierarchical shape retrieval can be represented as following.

```
if ( d(outer shape of A, outer shape of B)
       <shape similarity threshold )
{
      for each inner shape ha_i in A
           for each inner shape hb_j in B
                 sd_{ij} = d(ha_i, hb_j)
      for each inner shape ha_i in A
      match ha_i with the nearest hb_j that is not yet
      matched with any ha_i
      if (∀i∈1,…,n   sd_i = sd_{ij}≤δ_s)
                        ∑_{i=1}^{n} sd_i
      then D(A,B) = ─────────────
                            n
}
```

D(A,B) is the hierarchical inner shape distance. The system shows the retrieval result image set in the most matched order. That is, the most similar image will be the first image in the result set.

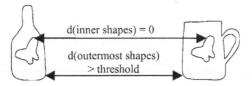

Fig. 3. The two objects are not matched even the inner shapes are the same

6   Experimental Results

6.1   Retrieval Example

This image retrieval system has been implemented and the shape similarity retrieval examples are presented in Fig. 4 and Fig. 5. There are 157 objects in the testing database and the objects are classified into three categories of bottles, cups and gun like objects. Fig. 6 shows the contents of the database used in the testing. The retrieval process can be made more flexible by varying the trust factor. The trust factor represents the minimum similarity threshold of shape distance. Higher trust factor value implies larger shape distance threshold. As the result, the number of retrieved objects can be increased.

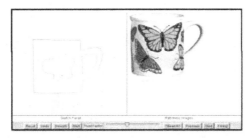

Fig. 4. A cup with hierarchical sketches (left) and its retrieval result (right)

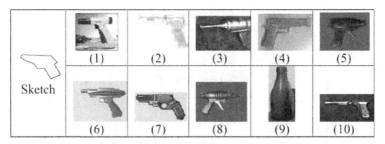

Fig. 5. Outer shape retrieval example with result set in most matched order

6.2   Retrieval Accuracy Analysis

Three different sets of retrieval performance have been analyzed. They are focused on outer boundary retrieval, hierarchical shape retrieval and retrieval with incomplete query.

Outer Boundary Retrieval. As the total number of similar shapes in the database corresponding to the query sketch is known, the analysis of precision(P) and recall(R) can be computed. The definition of precision and recall are defined as follow:

$$precision = \frac{number\ of\ relevant\ items\ retrieved}{total\ number\ of\ retrieved\ items} \tag{4}$$

$$recall = \frac{number\ of\ relevant\ items\ retrieved}{total\ number\ of\ relevant\ items\ available} \tag{5}$$

Fig. 6. The objects contained in the database

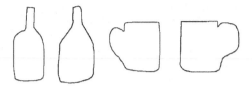

Fig. 7. Sketched shapes used in the test

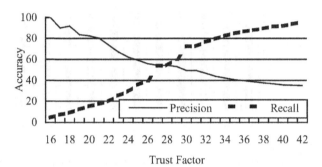

Fig. 8. Recall and precision as a function of the trust factor

The test shapes (Fig. 7), two bottle sketches and two cup sketches, are used. The plotted result (Fig. 8) is the average of the four test sketches. The precision remains high at the trust factor from 16 to 23 and the recall quickly reaches 70% at the trust factor of 29.

Hierarchical Shape Retrieval.   The comparison of precisions of two queries is shown here. The first query sketch has only the outer boundary defined which is the method of Berretti et al [5]. The second one is the hierarchical shape described query in which the outer boundary is exactly the same as the first one. In Fig. 9, a hierarchical shape retrieval result set is shown and the comparison result with outer shape retrieval is shown in Fig. 10. The hierarchical shape descriptor has a better performance than the outer boundary retrieval does. In addition, it gives a high stability because the accuracy is always above 70%.

Fig. 9. Hierarchical shape retrieval example with result set in most matched order

Retrieval with Incomplete Sketch.   The same test sketches as in the outer boundary retrieval experiment are used. For each query sketch, different degrees of missed sketch of 0%, 30% and 60% of the sketched shapes are defined. In addition, incomplete sketches, either of 30% or 60%, in three different directions are experimented. Fig. 11 shows the idea of incomplete sketch of 30% and 60%. The analysis of precision and recall as a function of missed sketch are shown in Fig. 12. There is only a small drop in both the precision and recall even 60% of the original shape are missing.

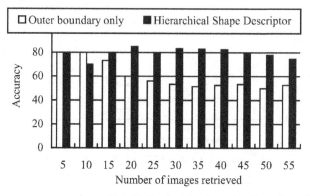

Fig. 10. The accuracy comparison of two kinds of retrieval queries: Outer boundary retrieval and Hierarchical Shape Descriptor

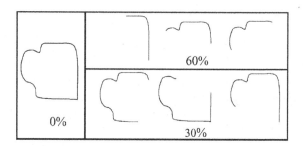

Fig. 11. Examples of incomplete queries of 0%, 30% and 60% in different directions

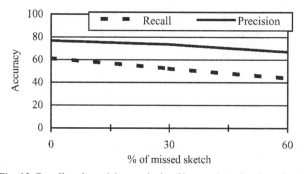

Fig. 12. Recall and precision analysis of incomplete sketch retrieval

7   Conclusion

An object-based image retrieval by user sketch system is developed. Furthermore, a new idea of Hierarchical Shape Descriptor is also proposed. Turn angle and orientation are the features for measuring the similarity of token and M-tree indexing helps to speed up the retrieval speed. The importance of retrieval accuracy and missing sketch has been considered in this project. Different kinds of retrieval analysis have been carried out. Our results show that the proposed system is fairly accurate and the Hierarchical Shape Descriptor is a better image content descriptor than the existing method using only the outer boundary.

References

1. A. W. M. Smeulders, M. Worring, S. Santini, A. Gupta, and R. Jain, "Content-based image retrieval at the end of the early years," IEEE Transactions on Pattern Analysis and Machine Intelligence, Vol. 22, No. 12, pp. 1349-1380, December 2000.
2. E. Vicario (Ed.), Image description and retrieval, Plenum Press, New York, 1998.
3. Y. Deng, B. S. Manjunath, C. Kenney, M. S. Moore, and H. Shin, "An efficient colour representation for image retrieval," IEEE Transactions on Image Processing, Vol. 10, No. 1, pp. 140-147, January 2001.
4. B. M. Mehtre, M. S. Kankanhalli, and W. F. Lee, "Shape measures for content based image retrieval: a comparison," Information Processing & Management, Vol. 33, No. 3, pp. 319-337, 1997.
5. S. Berretti, A. del Bimbo, and P. Pala, "Retrieval by shape similarity with perceptual distance and effective indexing," IEEE Transactions on Multimedia, Vol. 2, No. 4, pp. 225-239, December 2000.
6. S. Abbasi, F. Mokhtarian, and J. Kittler, "Enhancing CSS-based shape retrieval for objects with shadow concavities," Image and Vision Computing, Vol. 18, pp. 199-211, 2000.
7. F. Liu and R. W. Picard, "Periodicity, directionality, and randomness Wold features for image modelling and retrieval," IEEE Transactions on Pattern Analysis and Machine Intelligence, Vol. 18, No. 7, pp. 722-733, July 1996.
8. B. S. Manjunath, J.-R. Ohm, V. V. Vasudevan, and A. Yamada, "Colour and texture descriptors," IEEE Transactions on Circuits and Systems for Video Technology, Vol. 11, No. 6, pp. 703-715, June 2001.
9. T. Gevers and A. W. M. Smeulders, "PicToSeek: combining colour and shape invariant features for image retrieval," IEEE Transactions on Image Processing, Vol. 9, No. 1, pp. 102-119, January 2000.
10. M. Flickner, H. Sawhney, W. Niblack, J. Ashley, Q. Huang, B. Dom, M. Gorkani, J. Hafner, D. Lee, D. Petkovic, D. Steele, and P. Yanker, "Query by image and video content: the QBIC system," IEEE Computer, pp. 23-32, September 1995.
11. S. Santini and R. Jain, "Similarity measures," IEEE Transactions on Pattern Analysis and Machine Intelligence, Vol. 21, No. 9, pp. 871-883, September 1999.

12. L. J. Latecki and R. Lakamper, "Convexity Rule for Shape Decomposition Based on Discrete Contour Evolution," Computer Vision and Image Understanding, Vol. 73, No. 3, March, pp. 441-454, 1999.
13. P. Zezula, P. Ciaccia, and F. Rabitti. M-tree: A dynamic index for similarity queries in multimedia databases. Technical Report 7, HERMES ESPRIT LTR Projects, 1996.

Multimodal Person Identification in Movies*

Jeroen Vendrig and Marcel Worring

MediaMill, University of Amsterdam
Kruislaan 403, NL-1098 SJ Amsterdam
{vendrig,worring}@science.uva.nl

Abstract. An important task for annotation of movies is finding out which characters are playing in a shot. Character identification is based on available information sources from various modalities. Fully automatic character identification is not feasible as the modalities are not semantically synchronized. As manual annotation is too time consuming, an interactive tool assisting the annotator is needed. We propose the WhoIsWho function for our interactive i-Notation system.

WhoIsWho relates visual content to names extracted from movie scripts, working in both ways. We present extensive evaluation of character identification on six hours of movies. Employment of a user model enables evaluation of interactivity in WhoIsWho. Quantitative results show that WhoIsWho is successful in helping annotators identify movie characters.

1   Introduction

The advance of digital movies promotes the movie experience from passive viewing to interactive entertainment. To accommodate the variety in viewer demands, we target at assisting the annotators for interactive movies in enriching the video data with semantic metadata. People are mostly interested in seeing people in movies, and consequently people dominate most shots in a movie. Therefore, we focus on annotating *who?* is in the video leading to a range of new applications.

The *who?* question is answered by describing for each shot all characters visible. A character is the name of a personage with a speaking part.

Character identification for a set of shots is an important component of an effective annotation method. An annotator can recognize Robert DeNiro easily. If DeNiro's fame is not sufficient, his picture is found in any movie encyclopedia. But most characters are played by unknown actors. Finding out who those actors and their characters are, is a time consuming task. A video annotation tool should efficiently assist the annotator in finding out who is in the shot.

For computer assisted annotation we use our interactive, adaptive annotation tool i-Notation [4]. It assists annotators in labeling shots in movies. In this paper, we describe i-Notation's character identification, where the system supports the user by making suggestions.

As annotation systems target at improving efficiency, objective evaluation is important. We employ a user model for evaluation of the i-Notation tool's

---

* Funded by NWO and supported by the Multimedia Information Analysis project.

M. S. Lew, N. Sebe, and J. P. Eakins (Eds.): CIVR 2002, LNCS 2383, pp. 175–185, 2002.

performance allowing to measure the effectiveness of character identification in an interactive session.

2 Information Sources

For annotation of movie characters, various information sources are available for automatic processing. The movie itself contains visual and aural information as well as textual closed captions, while scripts carry information about movie content. Visual information conveys who appears in the video. The audio signal discloses what is said and by whom. Closed captions too describe what is said, but with a smaller error rate [1]. Script text describes what is said and by whom, but unrelated to time codes. Thus, scripts and closed captions are supplementary in the textual modality. In this section we describe what features of the available information sources are used to find the name of the character playing in a shot.

To identify who is visible, detection of faces is important. Face detection as used in Name-It [3] for news videos can be used in movies as well, but results in lower performance. Frontal faces are rare in movies, as actors are usually filmed sideways. Also, people move around more than in news videos. A movie annotation system making use of face detection has to deal with large uncertainty in detection results.

Visual coherence making use of repetition in movies is important for annotation as well. Narrative movies are divided into parts with semantic coherence, similar to chapters and sections in a book. Visually the coherence translates to a structure based on the repetition of shots that look similar. For example, the camera can alternate between the faces of two persons talking to one another. The coherence has been exploited successfully to detect movie scenes, as evaluated in [5]. The coherence can be used to support effective annotation in a similar way. Since generally a character remains in the same setting for a while, the background can be used to relate shots with the same character. An example is the Hue-Saturation histogram as used for movie scene segmentation in [2]. For finding a character in several shots, the use of visual features to compare backgrounds is more realistic than face recognition of unknown characters appearing in a non frontal pose.

As aural information is duplicated in closed captions and scripts, textual information is used to identify speakers. Scripts contain the information about speakers, while closed captions have the time codes for the video stream. Therefore, scripts are synchronized with the visual content by using closed captions. Based on the time code aligned script information, to each shot σ_k in the video a potential label $\tilde{L}_k$ is assigned. Potential label $\tilde{L}_k$ contains all character names in the script for the duration of σ_k. Multiple occurrences of the same name are included in $\tilde{L}_k$ so that information about speaker frequency is maintained.

In conclusion, processing of the available information sources results in detection of visual coherence between shots and in generation of potential labels.

3   Alignment of Textual Sources

Alignment of visual and aural presence of a character has to deal with a funda-mental and a practical problem. The fundamental problem is that when a char-acter speaks he does not necessarily have to be visible in the shot. In movies, often characters other than the speaker are visible to show the reaction on what is said. In our data set, we observed that approximately 33% of all potential la-bels $\tilde{L}_k$ are correct. If just shots that contain a non-empty label are considered, performance is even lower. Hence, alignment of shots and speech is not sufficient for correct annotation.

The practical problem for alignment is its dependence on the quality of in-formation sources. There are three types of problems, viz. transcription errors, transcript deviation and script deviation.

Transcription errors are contributed to human error, e.g. mistakes in typing. As a consequence, exact match alignment of text lines is not feasible.

Transcript deviation occurs when text lines in both sources contain the same semantic content, but in a different representation. For example, the movie "Sneakers" starts with the naming of numbers. In the script this is transcribed as "Four six nine", while the closed captions represent it as "46-9". Another exam-ple of transcript deviation is the use of words that sound similar, such as "yes" and "yeah" [6]. Hence, even in the case of error free textual sources, alignment can not be done via an exact match of text lines.

Script deviation occurs when the movie differs from the production script. A few words can be changed in a text line, e.g. because it sounds more natural to the actor. But also changes with major impact on the movie content occur, such as the addition or deletion of scenes. As a consequence, there is no guaranteed one-to-one correspondence between text lines in the closed captions from the actual movie and text lines in the script. In addition, the order of text lines in closed captions and script differ.

To cope with the three problems, we use fuzzy string matching in a context window. For each closed caption line L_i, F_i^1 is the top ranked fuzzy matched line in the script. Using fuzzy matching instead of exact matching solves transcript errors directly as in general few characters are wrong in a misspelled word. Tran-script deviation is solved indirectly. If single words are represented differently, as in the case of the number example, the other words in a sentence compensate for the deviation. Text lines that differ in details only are aligned by the fuzzy string matching technique.

The context window targets at synchronizing the closed captions and script when the order of scenes has been changed. Alignment cannot be done based on one line, because short, general lines cannot be distinguished from one another. An example is the one word text line "Yes.". The context window takes lines before and after the current line into consideration, so that the correct instance of the text line is aligned.

The context window is used as follows. Let F_i^n be the top ranked fuzzy matched lines in the script for line L_i. We use $n = 5$. The same procedure is followed for a window around the closed caption line being aligned. We choose a window of size 4, i.e. the closed caption lines $\{L_{i-2}, L_{i-1}, L_{i+1}, L_{i+2}\}$. Then the script line in F_i^n is selected that is temporally closest to the script lines in $\{F_{i-2}^n, F_{i-1}^n, F_{i+1}^n, F_{i+2}^n\}$. Temporal distance is based on the ordered index numbers of script lines.

The quality of the alignment resulting from the above procedure depends on the quality of the information sources in terms of the three problems identified. The alignment output is a starting point for judgement by annotators, allowing for more efficient interaction.

4   Character Identification

Character identification provides the user with names of persons appearing in selected shots. As described in section 3, fully reliable answers cannot be expected from automatic character identification. Rather than letting users do post-processing of results, we propose to put the human in the loop in an interactive method combining user knowledge and automatic data processing.

For interactive character identification, we employ the WhoIsWho function in the i-Notation system. The WhoIsWho function works in two directions. *WhoIsPerson* produces potential character names given a set of shots. Given a character name, *WhoIsName* produces a set of shots in which the character is likely to appear. The former is used to identify characters, while the latter is used for verification of potential character names.

The shot-character matching function δ is used in both cases of WhoIsWho. Function δ expresses the likelihood that character p appears in shot σ_k, where k is the index of the shot. For each σ_k a potential label $\tilde{L}_k$ is used, based on automatic script alignment.

$$\delta(p, \tilde{L}_s) = \frac{\omega(p, \tilde{L}_s)}{\#(\tilde{L}_s)} \tag{1}$$

where $\omega(p, \tilde{L}_s)$ counts the number of occurrences of p in $\tilde{L}_s$. Count operator $\#$ gives the number of elements in the set.

4.1   WhoIsPerson

Character identification based on visual input can be formulated as follows: given a set of shots S, where each shot contains character p, find out what the value of p is. The shots in S are not restricted to containing character p only. Note, that when more than one character appears in all shots in S, results cannot be used to identify p exclusively.

Character names are determined as follows. Let C_s be the set of all character names occurring in the script. For all names $p \in C_s$, a character score γ_p is computed, expressing the extent to which p matches S.

$$\gamma_p(S) = \sum_{\sigma_k \in S} \delta(p, \tilde{L}_k) \cdot \frac{1}{\#(S)} \tag{2}$$

Finally, the scores $\gamma_p > 0$ for each $p \in C_s$ are ranked. The p having the highest score is expected to be the character name. However, when the top ranked names have the same score, the character remains unidentified due to ambiguity. User interaction with WhoIsPerson is then necessary.

4.2   WhoIsName

Character identification based on textual input can be formulated as follows: given a name p, find a set of shots in which p is visible. WhoIsName is used to verify whether names found are correct.

Given scoring function δ, WhoIsName is trivial. For each unlabeled shot, a score is computed by δ. The character names of labeled shots are known already from previous interaction.

After computing δ for the shots, the scores are ranked. The highest ranking non-zero scores are displayed. The user evaluates whether the faces shown match the initially selected shots in S. If not, the selected p cannot be correct. Otherwise, although guarantees cannot be given, all evidence is pointing towards p being the name sought.

4.3   WhoIsWho Interaction

The two forms of WhoIsWho are used in an interactive process to identify characters. The process is as follows:

1. User selects shots S containing character p.
2. If desirable, user asks for shots visually coherent with S and goes to step 1.
3. System displays ranking of potential characters for S, based on WhoIsPerson.
4. If top ranked score is higher than 2nd ranked score, p is identified. User exits.
5. The top ranked characters with the same score form P.
6. WhoIsName verification: for each $p' \in P$, show matching shots $S_{p'}$
7. If $S_{p'}$ shows p, the character is identified: $p = p'$. User exits.
8. p remains unidentified, as there is no $S_{p'}$ showing p.

The user's knowledge about which persons are the same combined with textual alignment and visual coherence information results in an interactive solution to the character identification problem.

5   Evaluation

In this section we establish what has to be evaluated and how it is done. First, we define requirements for a ground truth. Next, the goals of character identification evaluation are described. Based on the goals, we introduce evaluation criteria.

5.1   Ground Truth

Evaluation concerns the character identification process. For the actual character names a ground truth annotation is needed. To minimize subjectivity in ground truths, we set up two strict rules. Firstly, we focus on the head of a person, as it is usually the only identifiable body part. We require the head of a person to be visible in the key frame presented. Secondly, the face of the person is required to be visible and recognizable somewhere in the shot. The first rule restricts the number of people to be identified based on the key frame. The second rule states that people appearing in a key frame must be recognized as well, at any time in the entire shot.

Since character identification targets mostly at unknown characters, celebrities can be assumed known. We define a celebrity as a character played by an actor whose picture is published in his or her Internet Movie Database biography (www.imdb.com). Use of a publicly available reference assures the term celebrity does not depend on the annotator's showbiz knowledge.

5.2   Evaluation Goals

In this section, we describe evaluation goals measuring WhoIsWho performance.

Both necessity and accuracy for WhoIsWho depend on the quality of the aligned script. In case of a perfect script, the WhoIsWho function is redundant. If the script deviates too much from the content of the movie, WhoIsWho will not be able to perform well.

Quality of aligned script. Determines the input quality for WhoIsWho.

For evaluation of the entire i-Notation system, it is assumed that celebrities are known. As celebrities appear in most of the shots, the assumption has two effects. Firstly, the celebrities are annotated first. Apart from the higher probabilities of appearing at random, the system targets at showing most common characters first. Secondly, weeding out the celebrities first, prevents the suggested names for unknown characters from being dominated by the most appearing names.

Assuming celebrities are known, is reasonable for many of the Hollywood movies. However, the assumption hampers extension of the proposed methods to other movies and other domains. We need to evaluate to which extent the celebrity assumption influences end results.

Necessity for knowing celebrities a priori. Determines whether celebrity characters can be identified without a priori information, i.e. assuming *no celebrities*.

Adding visually similar shots generates additional information to identify characters. The shots are added *interactively*, or *automatically*. Especially when a character has few lines in the movie, this could be a helpful strategy. However, it could as well add confusion.

Impact of visually similar shots addition. Determines whether addition of shots pays off overall.

The WhoIsName function is used for validation of proposed character names. It is used at the user's discretion. Therefore, it is not possible to predict when it will be used. Still, evaluation is imperative, as the function is an important supporting tool for user interaction.

WhoIsName quality. Determines whether names can be associated with visual appearances of the character.

The first screen shown in the i-Notation system is a WhoIsName query based on the character name found most in the script. The more shots on the start screen have the same label, the more efficient shot selection is performed.

Start screen quality. Determines success of first annotation interaction.

5.3   Evaluation Criteria

Quality of the aligned script is measured by two criteria. The first criterion measures completeness of the script, i.e. the percentage of shots that is annotated correct automatically. Let L_k be the ground truth annotation for shot σ_k.

$$S_c = \frac{\#(\{\sigma_k \mid L_k = \tilde{L}_k \wedge L_k \neq \emptyset\})}{\#(\{\sigma_k \mid L_k \neq \tilde{L}_k\})} \cdot 100\% \tag{3}$$

where $L_k = \emptyset$ denotes the label contains no characters.

The second script quality criterion measures character overlap between ground truth annotation and the original script. Let C_g be the set of characters in the ground truth and C_s the set of characters in the script.

$$S_o = \frac{\#(C_g \cap C_s)}{\#(C_g \cup C_s)} \cdot 100\% \tag{4}$$

For evaluation of interactive sessions, we employ user modeling in which rules are defined for user choices during the interaction process. Hence, the a priori knowledge of the user is controlled. The user model provides full control over experiment parameters, resulting in consistent and objective evaluation.

Overall character identification results are evaluated by measuring how often the system's proposed character name was correct. The user is modeled as follows. First, the user selects S and performs the WhoIsPerson function. The user evaluates the names proposed. By default, we assume the user knows the celebrities in the movie, so that they are dismissed from the names list. Names of characters that have been used in a label already are dismissed as well. The top ranked name remaining is processed to result in a global evaluation measure W_g for the WhoIsPerson function.

$$W_g = \frac{\#\text{correct identifications}}{\#\bar{C}_g} \cdot 100\% \tag{5}$$

where $\bar{C}$ denotes set C without celebrities.

By lack of other systematic approaches to the character identification problem, success of WhoIsPerson is measured by comparison to random selection of

Table 1. Characteristics of the movie collection

Movie	#shots	#characters C_g	#celebrities	#unknown characters
Shakespeare in Love	1928	45	7	38
Sneakers	1545	30	7	23
LA Confidential	2426	51	10	41

Table 2. Script quality for the movie collection

Movie	S_c	S_o
Shakespeare in Love	16%	79%
Sneakers	27%	72%
LA Confidential	11%	54%

a name from the set of unidentified names $\bar{C}_s$. We use the probability that the correct name is selected. The *random comparison* measure is defined as follows:

$$W_r = \sum_{i=0}^{\bar{C}_g} \frac{1}{\bar{C}_s - i} \cdot \frac{1}{\bar{C}_g} \cdot 100\% \tag{6}$$

where counter i equals the amount of names in $\bar{C}_g$ and $\bar{C}_s$ that has been identified during the interaction process.

In addition to the overall criteria, we propose two criteria measuring performance at intermediate levels of the annotation process. For evaluation of the start screen quality we propose character similarity. Character similarity W_s measures the usefulness of the start screen for shot selection. To this end, the label L is found that describes the shots S in the start screen best by measuring the percentage of shots in the start screen with the same label.

$$W_s = \frac{\#(\{\sigma_k \in S \mid L_k = L\})}{\#(S)} \tag{7}$$

For evaluation of WhoIsName, we measure performance of the function at various points in the interaction session, e.g. each time when another n shots have been annotated. Let P be all characters yet unidentified, and S the shots found by the WhoIsName function. The number of shots in S is bounded by the number of shots displayed on screen. Criterion W_t^n then measures how often the WhoIsName function is correct for all names in P:

$$W_t^n = \sum_{p \in P} (\sum_{\sigma_k \in S} \#(\delta(p, L_k) > 0) \cdot \frac{1}{\#(S)}) \cdot \frac{1}{\#(P)} \cdot 100\% \tag{8}$$

6   Results

Performance of the WhoIsWho function is evaluated for three well known Hollywood movies. Characteristics of the movies are given in table 1.

Results for the two script quality criteria are given in table 2. Sixty to eighty percent of the shots have a non-empty label. Detailed inspection of results shows that the scripts of "Sneakers" and "Shakespeare in Love" are relatively accurate. However, in the latter script confusion is caused by the use of various names for the same character. For example, the name of a rascal boy changes from the general term "urchin" into his actual name "Webster" half way the movie. The script for "L.A. Confidential" contains scenes not part of the movie or changed significantly. For other scenes, quality is comparable to the other scripts.

The impact of changing settings of the *default* user model is an important part of evaluation. The default user model has the following settings: celebrities are assumed known; nine shots are shown on screen during each interaction.

In table 3, results are shown for the default user model and three variants to measure the various evaluation goals from section 5.2. Treating every character as a non celebrity influences end results slightly. The better value of W_g for "L.A. Confidential" is explained by the good results for identification of celebrities. Celebrity characters in the movie are relatively easily identifiable, as they appear alone often, in contrast with celebrities in the other two movies.

Addition of visually similar shots to the set on which a WhoIsPerson function is performed, does improve performance. However, results show interactive selection is necessary. The weak performance of automatic addition for "Shakespeare in Love" and "L.A. Confidential" is explained by their poor script quality.

Results for start screen quality vary from $W_s = 0.33$ for "Shakespeare in Love" to $W_s = 0.67$ for "Sneakers". Experiments with additional movies show that results fall in the same range. As a large set of shots is available for generation of the start screen, results are not satisfactory. This can be contributed to dialog scenes in which the speaker is not visually present. Also, appearance of other characters in the same shot causes a performance decrease.

An example of WhoIsName results is given in figure 1. In all movies, W_t is stable until approximately 30%-40% of the characters is left unidentified. After 40%, W_t performance increases. Note, that the number of measurements decreases with the percentage of unidentified characters, thereby causing higher volatility in the values of W_t. For example, the last value in figure 1 is 83%, but it is based on identifying three characters only.

7   Conclusion

For identification of characters in a movie, automatic alignment of scripts with visual appearances is not sufficient. The interactive method proposed allows annotators to quickly find out who is in selected shots. The WhoIsWho function in our i-Notation system matches names to shots and vice versa.

Table 3. Evaluation of character identification for various settings

	Default	Random comparison	No celebrities	Automatic addition	Interactive addition
Movie	W_g	W_r	W_g	W_g	W_g
Shakespeare in Love	49%	5%	45%	43%	54%
Sneakers	73%	9%	71%	77%	82%
LA Confidential	40%	4%	42%	38%	43%

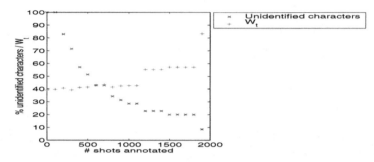

Fig. 1. Number of shots annotated set out against percentage of unidentified characters and W_t for the movie "Shakespeare in Love"

We employed a user model to evaluate the WhoIsWho function and various settings for its use. Assuming celebrities are known in advance reduces complexity, so that unknown characters are identified better. However, the function does not depend on the assumption. Hence, it can be used for movies without Hollywood stars as well.

An improvement of the identification process is the use of additional evidence, i.e. shots visually similar to the initially selected shots. When shots are added automatically, impact on the results depends on the quality of the movie script. Interactive addition as proposed in this paper, shows impact on results that is both positive and stable.

Use of the WhoIsWho function to create a good start screen for an interactive session does not yet live up to expectations. Future research focuses on predicting the reliability of textual information based on visual information to improve the start screen quality. Also, in future research the character identification methods will be extended to other domains where identification of people plays an important role.

References

1. P. J. Jang and A. G. Hauptmann. Learning to recognize speech by watching television. *IEEE Intelligent Systems*, 14(5):51–58, September/October 1999. 176
2. Y. Rui, T. S. Huang, and S. Mehrotra. Constructing table-of-content for videos. *Multimedia Systems, Special section on Video Libraries*, 7(5):359–368, 1999. 176
3. S. Satoh, Y. Nakamura, and Takeo Kanade. Name-it: Naming and detecting faces in news videos. *IEEE Multimedia*, 6(1):22–35, 1999. 176
4. J. Vendrig and M. Worring. Interactive adaptive movie annotation. Technical Report 2002-05, ISIS Group, University of Amsterdam, 2002. 175
5. J. Vendrig and M. Worring. Systematic evaluation of logical story unit segmentation. *IEEE Transactions on Multimedia*, to appear, June, 2002. 176
6. J. S. Wachman and R. W. Picard. Tools for browsing a tv situation comedy based on content specific attributes. *Multimedia Tools and Applications*, 13(3):255–284, March 2001. 177

Automated Scene Matching in Movies

F. Schaffalitzky and A. Zisserman

Robotics Research Group, Department of Engineering Science
University of Oxford, Oxford, OX1 3PJ
{fsm,az}@robots.ox.ac.uk

Abstract. We describe progress in matching shots which are images of the same 3D scene in a film. The problem is hard because the camera viewpoint may change substantially between shots, with consequent changes in the imaged appearance of the scene due to foreshortening, scale changes and partial occlusion.

We demonstrate that wide baseline matching techniques can be successfully employed for this task by matching key frames between shots. The wide baseline method represents each frame by a set of viewpoint invariant local feature vectors. The local spatial support of the features means that segmentation of the frame (e.g. into foreground/background) is not required, and partial occlusion is tolerated.

Results of matching shots for a number of different scene types are illustrated on a commercial film.

1   Introduction

Our objective is to identify the same rigid object or 3D scene in different shots of a film. This is to enable intelligent navigation through a film [2,4] where a viewer can choose to move from shot to shot of the same 3D scene, for example to be able to view all the shots which take place inside the casino in 'Casablanca'. This requires that the video be parsed into shots and that the 3D scenes depicted in these shots be matched throughout the film.

There has been considerable success in the Computer Vision literature in automatically computing matches between images of the same 3D scene for nearby viewpoints [18,21]. The methods are based on robust estimation of geometric multi-view relationships such as planar homographies and epipolar/trifocal geometry (see [3] for a review).

The difficulty here is that between different shots of the same 3D scene, camera viewpoints and image scales may differ widely. This is illustrated in figure 1. For such cases a plethora of so called "wide baseline" methods have been developed, and this is still an area of active research [1,7,8,9,11,12,13,14,15,17,19,20].

Here we demonstrate that 3D scene based shot matching can be achieved by applying wide baseline techniques to key frames. A film is partitioned into shots using standard methods (colour histograms and motion compensated cross-correlation [5]). Invariant descriptors are computed for individual key frames within the shots (section 2). These descriptors are then matched between key

M. S. Lew, N. Sebe, and J. P. Eakins (Eds.): CIVR 2002, LNCS 2383, pp. 186–197, 2002.

Fig. 1. These three images are acquired at the same 3D scene but from very different viewpoints. The affine distortion between the imaged sides of the tower is evident, as is the difference in brightness. There is considerable foreground occlusion of the church, plus image rotation ...

frames using a set of progressively stronger multiview constraints (section 3). The method is demonstrated on a selection of shots from the feature film 'Run Lola Run' (Lola Rent).

2   Invariant Descriptors for Multiview Matching

In this section we describe the invariant descriptors which facilitate multiple view matches, i.e. point correspondences over multiple images.

We follow the, now standard, approach in the wide baseline literature and start from features from which we can compute viewpoint invariant descriptors. The viewpoint transformations we consider are an affine geometric transformation (which models viewpoint change locally) and an affine photometric transformation (which models lighting change locally). The descriptors are constructed to be unaffected by these classes of geometric and photometric transformation; this is the meaning of invariance.

Features are determined in two stages: first, image regions which transform covariantly with viewpoint are detected in each frame, second, a vector of invariant descriptors is computed for each region. The invariant vector is a label for that region, and will be used as an index into an indexing structure for matching between frames — the corresponding region in other frames will (ideally) have an identical vector.

We use two types of features: one based on interest point neighbourhoods, the other based on the "Maximally Stable Extremal" (MSE) regions of Matas *et al.* [8]. In both types an elliptical image region is used to compute the invariant descriptor. Both features are described in more detail below. It is necessary to have (at least) two types of feature because in some imaged scenes one particular type of feature may not occur. The interest point features are a generalization of the descriptors of Schmid and Mohr [16] (which are invariant only to image rotation).

Fig. 2. Covariant region I. Invariant neighbourhood process, illustrated on details from the first and last images from figure 1. In each case, the left image shows the original image and the right image shows one of the detected feature points with its associated neighbourhood. Note that the ellipses deform covariantly with the viewpoint to cover the same surface region in both images

Invariant Interest Point Neighbourhoods: Interest points are computed in each frame, and a characteristic scale associated with each point using the method of [10], necessary to handle scale changes between the views. For each point we then compute an affine invariant neighbourhood using the adaptive method, proposed by Baumberg [1], which is based on isotropy of the gradient second moment matrix [6]. If successful, the output is an image point with an elliptical neighbourhood which transforms co-variantly with viewpoint. Figure 2 shows an example. Similar neighbourhoods have been developed by Mikolajczyk and Schmid [11].

For a 720 × 576 pixel video frame the number of neighbourhoods computed is typically 1600, but the number depends of course on the visual richness of the image. The computation of the neighbourhood generally succeeds at points where there is signal variation in more than one direction (e.g. near "blobs" or "corners").

MSE Regions: The regions are obtained by thresholding the intensity image and tracking the connected components as the threshold value changes. A MSE region is declared when the area of a component being tracked is approximately stationary. See figure 3 for an example. The idea (and implementation used here) is due to Matas *et al.* [8]. Typically the regions correspond to blobs of high contrast with respect to their surroundings such as a dark window on a grey wall. Once the regions have been detected we construct ellipses by replacing each region by an ellipse with the same 2nd moments.

Size of Elliptical Regions: In forming invariants from a feature, there is always a tradeoff between using a small intensity neighbourhood of the feature (which gives tolerance to occlusion) and using a large neighbourhood (which gives discrimination). To address this, we take three neighbourhoods (of relative

Fig. 3. Covariant regions II. MSE (see main text) regions (outline shown in white) detected in images from the data set illustrated by figure 1. The change of view point and difference in illumination are evident but the same region has been detected in both images independently

sizes $s = 1, 2, 3$) of each feature and use all three in our image representation. This idea has been formalized by Matas [9], who makes a distinction between the region that a feature occupies in the image and the region (the *measurement region*) which one derives from the feature in order to describe it. In our case, this means that the scale of detection of a feature needn't coincide with the scale of description.

Invariant Descriptor: Given an elliptical image region which is co-variant with 2D affine transformations of the image, we wish to compute a description which is *invariant* to such geometric transformations *and* to 1D affine intensity transformations.

Invariance to affine lighting changes is achieved simply by shifting the signal's mean (taken over the invariant neighbourhood) to zero and then normalizing its variance to unity.

The first step in obtaining invariance to the geometric image transformation is to affinely transform each neighbourhood by mapping it onto the unit disk. The process is canonical except for a choice of rotation of the unit disk, so this

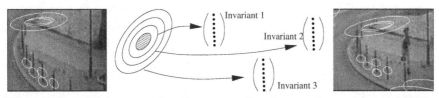

Fig. 4. Left and right: examples of corresponding features in two images. Middle: Each feature (shaded ellipse) gives rise to a set of derived covariant regions. By choosing three sizes of derived region we can tradeoff the distinctiveness of the regions against the risk of hitting an occlusion boundary. Each size of region gives an invariant vector per feature

device has reduced the problem from computing affine invariants to computing rotational invariants. The idea was introduced by Baumberg in [1].

Here we use a bank of orthogonal complex linear filters compatible with the rotational group action. It is described in detail in [15] and has the advantage over the descriptors used by Schmid and Mohr [16] and Baumberg [1] that in our case Euclidean distance between invariant vectors provides a lower bound on the Squared Sum of Differences (SSD) between registered image patches. This allows a meaningful threshold for distance to be chosen in a domain independent way. The dimension of the invariant space is 32.

Fig. 5. Ten key frames from the film "Run Lola Run". Each pair of frames was taken at the same 3D scene but comes from two different shots

3   Matching between Shots

Our objective here is to match shots of the same 3D scene given the invariant descriptors computed in the previous section. Our measure of success is that we match shots of the same scene but not shots of different 3D scenes.

Shots are represented here by key frames. The invariant descriptors are first computed in all frames independently. The matching method then proceeds in four stages (described in more detail below) with each stage using progressively stronger constraints:

(1) Matching using the invariant descriptors alone. This generates putative matches between the key frames but many of these matches may be incorrect.
(2) Matching using "neighbourhood consensus". This is a semi-local, and computationally cheap, method for pruning out false matches.
(3) *Local* verification of putative matches using intensity
(4) *Semi-local and global* verification where additional matches are grown using a spatially guided search, and those consistent with views of a rigid scene are selected by robustly fitting epipolar geometry.

Table 1. Tables showing the number of matches found between the key frames of figure 5 at various stages of the matching algorithm. The image represents the table in each row with intensity coding the number of matches (darker indicates more matches). Frames n and $n + 5$ correspond. The diagonal entries are not included. (a) matches from invariant indexing alone. (b) matches after neighbourhood consensus. (c) matches after local correlation/registration verification. (d) matches after guided search and global verification by robustly computing epipolar geometry. Note how the stripe corresponding to the correct entries becomes progressively clearer

(a)

	0	1	2	3	4	5	6	7	8	9
0	-	219	223	134	195	**266**	187	275	206	287
1	219	-	288	134	252	320	**246**	345	189	251
2	223	288	-	178	215	232	208	**341**	190	231
3	134	134	178	-	143	158	130	173	**169**	172
4	195	252	215	143	-	228	210	259	174	**270**
5	**266**	320	232	158	228	-	189	338	210	295
6	187	**246**	208	130	210	189	-	278	171	199
7	275	345	**341**	173	259	338	278	-	231	337
8	206	189	190	**169**	174	210	171	231	-	204
9	287	251	231	172	**270**	295	199	337	204	-

(b)

	0	1	2	3	4	5	6	7	8	9
0	-	2	3	6	4	**13**	1	6	2	3
1	2	-	5	3	4	11	**34**	3	5	3
2	3	5	-	14	2	6	10	**10**	8	4
3	6	3	14	-	6	0	0	1	**28**	5
4	4	4	2	6	-	1	7	0	3	**23**
5	**13**	11	6	0	1	-	2	2	8	6
6	1	**34**	10	0	7	2	-	2	4	3
7	6	3	**10**	1	0	2	2	-	5	11
8	2	5	8	**28**	3	8	4	5	-	1
9	3	3	4	5	**23**	6	3	11	1	-

(c)

	0	1	2	3	4	5	6	7	8	9
0	-	0	1	1	0	**8**	0	0	0	0
1	0	-	0	0	0	0	**16**	0	0	0
2	1	0	-	1	0	0	0	**5**	1	1
3	1	0	1	-	4	0	0	0	**11**	0
4	0	0	0	4	-	0	0	0	2	**14**
5	**8**	0	0	0	0	-	0	0	0	2
6	0	**16**	0	0	0	0	-	0	0	0
7	0	0	**5**	0	0	0	0	-	0	2
8	0	0	1	**11**	2	0	0	0	-	0
9	0	0	1	0	**14**	2	0	2	0	-

(d)

	0	1	2	3	4	5	6	7	8	9
0	-	0	0	0	0	**163**	0	0	0	0
1	0	-	0	0	0	0	**328**	0	0	0
2	0	0	-	0	0	0	0	**137**	0	0
3	0	0	0	-	0	0	0	0	**88**	0
4	0	0	0	0	-	0	0	0	9	**290**
5	**163**	0	0	0	0	-	0	0	0	0
6	0	**328**	0	0	0	0	-	0	0	0
7	0	0	**137**	0	0	0	0	-	0	0
8	0	0	0	**88**	9	0	0	0	-	0
9	0	0	0	0	**290**	0	0	0	0	-

Fig. 6. Verified feature matches after fitting epipolar geometry. It is hard to tell in these small images, but each feature is indicated by an ellipse and lines indicate the image motion of the matched features between frames. See figure 4 for a close-up of features. The spatial distribution of matched features also indicates the extent to which the images overlap

The method will be illustrated on 10 key frames from the movie 'Run Lola Run', one key frame per shot, shown in figure 5. Statistics on the matching are given in table 1. 'Run Lola Run' is a time movie where there are three repeats of a basic sequence. Thus scenes typically appear three times, at least once in each sequence, and shots from two sequences are used here.

Stage (1): Invariant Indexing: By comparing the invariant vectors for each point over all frames, potential matches may be hypothesized: i.e. a match is hypothesized if the invariant vectors of two points are within a threshold distance. The basic query that the indexing structure must support is "find all points within distance ε of this given point". We take ε to be 0.2 times the image dynamic range (recall this is an image intensity SSD threshold).

For the experiments in this paper we used a binary space partition tree, found to be more time efficient than a k-d tree, despite the extra overhead. The high dimensionality of the invariant space (and it is generally the case that performance increases with dimension) rules out many indexing structures, such as R-trees, whose performances do not scale well with dimension.

In practice, the invariant indexing produces many false putative matches. The fundamental problem is that using only local image appearance is not sufficiently discriminating and each feature can potentially match many other features. There is no way to resolve these mismatches using local reasoning alone. However, before resorting to the non-local stages below, two steps are taken. First, as a result of using several (three in this case) sizes of elliptical region for each feature it is possible to only choose the most discriminating match. Indexing tables are constructed for each size separately (so for example the largest elliptical neighbourhood can only match that corresponding size), and if a particular feature matches another at more than one region size then only the most discriminating (i.e. larger) is retained. Second, some features are very common and some are rare. This is illustrated in figure 7 which shows the frequency of the

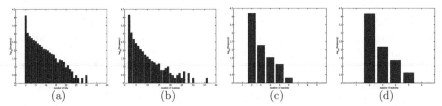

Fig. 7. The frequency of the number of hits for features in the invariant indexing structure for 10 key frames. (a): Histogram for all features taken together. The highest frequency corresponds to features with no matches, whilst some features have up to 30 matches. This illustrates that the density of invariant vectors varies from point to point in invariant space. (b), (c), (d): Histograms for intra-image matching for each of the three scales $s = 1, 2, 3$. It is clear that features described at the smallest scale are strikingly less distinctive than those described at higher scales. On the other hand, these have fewer matches as well

number of hits that individual features find in the indexing structure. Features that are common aren't very useful for matching because of the combinatorial cost of exploring all the possibilities, so we want to exclude such features from inclusion in the indexing structure. Our method for identifying such features is to note that a feature is ambiguous *for a particular image* if there are many similar-looking features in that image. Thus intra-image indexing is first applied to each image separately, and features with five or more intra-image matches are suppressed.

Stage (2): Neighbourhood Consensus: This stage measures the consistency of matches of spatially neighbouring features as a means of verifying or refuting a particular match. For each putative match between two images the K ($= 10$) spatially closest features are determined in each image. These K features define the neighbourhood set in each image. At this stage only features which are matched to some other features are included in the set, but the particular feature they are matched to is not yet used. The number of features which actually match between the neighbourhood sets is then counted. If at least N ($= 1$) neighbours have been matched, the original putative match is retained, otherwise it is discarded.

This scheme for suppressing putative matches that are not consistent with nearby matches was originally used in [16,21]. It is, of course, a heuristic but it is quite effective at removing mismatches without discarding correct matches; this can be seen from table 1.

Stage (3): Local Verification: Since two different patches may have similar invariant vectors, a "hit" match does not mean that the image regions are affine related. For our purposes two points are deemed matched if there exists an affine geometric and photometric transformation which registers the intensities of the elliptical neighbourhood within some tolerance. However, it is too expensive, and unnecessary, to search exhaustively over affine transformations in order to verify every match. Instead an estimate of the local affine transformation between the neighbourhoods is computed from the linear filter responses. If after this approximate registration the intensity at corresponding points in the neighbourhood differ by more than a threshold, or if the implied affine intensity change between the patches is outside a certain range, then the match can be rejected.

Stage (4): Semi-local and Global Verification: This stage has two steps: a spatially guided search for new matches, and a robust verification of all matches using epipolar geometry. In the first step new matches are *grown* using a locally verified match as a seed. The objective is to obtain other verified matches in the neighbourhood, and then use these to grow still further matches etc. Given a verified match between two views, the affine transformation between the corresponding regions is now known and provides information about the local orientation of the scene near the match. The local affine transformation can thus be used to guide the search for further matches which have been missed as hits, perhaps due to feature localization errors, to be recovered and is crucial in increasing the number of correspondences found to a sufficient level. This idea of growing matches was introduced in [13]. Further details are given in [15].

The second, and final, step is to fit epipolar geometry to all the locally verified matches between the pair of frames. If the two frames are images of the same 3D scene then the matches will be consistent with this two view relation (which is a consequence of scene rigidity). This is a global relationship, valid across the whole image plane. It is computed here using the robust RANSAC algorithm [18,21]. Matches which are inliers to the computed epipolar geometry are deemed to be globally verified.

Discussion: The number of matches for each of the four stages is given in table 1. Matching on invariant vectors alone (table 1a), which would be equivalent to simply voting for the key frame with the greatest number of similar features, is not sufficient. This is because, as discussed above, the invariant features alone are not sufficiently discriminating, and there are many mismatches. The neighbourhood consensus (table 1b) gives a significant improvement, with the stripe of correct matches now appearing. Local verification, (table 1c), removes most of the remaining mismatches, but the number of feature matches between the corresponding frames is also reduced. Finally, growing matches and verifying on epipolar geometry, (table 1d), clearly identifies the corresponding frames.

The cost of the various stages is as follows: stage 1 takes $5 + 10$ seconds (intra+inter image matching); stage 2 takes 0.4 seconds; stage 3 takes less than one millisecond; stage 4 takes $15 + 4$ seconds (growing+epipolar geometry). In comparison feature detection takes a longer time by far than all the matching stages.

The matches between the key frames 4 and 9 demonstrate well the invariance to change of viewpoints. Standard small baseline algorithms fail on such image

Fig. 8. Matching results using three keyframes per shot. The images represent the normalized 10×10 matching matrix for the test shots under the four stages of the matching scheme

pairs. Strictly speaking, we have not yet matched up corresponding frames because we have not made a formal *decision*, e.g. by choosing a threshold on the number of matches required before we declare that two shots match. In the example shown here, any threshold between 9 and 88 would do but in general a threshold on match number is perhaps too simplistic for this type of task. As can be seen in figure 6 the reason why so few matches are found for frames 2 and 7 is that there is only a small region of the images which do actually overlap. A more sensible threshold would also consider this restriction.

Using Several Key Frames per Shot: One way to address the problem of small image overlap is to aggregate the information present in each shot before trying to match. As an example, we chose three frames (30 frames apart) from each of the ten shots and ran the two-view matching algorithm on the resulting set of $3 \times 10 = 30$ frames. In the matrix containing number of matches found, one would then expect to see a distinct 3×3 block structure. Firstly, along the diagonal, the blocks represent the matches that can be found between nearby frames in each shot. Secondly, off the diagonal, the blocks represent the matches that can be found between frames from different shots. We coarsen the block matrix by summing the entries in each 3×3 block and arrive at a new and smaller 10×10 matrix M_{ij}; the diagonal entries now reflect how easy it is to match within each shot and the off-diagonal entries how easy it is to match across shots. Thus, the diagonal entries are used to normalize the other entries in the matrix by forming a new matrix with entries given by $M_{ii}^{-1/2} M_{ij} M_{jj}^{-1/2}$ (and zeroing its diagonal). Figure 8 shows these normalized matrices as intensity images, for the various stages of matching.

Note that although one would expect the entries within each 3×3 block between matching shots to be large, they can sometimes be zero if there is no spatial overlap (e.g. in a tracking shot). However, so long as the three frames chosen for each shot cover most of the shot, there is a strong chance that some pair of frames will be matched. Consequently, using more than one key-frame per shot extends the range over which the wide baseline matching can be leveraged.

4   Conclusions and Future Work

We have shown in this preliminary study that wide baseline techniques can be used to match shots based on descriptors which are ultimately linked to the shape of the 3D scene but measured from its image appearance. Using many local descriptors distributed across the image enables matching despite partial occlusion, and without requiring prior segmentation (e.g. into foreground/background layers).

Clearly, this study must be extended to the 1000 or so shots of a complete movie, and a comparison made with methods that are not directly based on 3D properties, such as colour histograms and image region texture descriptions.

In terms of improving the current method, there are two immediate issues: first, the complexity is $O(N^2)$ in the number of key frames (or shots) used. This must be reduced to closer to $O(N)$, and one way to achieve this is to use more discriminating (semi-)local descriptors (e.g. configurations of other descriptors). Second, no use is yet made of the contiguity of frames within a shot. Since frame-to-frame feature tracking is a mature technique there is a wealth of information that could be obtained from putting entire feature tracks (instead of isolated features) into the indexing structure. For example, the measurement uncertainty, or the temporal stability, of a feature can be estimated and these measures used to guide the expenditure of computational effort; also, 3D structure can be used for indexing and verification. In this way the shot-with-tracks will become the basic video matching unit, rather than the frames-with-features.

Acknowledgements

We are very grateful to Jiri Matas and Jan Paleček for supplying the MSE region code. Funding was provided by Balliol College, Oxford, and EC project Vibes.

References

1. A. Baumberg. Reliable feature matching across widely separated views. In *Proc. CVPR*, pages 774–781, 2000. 186, 188, 190
2. M. Gelgon and P. Bouthemy. Determining a structured spatio-temporal representation of video content for efficient visualization and indexing. In *Proc. ECCV*, pages 595–609, 1998. 186
3. R. I. Hartley and A. Zisserman. *Multiple View Geometry in Computer Vision*. Cambridge University Press, ISBN: 0521623049, 2000. 186
4. H. Lee, A. Smeaton, N. Murphy, N. O'Conner, and S. Marlow. User interface design for keyframe-based browsing of digital video. In *Workshop on Image Analysis for Multimedia Interactive Services. Tampere, Finland, 16-17 May*, 2001. 186
5. R. Lienhart. Reliable transition detection in videos: A survey and practitioner's guide. *International Journal of Image and Graphics*, Aug 2001. 186
6. T. Lindeberg and J. Gårding. Shape-adapted smoothing in estimation of 3-d depth cues from affine distortions of local 2-d brightness structure. In *Proc. ECCV*, LNCS 800, pages 389–400, May 1994. 188

7. J. Matas, J. Burianek, and J. Kittler. Object recognition using the invariant pixel-set signature. In *Proc. BMVC.*, pages 606–615, 2000. 186

8. J. Matas, O. Chum, M. Urban, and T. Pajdla. Distinguished regions for wide-baseline stereo. Research Report CTU–CMP–2001–33, Center for Machine Perception, K333 FEE Czech Technical University, Prague, Czech Republic, November 2001. 186, 187, 188

9. J. Matas, M. Urban, and T. Pajdla. Unifying view for wide-baseline stereo. In B Likar, editor, *Proc. Computer Vision Winter Workshop*, pages 214–222, Ljubljana, Sloveni, February 2001. Slovenian Pattern Recorgnition Society. 186, 189

10. K. Mikolajczyk and C. Schmid. Indexing based on scale invariant interest points. In *Proc. ICCV*, 2001. 188

11. K. Mikolajczyk and C. Schmid. An affine invariant interest point detector. In *Proc. ECCV*. Springer-Verlag, 2002. 186, 188

12. P. Pritchett and A. Zisserman. Matching and reconstruction from widely separated views. In R. Koch and L. Van Gool, editors, *3D Structure from Multiple Images of Large-Scale Environments, LNCS 1506*, pages 78–92. Springer-Verlag, Jun 1998. 186

13. P. Pritchett and A. Zisserman. Wide baseline stereo matching. In *Proc. ICCV*, pages 754–760, Jan 1998. 186, 194

14. F. Schaffalitzky and A. Zisserman. Viewpoint invariant texture matching and wide baseline stereo. In *Proc. ICCV*, Jul 2001. 186

15. F. Schaffalitzky and A. Zisserman. Multi-view matching for unordered image sets, or "How do I organize my holiday snaps?" . In *Proc. ECCV*. Springer-Verlag, 2002. 186, 190, 194

16. C. Schmid and R. Mohr. Local greyvalue invariants for image retrieval. *IEEE PAMI*, 19(5):530–534, May 1997. 187, 190, 193

17. D. Tell and S. Carlsson. Wide baseline point matching using affine invariants computed from intensity profiles. In *Proc. ECCV*, LNCS 1842-1843, pages 814–828. Springer-Verlag, Jun 2000. 186

18. P. H. S. Torr and D. W. Murray. The development and comparison of robust methods for estimating the fundamental matrix. *IJCV*, 24(3):271–300, 1997. 186, 194

19. T. Tuytelaars and L. Van Gool. Content-based image retrieval based on local affinely invariant regions. In *Int. Conf. on Visual Information Systems*, pages 493–500, 1999. 186

20. T. Tuytelaars and L. Van Gool. Wide baseline stereo matching based on local, affinely invariant regions. In *Proc. BMVC.*, pages 412–425, 2000. 186

21. Z. Zhang, R. Deriche, O. D. Faugeras, and Q.-T. Luong. A robust technique for matching two uncalibrated images through the recovery of the unknown epipolar geometry. *Artificial Intelligence*, 78:87–119, 1995. 186, 193, 194

Content Based Analysis for Video from Snooker Broadcasts*

H. Denman, N. Rea, and A. Kokaram

Electronic and Electrical Engineering Department
University of Dublin, Trinity College, Dublin, Ireland
hdenman@cantab.net
{oriabhan,anil.kokaram}@tcd.ie

Abstract. This paper presents the tools in a system for creating a semantically meaningful summary of broadcast Snooker footage. The system parses the sequence, identifies relevant camera views and tracks ball movement. Two new algorithms for video retrieval are presented. The first is the use of the 2nd order moment of the Hough Transform of the image for sequence parsing. The second method is the use of enclosed regions for detecting the disappearance of objects. The moment feature captures the geometry of the scene *without* extracting the 3D scene geometry. It is expected that this new feature is applicable to most sports in which a playing area is delineated by field lines.

1 Introduction

The problem of content-level processing for multimedia has much exercised the research community in recent years [3]. Video, in particular, is the focus of much research; the high volume of content makes content-level video processing difficult. Content-based retrieval for video involves developing search-engines capable of running queries against a corpus of video footage.

Common to most content-level multimedia applications is a feature extraction stage, in which the video footage is annotated with derived information or metadata. For example, simple summary generation is often based on the extraction of the first frame of each scene; in this instance scene cuts are the feature to be extracted. Automated feature extraction has been implemented in numerous research and commercial systems. Some systems augment the stored corpus with extracted features such as global colour, texture and motion [6], and dominant camera motion [2]. Systems such as QBIC, Virage, VisualGREP rely on distance measures against these features to enable query by example.

However, any domain-agnostic system relying on automated feature extraction will be limited in the kinds of query that it can support. Specifically, semantic level queries against the 'meaning' of the video are not possible at present;

---

* Work sponsored by Enterprise Ireland Project MUSE-DTV (**M**achine **U**nderstanding of **S**ports **E**vents for **D**igital **T**elevision)

M. S. Lew, N. Sebe, and J. P. Eakins (Eds.): CIVR 2002, LNCS 2383, pp. 198–205, 2002.

supporting such queries in a domain-agnostic system is an AI-complete problem, as the machine must understand the video presented. To enable high-level queries against the corpus, it is necessary to restrict the domain addressed [1]. Previously successful domain-specific systems include the tennis classification system developed by Jain [4]. This was capable of classifying footage of tennis shots into passing shots, volleys, etc.

This paper concentrates on televised Snooker coverage. The focus of this work is the creation of a semantically meaningful summary of the game. This implies identifying meaningful table views and ball movement events. We illustrate that unlike previous work in sports retrieval [4], the content analysis engine *does not need to extract the 3D scene geometry* to parse the video effectively. In particular, a new geometric feature for parsing this kind of video sequence is presented. In the next section, the initial parsing of the footage according to scene geometry is presented. There then follows a discussion about a novel algorithm that allows the detection of the most important event in the game: Ball potting; and finally a brief summary of our ball tracking algorithm.

2   Parsing Sport Video Using Implicit Geometry

For Snooker, the playing area is a green table. The main view that is important for content summary is shown in Figure 1 and Figure 3. It is sensible then that the video should be parsed according to the geometry of the table view. This method is an alternative to the use of Histograms [3] (for instance) to detect shot cuts since, the shot cuts do not immediately yield semantic episodes and it is also very difficult to differentiate between cuts, wipes, zooms etc, using Histograms. Rather than get involved with calculating 3D camera view orientations from edge maps of the scene, a much simpler process yields very good results.

A segmentation of the green area of each image frame is created using a threshold on the two colour planes as follows

$$b(i,j) = \{(G(i,j) - R(i,j)) > T\} \wedge$$
$$\{(G(i,j) - B(i,j)) > T\} \tag{1}$$

where $b(i,j)$ is set to 1 at sites where both colour differences exceed T and set to 0 otherwise. The difference signals measure the *greenness* of the images. See Figure 1 for a typical result. All analysis here was generated using $T = 25$.

In order to parse the images based on view geometry, the edges of these maps are then transformed using the Hough Transform. Significant peaks in this space indicate significant straight lines in the image. Here we use the ρ, θ form of the Hough Transform with $\theta = [0 : 180°]$; $\rho = [0 : 720]$ for Standard Definition PAL frames. Because each camera view has significantly different geometry, the orientation of the edges of the table area create substantially different Hough spaces when transformed. We have discovered that calculating the second order moment of that Hough space allows a single feature to be generated that summarises the *surface* in Hough space. Figure 1 shows the Hough Transform of a typical edge map in the main view. Figure 2 shows the 2nd order image moment calculated on the Hough space for each of 3500 frames (about 2.5 mins) from a broadcasted sequence. The colour bar along the bottom of the plot indicates the events taking place in the game. The amazing result is that the different camera views and *semantic* events are clearly delineated as plateaus with significantly different values of 2nd order moments. Furthermore, crowd scenes and close ups of players show significantly lower plateaus in the moment signal since they contain significantly less straight line geometry. Wipes, fades and dissolves can all be detected as impulsive transitions in the moment signal, and shot cuts show up clearly as step edges.

This new feature is rich with information about the video footage, but we are only interested in the "full-table" view as far as semantic summarisation is concerned (Figures 1,3). From the Hough transform of each image therefore, we detect the presence of peaks corresponding to the view geometry in question. This implies restricting the transform to the ranges [3..25], [89..90], and [155..177]. These ranges correspond to the regions of parameter space where the table edges are likely to be found, and restricting θ in this way increases computation speed considerably. Peaks in the parameter space correspond to lines in the image. Lines in the configuration corresponding to a Snooker table create a characteristic pattern in the transform space (figure 1): one peak each in the ranges $\theta \in [3..25]$, and $\theta \in [155..177]$, and two in the range $\theta \in [89..90]$. If peaks in this configuration are found, the shot is selected as a full-table footage shot which contributes to shot segmentation. Figure 1 shows the lines extracted in a correctly detected table view.

We find that using the 2nd order moment of the Hough transform, we can detect relevant table shots with 100% accuracy and zero false alarm.

2.1   Table Geometry

Once the edges of the table have been detected, the table geometry must be inferred, i.e. the locations of the ball spots and the pockets. As the table is distorted by a perspective projection resulting from the camera angle in full-table footage, attempting the inverse perspective transform is an obvious candidate.

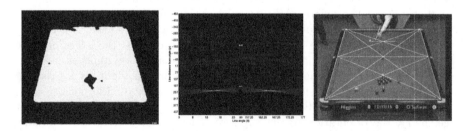

Fig. 1. Extracting geometry. Left to right: Binary map created from colour differences; Hough Transform of the full table shot. Table geometry, spots and pockets recovered

However, this is not practical: the inverse perspective transform is an ill-posed computation, especially when only one view is available. Sudhir, Jain, and Lee described in [4] a method that uses basis vectors, spanning the playing area, to find points of interest in the playing area.

Here we introduce the novel approach of using the fact that the intersection of the diagonals of a trapezoid are invariant under perspective distortion. Having located the table edges, therefore, we can calibrate the table surface *within the view geometry* without actually extracting the 3D geometry of the scene. Once the corner pockets have been found, the table dimensions assist in locating the other points of interest.

Because the diagonals of a trapezoid intersect at the midpoint of the trapezoid, irrespective of perspective transformation; the table can be repeatedly subdivided vertically to find any vertical position. For example, the black spot is 320 mm up from the bottom edge of the table, whose total length is 3500mm. 320/3500 expressed in binary is approximately 0.0001011101, so $320 = 3500 * (2^{-4} + 2^{-6} + 2^{-7} + 2^{-8} + 2^{-10})$. Each of these negative powers of two can be found by division along the vertical dimension of the table. In practice, the position of the black ball is adequately approximated with 0.00011, or 3/32 along the vertical. Repeated subdivision along the vertical yields the positions 1/8 and 1/16 of the way up the table; subdivision between these two points yields the black spot approximation, 3/32 of the way up the table. The perspective projection does not affect horizontal proportion, so to localise horizontal position it is sufficient to use the table dimensions directly. Most ball spots are exactly half-way between the table edges. Figure 1 shows the full table view with all relevent spots and pockets recovered.

3   Event Detection: Ball Potting

Having parsed the footage into the important table views, it is then necessary to locate the instants in which balls are potted. This makes possible event-driven state tracking, as distinct from state tracking via continuous monitoring of the entire table. We describe here a new histogram based approach for detecting the disappearance of objects (the ball) from a scene.

Regions around each pocket are monitored. The pockets have been located as described previously. Two regions are monitored, a small region and a larger one encompassing the smaller one. The small regions are 1/10 of the table width on a side, and the large regions 1/7 of the table width on a side. The colour distribution within each area will change when a ball enters the area. If a ball enters the large, surrounding area, then enters the smaller area near the hole, and then leaves both areas simultaneously, it has been potted. If it leaves the small area first, and then the large, it has bounced back from the cushion and has not been potted. If it enters the small area but does not leave, it has stopped moving in close proximity to the hole. These latter two possibilities are 'near miss' events, of interest in themselves in summary generation.

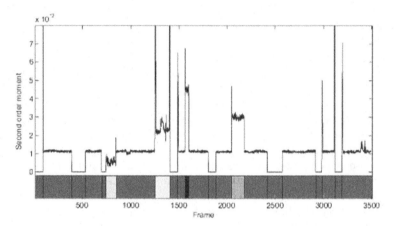

Fig. 2. Illustrating the effectiveness of the new Moment feature. The lowest second order moment values (red) signify crowd/player shots. Green indicates a full table shot (Moment $\approx 0.1 \times 10^{13}$). Editing effects such as wipes and dissolves yield impulses in the Moment. The yellow segment corresponds to a shot from the black (bottom right) pocket, orange corresponds to a close up of the player leaning over the table and the blue and light blue are shots over the yellow (top left) and green (top right) pockets respectively

Initial efforts to develop an event-driven state tracker used explicit colour distribution bins for each ball. Thus an increase in the number of pixels with a yellow hue in a region was interpreted as 'yellow ball has entered the region'. The difficulty with this approach is that the particular colour values for the various balls vary considerably between footage sources. A refined approach consisted of initialising the model of each ball's colour distribution from a known segmentation, computed from the first frame of the clip. However, this requires a reliable initial segmentation, which is at best a time-consuming computation. A further drawback is that the colour distribution of each ball evolves over time, to a considerable extent if the ball is in motion but with noticeable effect even on stationary balls.

The approach currently in use takes a snapshot of each area at the start of each clip and compares the area to the snapshot in each successive frame using the sum of histogram differences. Detecting peaks in this trace offers suitable event detection. Figure 3 shows the trace of a clip when a ball is potted. The ball enters the large area in frame 14 and the small area at frame 16 and leaves both at frame 21 which indicates that the ball has disappeared i.e. has been potted.

4   Initial Ball Localisation

The initial localisation of ball positions based on analysis of the first frame of a shot, is important for ball tracking during shot episodes. The first step in the initial localisation is the discovery of non-table objects in the frame. We wish to exclude the player from consideration, so the player is first removed using a simple thresholding scheme based on the a-priori knowledge that the players wear Black and White coloured clothes. A watershed algorithm [6] is then used to generate a first approximation to the object segmentation (Figure 3). The Hue and Saturation planes are independently segmented, and objects that are smaller than 10×10 pixels (in size) are selected. This thresholding on the basis of size is feasible because the retrieved table shot is always from the same camera view. The logical AND of the two segmentation maps then allows good ball location.

It is then necessary to identify the colour of the ball objects that have been segmented. Rather than attempt to classify directly using colour spectra, the relative colour properties are used. For example, the hue of green is higher than pink, red, and yellow—all of which have very similar values of hue. First, the three brightest balls are found, using the average luminance within an object as a measure of brightness. These are the pink, yellow, and white balls. Of these balls, the yellow ball will have the highest saturation. The pink and white can be difficult to distinguish; the hue of the pink ball is usually less than the white.

The remaining ball with the lowest median brightness, is the black ball. Median brightness allows robustness to specular reflection. The balls with the highest values of hue are the green and the blue; generally the highest hue is that of the blue. This leaves the brown and red balls, which are quite difficult to tell apart. Figure 3 shows several colours located and properly classified (using this algorithm) on the playing area. Correct disambiguation between brown/red; white/pink pairs remains a difficulty when using poor video recordings.

Further cues for object localisation become available when the systems described are integrated into a whole. For example, when player masking is being used, the white ball will frequently be a bright object close to the player mask, which can aid in white/pink disambiguating.

5   Tracking

To maintain a complete model of the game state from clip to clip, it is necessary to explicitly track the balls from frame to frame. The algorithm used to track balls in this work is condensation (CONditional DENSity propagATION), as in [5]. Both the evolution of the ball's colour distribution and the changes in the ball's position are tracked within the condensation framework.

The position of the ball is represented by a number of samples drawn from a Gaussian distribution centred on the region known to contain the ball being tracked. By assuming constant velocity this region can be roughly found in consecutive frames. Each position sample is initially assigned an equal weight. Means ($\bar{c}$) and standard deviations, (σ_c) of the hue, saturation, and brightness

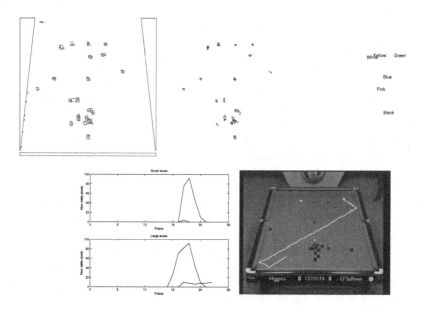

Fig. 3. Top Row: Implementation of the watershed algorithm showing the object map; Superfluous objects removed from the object map; Several balls found and labelled; Bottom Row: (Left) The top figure shows the small region around the pocket and the bottom one is the surrounding region. The ball enters the larger region first and subsequently the smaller one. The peaks decrease simultaneously in both indicating that the ball has been potted. The smaller peak in both traces is due to player occlusion. (Right) Ball tracking via CONDENSATION

(collectively denoted **c**) of the sample sets are calculated. The video is advanced one frame, and the position of each sample is perturbed by a small, random amount. The likelihood attached to each sample is computed according to $(c - \bar{c})/\sigma_c^2$. If the sum of the sample likelihoods is below a threshold, the ball is deemed to have been found in this frame. In this case, $\bar{c}$ and σ_c are updated according to the current sample set and the video is advanced to the next frame. If the sum of the likelihoods exceeds the threshold, the ball has not been found; in this case the points are further perturbed—points with a low likelihood are perturbed according to $\mathcal{N}(0,6)$, while points with a high likelihood are perturbed using $\mathcal{N}(0,2)$. The samples with a high likelihood must be perturbed as otherwise σ_c become zero. An example of the tracking performance is shown in Figure 3.

It is in the nature of the condensation algorithm that the modelled hue, saturation and brightness of the ball being tracked will lock on to a false object if the ball cannot be found—for example, if the ball has been potted. When this happens, a considerable change in one of these parameters will be observed. In the implementation considered here, a 25% change in any one of the parameters is taken to indicate that the ball has been potted.

6   Final Comments

This paper has introduced a system that allows semantic level parsing of video footage of Snooker. It shows that some level of domain specific information could yield a massive jump in the execution of a relevant analysis. Results show that we can successfully detect (and cluster) each camera view, detect Balls being potted and create Ball tracks. We have also presented two new processes that could be used generically for video retrieval. The first is the use of the 2nd order moment of Hough Space as an indicator of scene geometry. The second is the use of enclosed regions for object disappearance. It is expected that the new moment feature could be useful in many different types of sports footage where a playing area is clearly delineated e.g. Tennis, Table Tennis and Badminton.

References

1. Di Zhong and Shih-Fu Chang: Structure Analysis of Sports Video Using Domain Models. IEEE ICME (2001)   199
2. W Xiong, J C.-M. Lee: Automatic dominant camera motion annotation for video retrieval. SPIE International Conference on Storage and Retrieval for Image and Video Databases VI, Vol. 3312 (1997)   198
3. Alberto Del Bimbo: Visual Information Retrieval. Morgan Kaufmann (1999)   198, 199
4. G. Sudhir and John C. M. Lee and Anil K. Jain: Automatic Classification of Tennis Video for High-level Content-based Retrieval. International Workshop on Content-Based Access of Image and Video Databases (CAIVD '98) (1998)   199, 201
5. Michael Isard and Andrew Blake: CONDENSATION – conditional density propagation for visual tracking. Int. J. Computer Vision, Vol. 29 (1998)   203
6. Y. Deng, D. Mukherjee, B. S. Manjunath: Netra-V: toward an object-based video representation. SPIE International Conference on Storage and Retrieval for Image and Video Databases VI, Vol. 3312 (1997)   198, 203

Retrieval of Archival Moving Imagery - CBIR Outside the Frame?

Peter G. B. Enser and Criss J. Sandom

School of Information Management, University of Brighton
Watts Building, Moulsecoomb, Brighton BN2 4GJ, U.K.
pgbe@bton.ac.uk
cs117@bton.ac.uk

Abstract. Digitization may offer a means of making archival film content more accessible by allowing moving imagery to be delivered via the Internet and by CBIR-enabled mitigation of the effects of archivists' inconsistent indexing practices. This paper reports on research which sought to gain an informed view on this potential. Working in collaboration with eleven representative film collections, evidence was gathered relating to subject access facilities and to client requirements in the form of 1,270 requests. The work revealed that the formulation and satisfaction of requests for archival footage places a heavy reliance on human intellectual input for which CBIR techniques offer little prospect of being an effective substitute. The conclusion is drawn that the combination of limited CBIR functionality and lack of adherence to cataloguing standards seriously limits the Internet's potential for providing enhanced access to film and video-based cultural resources.

1   Introduction

Archival film and video material is an important contributor to our cultural heritage. With the enhanced interest evoked in that heritage among our increasingly leisured society the demand to gain subject access to such material has become more pronounced. Digitization appears to provide a means of making historic footage more accessible by allowing moving imagery to be displayed via the Internet. Moreover, digitization opens up the possibility of using automated indexing and retrieval techniques to assist in the transition from human-mediated to computer-mediated subject access to film and video material. Against such a background this paper reports on research into the characteristics of expressed information need for such material, as a means of gaining an informed view on the likely efficacy of content-based image retrieval (CBIR) techniques and Web-enabled access in this context.

The intellectual and practical challenges posed by the semantic indexing of still image material have been widely reported [1] - [6]. In the case of moving image material the challenges are accentuated, but less often encountered in the literature.

M. S. Lew, N. Sebe, and J. P. Eakins (Eds.): CIVR 2002, LNCS 2383, pp. 206-214, 2002.
© Springer-Verlag Berlin Heidelberg 2002

For the practitioner, however, the rate at which video material is being generated and submitted for archiving presents some organisations – national television broadcasting agencies being an obvious example – with a huge practical problem in the cataloguing and semantic annotation of such material.

Among film/video librarians and archivists there are widely varying responses to these challenges. In the U.K., at least, cataloguing practice tends to reflect local rather than (inter)national standards, and content annotation is effected by means of synopses or summaries of greatly varying levels of detail, or by exhaustive shot listing [7]. Synopses and summaries seek to represent the storyboard or continuity of action, and inevitably leave unreported the great majority of the actual visual content of a film or video. The example in Figure 1 shows a synopsis of the famous 1955 Ealing comedy *The Ladykillers*. The film runs for 87 minutes, provides a rich vein of mid-20th century English cultural and social content, all of which this 40 word summary allows to escape.

A gang planning a 'job' find themselves living with a little old lady, who thinks they are musicians. When the gang set out to kill Mrs Wilberforce, they run into one problem after another, and they get what they deserve.

Fig. 1. Example synopsis [8]

The second example is an English regional film archive's content description of an amateur film of some 10 minutes' duration.

Group visiting the Paris Exhibition in 1937.

Fig. 2. Example synopsis (South East Film & Video Archive)

A shotlist, by contrast, provides a detailed – and often time-coded – representation of the content of each shot. Figure 3 contains the shotlist corresponding with the synopsis shown in Figure 2. Clearly, shotlist compilation can be extremely time consuming: up to 30 hours' cataloguing time to 1 hour of transmission time has been quoted to us by one major national television broadcaster [7].

The different characteristics of shotlists and synopses reflect the distinction which has been drawn between the 'of-ness' and 'about-ness' of visual image content [4]. Shots are 'of' visible entities, whereas synopses tend to summarise what a particular length of footage is 'about'. Whichever type of content representative is used, however, any features within a film which are not included in the shotlist or scenario represent visually encoded knowledge which (in the context of retrieval transactions adopting the ubiquitous concept-based, rather than content-based paradigm [2], [9]) is effectively unretrievable except by sequential scanning of the entire footage.

Group walking under the Arc de Triomphe.
Champs Elysees seen from a car.
Fountain.
River boat "Sandettie" on the Seine, near the Eiffel tower.
Panorama following the river.
Exhibition site.
Beehive type of architectural towers.
Sculpture of a nude.
"L'Ocean" bridge over the Seine to reach the exhibition area.
Exhibition sites on the waterside.
Sculpture (boy + girl raising their arms in Victory) seen against the setting sunlight.
Water ballet.
Eiffel tower from nearby.
High exhibition building.
Water ballet, seen from the Eiffel tower.
Exhibition buildings floating on the Seine.
Streets and sites seen from a high spot.
Advertisement: 'Huile Lesieur'.

Fig. 3. Example shotlist (South East Film & Video Archive)

2   The VIRAMI Project

2.1 The CBIR Potential: Analysis of Archival Footage Requests

It is against such a background that the VIRAMI (Visual Information Retrieval for Archival Moving Imagery) project was undertaken at the University of Brighton. Funded by Resource, the Council for Museums, Archives and Libraries in the U.K., the project researched *inter alia* the information needs and retrieval strategies of users of moving image archives. One of the objectives of this project was to provide an informed view on the role, if any, which Content Based Image Retrieval (CBIR) techniques might play in answering user requirements, and thereby alleviating some of the moving image archives' dependency on the time-consuming and costly process of subject and content descriptive cataloguing.

The work was undertaken in collaboration with eleven film archives of different types: commercial footage companies, national and regional public archives, collections associated with museums, corporate collections, news archives, and television libraries. As part of the evaluation of current practice and the analysis of content representation and retrieval within these archives, a sample of users' requests for film footage was collected from each, resulting in a total of 1,270 individual requests from the eleven case study archives.

As Figure 4 indicates, the users' footage requests ranged from simple uniterms to highly complex requests that stipulated specific people, places, dates, and times – often for unique events, and sometimes also including technical shot or film format requests.

Figure 5 gives examples of requests which required some subjective judgement on the part of the researcher or cataloguer, whilst Figure 6 illustrates some requests which demonstrated a lack of understanding of moving imagery within the historical context, and which thus could not be resolved with documentary footage; only feature or other non-documentary footage could be offered to resolve the queries.

The requests were considered in a number of ways, particular use being made of the Panofsky-Shatford mode/facet matrix as a tool of analysis. Founded on the work of the art historian, Erwin Panofsky, three levels at which the subjects or concepts of pictures can be analysed are identified; in broad terms these are pre-iconographic, which addresses what an image shows in generic terms, e.g. a woman, a baby, a building; iconographic, which addresses the specific, interpreted subject matter of an image, with naming of features, e.g. Madonna and child, Taj Mahal; and iconological, which addresses the symbolism or intrinsic meaning of the image – its abstract concepts, e.g. hope, salvation, love [10]. Although conceived as an analytic tool in the particular context of Western Renaissance art, the approach has been generalised by Shatford [4] for application to any representational pictorial work. The resulting Panofsky-Shatford mode/facet matrix was used in an earlier research project, relating to user information needs in still image archives, [11] , and appeared to offer an equally convenient formulation for moving images.

Simple single term queries
- Transport
- Meercats
- Portillo
- The supernatural

Multi-faceted requests
- Ferries departing at night
- Animal rights protestors outside Hillgrove cat farm, wearing masks
- side pov of German countryside from train window and pov through stations
- Delegation of Australians arrive to discuss new constitution - Alfred Deakin, Edmund Baston, Robert Dickson in London with Joseph Chamberlain (Secretary of State for the Colonies 1900)
- "Famous" b & w shot circa 40-50 of a couple embracing in silhouette in an alleyway at night

Highly specific and complex requests
- John Lennon/Yoko Ono sleep-in, Amsterdam, March 21 in 1969
- Newsreel of the Britannia Shield (Boxing) which took place at the Empire Pool, Wembley, on Wednesday 5th October 1955.
- Mother [of the enquirer] achieving world athletics record for 880 yards at the White City, on 17 September 1952 of 2 minutes 14.5 seconds.

Fig. 4. Examples of the different levels of complexity in requests

- women at work and at play - must look 'British'
- Quirky - light hearted view of transport - cars - 20s to 70s
- Unusual people wearing Scottish kilts
- Sea lions - anything clever

Fig. 5. Examples of requests the resolution of which demands subjective judgement

- First phonograph (cylinder) Edison 1877
- British settlers arriving in Cape Harbour, South Africa, in the early 1800s
- Ancient Britain - Queen Bodicea with chariot and blades
- Livingstone and bearer

Fig. 6. Examples of requests which could not be satisfied by documentary footage

This analytical device, which is illustrated in Figure 7, allows footage request subjects – people or objects, events, places or times – to be stratified in terms of their specific, generic or abstract nature. It also allows enquiries comprising a number of different facets to be recognized and used as an measure of complexity.

Of the sample of 1,270 footage requests, 1,148 requests were subject requests: the remainder were requests for particular titles, directors, actors, shot types and the like. These non-subject requests were omitted from the mode/facet analysis, the results of which are shown in Table 1. It can be seen from the table that there was a large number of requests for footage of specific subjects; for example, 373 requests included a named person, group or object, 100 requests included a specific action or event, and 360 a specific place. In all, the footage requests included 1,143 named people, events, places or times. It should be noted that the total number of facets is more than the total number of requests, because many of the requests comprised more than one facet. Requests for specific subjects were more evident than for generic subjects, and requests for abstract subjects were unusual.

	Specific (Iconographic)	Generic (Pre-Iconographic)	Abstract (Iconological)
Who	Individually named person, group, thing	kind of person or thing	mythical or fictitious being
What	Individually named event, action	kind of event, action, condition	emotion or abstraction
Where	Individually named geographical location	Kind of place geographical architectural	place symbolized
When	linear time: date or period	Cyclical time: season time of day	emotion, abstraction symbolized by time

Fig. 7. Panofsky/Shatford mode/facet matrix

Table 1. Summary stratification of subject requests

	Specific (Iconographic)	Generic (Pre-Iconographic)	Abstract (Iconological)
Who	373	409	4
What	100	310	16
Where	360	120	1
When	310	13	1
Total	1143	852	22

In Table 2 the number of requests is stratified by the number of facets of each category and in total. This gives an indication of the complexity of the requests. The table shows that 780 of the 1148 subject requests (68%) contained one or more specific facets; i.e. requests for footage of a named person, group, event, place or time, and that 645 (56%) included one or more generic facets. More than half of the subject requests comprised two or more facets.

Table 2. Analysis of request complexity

Number of facets in request. (n)	Requests with (n) Specific facets	Requests with (n) Generic facets	Requests with (n) Abstract facets	Total number of requests with (n) facets
0	368	503	1126	0
1	519	456	22	542
2	171	172	0	390
3	79	16	0	173
4	11	1	0	40
5	-	-	-	3
Total no of requests n ≥ 1	780	645	22	1148

The high proportion of requests for specific, named items found in this study replicates the findings of earlier studies into user needs for still images [11], [12]. The use of CBIR techniques is effectively precluded in such information retrieval transactions. Staff within all of the archives which participated in the VIRAMI study shared the view that such techniques offered very little prospect of reducing the heavy reliance on human intellectual input for the formulation and satisfaction of requests for archival footage.

Although CBIR would appear to have very limited potential as a retrieval tool within film/video archives and libraries, various advances in CBIR techniques do have the potential to make the process of subject description and cataloguing considerably simpler. Descriptions of several experimental systems can be found via the Internet. These include: shot level video segmentation and key frame detection, which can be used to provide a storyboard which can then be used by a cataloguer to describe, in words, what each shot depicts [13], [14]; speech recognition to provide automatic transcription of video soundtracks, which can be used with time codes as a textual index for the footage [15],[16]; video skimming, which creates a video abstract that can be viewed in considerably less time than the original (*ibid*).

2.2 The Web-Enabled Delivery Potential

Having found little evidence that CBIR techniques were known to or used by film researchers it was interesting to establish the extent to which CBIR functionality was embedded within Web image search engines. A survey was undertaken, both by using a number of gateways [17], [18], [19] and by searching via Google [20] using the term "image search engine". A total of 43 image search engines or metasearch engines was found, 15 of which searched the entire web for images, the remainder searched within their own or other named image databases.

Four of the search engines offered some CBIR functionality: Cobion [21], Diggit [22], QBIC [23], and Webseek [24]. One additional search engine, AltaVista [25] had, in the past, provided a query by picture example facility, but this is no longer available.

However, it must be remembered that CBIR techniques are only applicable to digitized media and there are issues that need to be considered before any moving image archive embarks on wholesale digitization. Digitization is just not a practical option for many archives, and only one of the project's case study collections was actively undertaking a large-scale digitization programme. The creation of digital copies is in itself a time consuming activity and although storage costs are decreasing year by year, it would not be practical to consider anything but a very limited digitization programme, in view of the size of some moving image collections. The U.K. Imperial War Museum, for example, has 120 million feet of film [26].

The prospect of digitization and image database construction not infrequently stimulates a reconsideration of an archive's cataloguing and indexing practices, itself a major undertaking, quite apart from the cataloguing backlog which film archives commonly face. And the real benefits of Web-enabled access to such cultural heritage artefacts are made more difficult to realise in the absence of adherence to common standards of cataloguing among film libraries and archives: of the eleven case study archives surveyed in the VIRAMI project, for example, only three catalogued according to any of the published standards.

Of concern from the archival point of view, moreover, is the perception that developments in computer hardware, especially digital storage technologies, and software mean that the current digital media may swiftly become obsolete. There is, also, uncertainty about the longevity of digital media.

Transmission speeds also present a problem if access to digitized archival moving imagery is to be made available via the Internet. Film clips that are currently available are generally poor quality, low resolution, and with a small display area. Even so, they create large files: a typical Quicktime movie clip has a file size of approximately 116 KBytes per second playing time, which equates to nearly 420 MBytes per hour. One hour of compressed colour video is estimated to require up to 2 Gbytes, dependant upon image quality [27]. To download even a low resolution, poor quality, movie clip is slow: with modem speeds of 56Kbits per second, a typical 20 second clip would currently take a minimum of five and a half minutes to download.

3 Conclusion

Digitization appears to provide a means of making important cultural artefacts in the form of historic footage more accessible by allowing moving imagery to be made available for unmediated retrieval transactions via the Internet. The study reported in this paper would suggest that a balanced view has to be taken of such technological thrusts.

The VIRAMI project has revealed the emphasis which clients place on retrieving footage within which specifically named persons, objects, places and events are featured. As other studies of clients' needs have established in the context of still images, the formulation and satisfaction of requests for archival footage places a heavy reliance on human intellectual input, and there would appear to be little prospect of CBIR techniques offering an effective substitute for the amalgam of collection, domain and tacit knowledge which the experienced film archivist/librarian brings to bear in response to real client requirements. More encouragingly, advances in techniques such as video skimming and shot segmentation can provide an invaluable tool for film cataloguers, while speech recognition techniques may further assist by enabling footage to be automatically indexed by the accompanying soundtrack.

There are a number of practical hurdles which act as deterrents to digitization programmes in large film archives. These, together with the lack of commonly implemented standards for the cataloguing and indexing of film and video material, must be seen as further limitations on the Internet's potential for providing enhanced access to film and video-based cultural resources.

References

1. Enser, P. G. B.: Pictorial Information Retrieval. (Progress in Documentation). Journal of Documentation, 51(2) (1995) 126-170
2. Rasmussen, E. M.: Indexing Images. In: Williams, M. E. (ed.): Annual Review of Information Science, 32. Information Today, Medford, New Jersey (1997), 169-196
3. Fidel, R.: The Image Retrieval Task: Implications for the Design and Evaluation of Image Databases. The New Review of Hypermedia and Multimedia, (1997), 181-199
4. Shatford, S.: Analyzing the Subject of a Picture: A Theoretical Approach. Cataloguing & Classification Quarterly, 5(3) (1986), 39-61
5. Shatford, S.: Describing a Picture: A Thousand Words are Seldom Cost Effective. Cataloging & Classification Quarterly, 4(4) (1984,) 13-30
6. Markey, K.: Interindexer Consistency Tests: A Literature Review and Report of a Test of Consistency in Indexing Visual Materials. Library & Information Science Research, 6(2) (1984) 155-177
7. Sandom, C. J; Enser, P. G. B.: VIRAMI – Visual Information Retrieval for Archival Moving Imagery. Library and Information Commission Research Report 129. Resource: The Council for Museums, Archives and Libraries, London (2002)

8.   The Internet Movie Database: http://us.imdb.com/Plot?0048281
9.   Enser, P.: Visual Image Retrieval: Seeking the Alliance of Concept-Based and Content-Based Paradigms. Journal of Information Science, 26(4) (2000), 199-210
10.   Panofsky, E.: Studies in Iconology: Humanistic Themes in the Art of the Renaissance. Harper & Rowe, New York, (1962)
11.   Armitage, L. H.; Enser, P. G. B.: Analysis of User Need in Image Archives. Journal of Information Science, 23(4) (1997), 287-299
12.   Enser, P. G. B.; McGregor, C. G.: Analysis of Visual Information Retrieval Queries. Report on Project G16412 to the British Library Research & Development Department (1992)
13.   Dublin City University, Centre for Digital Video Processing - DCU: Overview. http://lorca.compapp.dcu.ie/Video/overview.html (2001). Accessed September 2001
14.   Lew, M. S.; Haas, M.; Touber, P.; Wentzler, D.: Leiden Video Segmentation and Classification. http://www.wi.leidenuniv.nl/home/lim/video.seg.classif.html (2001). Accessed September 2001
15.   Savino, P.: Searching Documentary Films On-line: the ECHO Project ERCIM News No.43. http://www.ercim.org/publication/Ercim_News/enw43/savino.html (2000). Accessed September 2001
16.   Carnegie Mellon University: Informedia Digital Video Library at CMU. http://www.informedia.cs.cmu.edu/ (2001). Accessed September 2001
17.   National Information Services and Systems. WWW - Search Engines. http://www.niss.ac.uk/lis/search-engines.html (2001). Accessed September 2001
18.   Meyer, H.: Image and sound search engine page - Allsearchengines.com http://www.allsearchengines.com/images.html (2001). Accessed September 2001
19.   Bernstein, Paula: Finding images online: links to image resources. http://www.berinsteinresearch.com/fiolinks.htm (2001). Accessed September 2001
20.   Google: Google http://www.google.com (2001). Accessed September 2001
21.   Cobion: Cobion Visual Content Search. www.cobion.com/services/portal.shtm - (2001). Accessed September 2001
22.   Bulldozer Software, Inc.: Diggit! Image search engine. http://www.diggit.com (2001) Accessed July 2001
23.   IBM: QBIC$^{(TM)}$–IBM's Query By Image Content. http://wwwqbic.almaden.ibm.com/~qbic/ (1997). Accessed September 2001
24.   Columbia University, WebSEEk - Content Based Image and Video Search and Catalog Tool for the Web. http://www.ctr.columbia.edu/webseek.
25.   AltaVista Company,. AltaVista Image Search. http://uk.altavista.com/searchimg?stype=simage (2001)
26.   Kirchner, Daniela. (ed.): The Researcher's Guide to British Film and Television Collections. British Universities Film and Video Council, London: (1997)
27.   Gilheany, S. Digital Image Sizes. http://www.ArchiveBuilders.com/whitepapers/27006v005p.pdf (2001). Accessed September 2001

Challenges for Content-Based Navigation of Digital Video in the Físchlár Digital Library

Alan F. Smeaton

Centre for Digital Video Processing, Dublin City University
Glasnevin, Dublin, 9, Ireland
Alan.Smeaton@dcu.ie

Abstract. Now that the engineering problems associated with creating, manipulating, storing, transmitting and playback of large volumes of digital video information are well on their way to being solved, attention is turning to content-based and other means to access video from large collections. In this paper we present an overview of the different ways in which video content can be used, directly, to support various ways of navigating within large video libraries. Some of these content-based mechanisms have been developed and implemented on video already and we use our own Físchlár system to illustrate many of these. Others remain beyond our current technological capabilities but by sketching out the possibilities and illustrating with examples where possible, as we do in this paper, we help to define what challenges still remain to be addressed in the area of content-based video navigation.

1 Introduction

The technical challenges associated with the capture, compression, storage, streaming, transmission and playback of digital video information have commanded most of the attention in the areas of digital video research and development over the last decade. This has been necessary in order to allow real video libraries which can deliver video content to users in an efficient manner, to be developed. The engineering problems and challenges associated with this are being solved or are close to being solved, thanks to the development and adoption of standards such as those from the MPEG family. Now that we can comfortably engineer and construct the actual video library systems, we can turn our attention to the problems faced by users who wish to use these systems, and foremost among those problems are the issues of how to effectively navigate through a video library. While the first video libraries were based around indexing and content navigation based on video metadata such as titles, actors, dates, etc. [1] such access does not fully leverage the advantages of having video content in digital format. What are also needed are techniques which operate *directly* on the video content, in the same way that text-based information retrieval searches are performed *directly* on the documents or web pages rather than on some metadata information.

M. S. Lew, N. Sebe, and J. P. Eakins (Eds.): CIVR 2002, LNCS 2383, pp. 215–224, 2002.
© Springer-Verlag Berlin Heidelberg 2002

As we begin to contemplate the possible ways in which digital video information can be manipulated, directly, based on its content, our natural inclination is to limit ourselves to the kinds of content based operations which we already perform on analogue video and to regard the benefits of being digital as simply being related to ease of access, playback, free ridership and replication. This, however, is under-exploiting the full advantages of video being digital and this is what we want to address in this paper.

Our aim in this paper is to introduce and summarise most of the functions which have been developed to allow content-based operations on text, image and audio, and to present these in a rough classification. The purpose of this is to then view digital video information as just another kind of media, and to examine the possibilities for content-based operations on video using the rough classification from content-based operations from other media. It is hoped that by examining video navigation from this more abstract position, we may gain some insights into what kinds of video navigation tools are possible, what is still missing based on current technology, and what remains to be discovered and built. We use our own video library system, Físchlár, to illustrate by example wherever possible.

The rest of this paper is organised as follows. In the next section we present – in an abstract way – the different kinds of content-based navigation operations that can be done on text, image and audio documents. Section 3 gives a brief summary of the Físchlár system while in section 4 we re-examine the content-based navigation tools which have been developed for video. For this section we illustrate from the Físchlár system wherever possible. Finally, a concluding section sets out the challenges which remain to be addressed in order to achieve the goal of truly flexible, effective, efficient, and scalable video navigation on large volumes of video content

2   Content-Based Navigation through Text, Image and Audio

In examining the broad amount of work done in content based retrieval we will deliberately exclude work done on retrieval using metadata such as descriptor captions or structured data fields, as we are interested on retrieval *directly* from information and we are also not concerned with issues of scale or size of the data being searched.

We have divided the whole gamut of content-based navigation operations into three broad categories, namely searching, grouping related objects, and summarising.

2.1   Searching

Searching through text, image or audio collections is based on matching a user's query against some set of units of information whose granularity can vary from an entire whole object (document, image, audio recording) to a small element within (a fact within a document, an object in an image or an utterance in an audio recording. Searching of text databases is well established and there are well-known techniques for retrieving whole documents (web searching for example), for retrieving parts of documents (such as passage retrieval) [2] and within recent years we've seen the emergence of question-answering [3], where a "factoid" such as a person name, monetary amount, a date or a location, is retrieved from a document database, in response to a query. In a sense this corresponds to retrieval of the "atomic" entities within a text document as retrieval of text units of information smaller than a factoid would simply not make sense.

For image retrieval, retrieval of entire whole images from a database in response to a query or sample image is the most common application [4]. Retrieval of parts of whole images or image fragments in response to a query image is generally restricted to specialist applications such as medicine, astronomy or geographic information systems where the original data source (the images in the database) consists of images which are much larger in size than that which is required by the user. An example from the medical domain can be seen in [5] where regions of an image in the database are retrieved in response to a query. Finally, retrieval of the "atomic" units making up an image is analogous to identifying and then retrieving objects from an image and this is possible only if the objects to be identified and retrieved, are homogenous. For example, it is possible to index a database of images of faces, or fingerprints or trademarks and to perform retrieval of the actual objects (faces, prints or trademarks) identified in those images. It is important to make the distinction between matching against and then retrieving the objects in an image, as opposed to retrieving the whole image which is easier.

If we restrict our analysis to audio recordings of speech then we can say that indexing and retrieval of recorded audio is less developed than on other media. This is because the analysis of the raw data into something meaningful requires speech recognition, or phone recognition at least. Systems such as the Taiscealaí system [6] which retrieve clips or segments of audio recordings, have been developed and these are analogous to retrieving passages from within text documents or fragments from within whole images. Retrieval of entire audio "documents" or recordings in response to a user query on content is generally not a useful operation, and retrieval of discrete and identified spoken "factoids" from an audio archive, is surely now within reach, although this should not be confused with retrieving arbitrary audio clips which just might contain "factoids".

2.2 Grouping Related Objects

The task of automatically grouping objects related by virtue of having related content and then using this as part of retrieval, broadly falls into three specific areas.

In the first case, object pairs (text documents, images) can be linked based on their direct similarity, independent of any superimposed or overarching structure. This would correspond to "find more like this" which uses a given whole text document or an image as a query and finds the documents or images which are most similar to the given one. Such links can be computed dynamically, or pre-computed and stored, and while this is reasonably commonplace for text and images it is not generally done for spoken audio. Following such links is akin to standard web browsing, following static hypertext links.

Another use for grouping of related objects is in automatically calculating a clustering of the set of objects in the collection. This is initially based on calculating pairwise object-object similarities and then applying a clustering algorithm to detect groups of related objects which form clusters. Sometimes outliers or small clusters of objects remain but most objects are generally brought into the overall classification. Like the calculation of independent pairwise links, clustering works well on text documents and on images but not audio data.

Somewhere in between independent pairwise linking and overall collection clustering lies the concept of automatically linking some related objects from the collection and superimposing some kind of higher level local structure. This is local in the sense that it is not applied to the entire collection but to some subset such as the output of a web search in order to impose some structure on the result of a query [7]. Another example of this would be a dynamically generated guided tour [8] and although we can find no examples of this applied to images there is no reason why it could not be applied.

2.3   Summarisation of Content

The summarisation task can be regarded as fairly straightforward. Given a single object (document, image, video program), automatically generate a smaller or shorter version which encapsulates the most important content from the original in such a way that an approximation of the meaning of the doucment, image or video, can still be obtained. Automatic summarisation of text documents is now an achievable task [9], although systems which do this are not yet robust and accurate enough to be in widespread use. Automatic summarisation of individual images is a trivial engineering task and involves generating a thumbnail or lower resolution and quality to the original. Automatic reduction of spoken audio information can involve detecting and eliminating silences and pauses in the original, and playback at a faster speed, and in this way the audio is reduced and takes less time to play. There is good potential for combining text summarisation with audio summarisation, and using audio features such as intonation and prosody, though this is still very much a research area.

The summarisation task may be applied to a set of objects and the task is much more difficult, to automatically generate a single, object, reduced in some way, which approximates the meaning of the set of objects. Such summarisation is only at the research stage, for all media.

3   Overview of the Físchlár Video Library System

Before we examine how searching, linking and summarisation can apply to video information, we will now introduce our own video digital library system, Físchlár, and use it for illustrating some content-based operations on video in the section to follow. Físchlár is a library of digital video information which records, analyses, indexes, stores and provides streamed playback as well as browsing and searching, on broadcast TV materials in a University campus environment. At any point in time there are between 300 and 400 hours of video content available to a userbase of over 1,500 registered users, though only about 1,000 of those are actually active and regular participants. The video streaming technology behind Físchlár is capable of supporting over 250 simultaneous streams of MPEG-1 encoded video and the interface to Físchlár is via a conventional web browser with a plug-in for video streaming.

There are two versions of the Físchlár system in use on our campus, one which allows recording, browsing and playback of TV programmes transmitted on any of 8 terrestrial TV channels and "Físchlár-News" which is the system of interest here. This automatically records the main evening news from the national broadcaster's main TV station and the archive of TV news has now grown to almost 1 year in size (c.170 hours of news content). Each program is digitised into MPEG-1 and then submitted to an analysis phase during which we automatically detect the boundaries between different camera shots. For shots over a threshold length we automatically select a single frame as the keyframe whose content is somehow indicative of the content of the whole shot. The entire News programme can then be presented to a user as a storyboard of keyframes, either the full set of keyframes (perhaps 200-250 such keyframes per broadcast) or an abstracted subset of 20 or 30. To support user browsing through these keyframes we have developed several keyframe browser interfaces [10] from which the user can choose. To compliment the keyframe browsing facilities in Físchlár, we have developed a text-based search tool for video digital libraries. We capture the closed captions associated with 6 TV channels, 24x7 and we link these to the archived broadcast. In this way we can support text-based searching through the text caption archive with relevant video clips as "answers" to queries.

Fig. 1. Screenshot from Físchlár showing search for "forged euro notes" and browsing through keyframes of relevant TV News program

An example of teletext search through a video news archive can be seen in Figure 1 where the user has requested clips about "forged euro notes", has chosen the TV News of 4 Jan 2002, is browsing the keyframes and associated teletext and is playing back a news clip of the anchorperson. However, teletext search is just part of our work on analysing video in order to support searching and browsing. We have also developed and tested techniques for news story segmentation, segmentation of programmes into scenes which are groups of shots, counting the number of human faces in a shot, speech-music discrimination, speaker identification, camera motion detection and detection of the amount of object motion in shots. We are also working on sophisticated object and shape detection and have been able to demonstrate this as a detector which recognises the appearance of either Bart or Homer Simpson in "The Simpsons", in real time. Further details of this work can be found on the project publications page at http://www.cdvp.dcu.ie/publications/

The Físchlár system has been deployed on campus for the last 2-and-a-half years and has allowed us to build up a userbase of real users with real video retrieval needs and we have been able to observe, first hand, how people want to do content access to digital video libraries.

4   Content-Based Operations on Digital Video

4.1   Searching Digital Video

Digital video information is structured from individual still frames into shots (single camera motion over time), then into logical scenes (groups of shots which make up a semantic whole), then into programmes and when we search video we may wish to retrieve from any of these levels of granularity. Retrieval of whole programmes can really only be done using metadata and this is not of interest to us here. Similarly, retrieval of individual frames is best done by treating each frame in the video as an image and using image retrieval techniques.

When it comes to searching for parts of video then a user's query can be either a text query – as in a web search engine – or a user may use an existing video shot or an image, as the query. In response to this, video shots and/or scenes can be retrieved. This requires shot boundaries and the boundaries between logical scenes to be detected, automatically. Shot bound detection is a well-established problem with many solutions and the recent TREC video track [11] benchmarked several different approaches to this problem. Detecting the boundaries between scenes can also be done, though this is more difficult. In Físchlár we can automatically detect the boundaries between news stories in a news broadcast and will shortly be able to retrieve an entire segmented news story in response to a user's query. Such segmentation is easier than detecting and retrieving logical scenes from other broadcast TV though this is something we are presently working on. To achieve this kind of scene segmentation we have developed techniques for speech/music discrimination, speaker identification and other audio analysis, face detection, anchorperson detection, etc. and we will incorporate these into the Físchlár system. Meanwhile a user's text query in Físchlár can be used to identify video clips based on the associated teletext which for the present is used to bring the user to a relevant part of the video, with no shot or scene segmentation for now. An example of this in action can be seen in Figure 1.

At Carnegie Mellon University, the Informedia project [12] has had much success in developing techniques for video indexing and retrieval. The CMU approach is based on searching speech transcripts, where the speech recognition has been performed by the Sphinx speech recognition engine. This is then coupled with automatic identification of faces from the video, naming of those faces when they are among a database of "VIPs", automatic identification of the occurrence of text on-screen in captions or as part of the image, OCR applied to such text, and automatic identification of named locations. These locations are then input into a mapping interface allowing a user to see whereabouts on the globe or on a smaller scale, that the video content refers to.

The IBM CueVideo project at IBM Almaden also does searching of video content and in this case it is based on speech recognition, smart browsing and other video analysis techniques [13].

Retrieval of video clips based on queries in media other than text was also supported in the TREC video track where user queries could include actual video shots and the task was to find other video shots similar to the query. Some exploratory approaches which have been taken include matching based on colour correlograms for a video shot (a multidimensional data structure incorporating colour histograms for frames spread throughout a shot) and automatic pre-classification of video shots into a semantic hierarchy [11]. While these, and other similar approaches represent the "high end" in terms of sophistication of approaches taken they do allow first attempts at automatic video retrieval based on video queries.

4.2   Grouping Related Video Clips

As we saw earlier, grouping pairs of objects on the basis of relatedness depends upon being able to compute similarity between objects – in the case of video, either shots or scenes. In the Físchlár system we are developing techniques to automatically link related news stories which have been automatically segmented. Our current work does this based on linking teletext associated with each story, which is effectively the same as link creation in text documents, but we are also working on incorporating image and audio analysis into this process. Automatically detecting anchorperson, image similarity and speaker identification are all being used to lead to a "higher level" of similarity matching between news stories, so we move away from a dependency on text (closed caption) similarity. Most of the difficulties associated with text-based information retrieval revolve around word polysemy and other language ambiguities [14] and adding in audio and image based analysis will help to redress some of these difficulties.

In our initial work, pairwise linking of news stories is done independently of all other stories but we hope to move on to more elaborate local clustering to sets of news clips, perhaps as part of a user's video navigation session in the same way that guided tours on text documents have been dynamically created in our previous work [8]. Following that we can address the possibilities of global clustering of an entire video archive, though we are not aware of any work reported to date which attempts to group related video clips. The hurdles to doing this are simply the ability to segment video accurately into meaningful units such as news stories, and the ability to calculate similarity values between such units. Our work on video searching, via text, plus image and audio analysis, gives us the foundation for such work.

4.3   Summarisation of Video

The most widespread type of video summarisation is based on generating keyframes from video shots and presenting these via some kind of keyframe browser interface. In Físchlár we have several keyframe browsing interfaces which allow a user to browse the several hundred keyframes from a TV program [10]. For example, a 30 minute news program may have between 300 and 400 such keyframes in one of these interface we also automatically reduce this large set to a summary of only about a dozen keyframes, one per news story.

The way in which one automatically summarising video into a shorter video clip is wholly dependent on the genre or type of the video. For some types, such as TV documentaries, soaps, etc., this is very difficult but for others – such as sports programmes – we have been able to make progress. In Físchlár we can generate summaries of certain types of sports programmes which have crowd noises and a rolling commentary based simply on audio analysis [15]. By detecting and measuring the level of crowd noise, and detecting and measuring the excitation level of a sports commentator, we have been able to identify the "most exciting" events in a sports event and bundle these together into a summary of whatever length a user requires. As a first approximation this is very simple, but effective, and we are working on identifying slow motion (which is also indicative of significant sports plays), camera long shots vs. close ups, as well as teletext analysis, each of which will further improve the quality of our summary.

Work on video summarisation and video skimming – which is like summarisation in that it involves identifying the most salient parts of a video programme and presenting them to a user, can be found as part of the VACE project, funded by the US ARDA programme [16].

While summarisation of sports and TV news is achievable, summarisation of other types will require the kind of discourse analysis which makes text summarisation so difficult, and we have a long way to go before we can do this effectively.

Finally, a set of online links to video information retrieval projects can be found at [17].

5   Challenges for Content-Based Navigation for Video Libraries

Current approaches to video navigation are based primarily around retrieval based on associated teletext but even moreso, browsing via selected keyframes. In comparison to text or other media we have not really started to explore grouping or linking together of related video clips and a summary of the different capabilities for different media can be seen in Table 1 . In Físchár, as in other projects, all our image and audio analysis techniques on video run automatically and as they are developed into robust and effective implementations they open up for us the possibility of automatically linking together related "chunks" of video, just as we have been linking between related pieces of text on the www. The pace of development of the www has shown us how comfortable and natural it is for us to browse from web page to page by following pre-constructed hypertext links, albeit that most of them are constructed manually. It is our belief that a similar kind of video navigation which seamlessly combines searching for objects, shots or scenes, browsing and following hyperlinks between related video elements, and summarisation based on generated summaries or sets of keyframes, will offer the most appropriate and effective type of navigation through video libraries. We have already achieved some of these technologies and now we know what remains to be done.

Table 1. Content operations on different media

	Text	Image	Audio	Video
Searching for …				
…whole objects	Web search	Image retrieval	N/A	Use metadata to retrieve programmes
..parts of objects	Passage retrieval	Medical, astronomy & other specialist	Retrieve audio clips	Use teletext to retrieve shots or scenes Use image/shot to retrieve shots or scenes
atomic elements / object retrieval	Factoids	Works for homogeneous collections	Not widespread	Use image retrieval on frames
Grouping / linking based on …				
…independently linked pairs	"more like this"	Image retrieval	N/A	Preliminary work reported
…overall local structure	Guided tours, cluster web search results	Not done !	N/A	Not done yet
…overall global structure	Text clustering	Image clustering	N/A	Not done yet
Summarisation	Not widespread	Thumbnails	Silence detection & FFwd.	Keyframes or audio-based and restricted to sports programs

Acknowledgement

The Support of the Informatics Research Initiative of Enterprise Ireland is gratefully acknowledged.

References

1. Little, T. D. C. and Venkatesh, D.: Prospects for Interactive Video-on-Demand *IEEE Multimedia, 1(3), 14-24, 1994.*
2. Salton, G., *et al.*: Approaches to passage retrieval in full text information systems. In Proceedings of ACM SIGIR, Pittsburgh, pp 49-58, 1993.

3. Flexible Query Answering Systems : Recent Advances : Proceedings of the 4th Intnl. Conf. on Flexible Query Answering Systems, H.L. Larsen (Ed), Physica Verlag, 2000.

4. Faloutsos, C., *et al.*: Efficient and Effective Querying by Image Content. Journal of Intelligent Information Systems, 3, 3/4, July 1994, pp. 231-262.

5. Wang, J. Z.: SIMPLIcity: A Region-based Image Retrieval System for Picture Libraries and Biomedical Image Databases.in *Proceedings of ACM Multimedia 2000,* Los Angeles, Ca., USA, October 30 to November 4, 2000.

6. Smeaton, A. F., Morony, M., Quinn G. and Scaife, R.: Taiscéalaí: Information Retrieval from an Archive of Spoken Radio News. in *Proceedings of the 2nd European Conference on Research and Advanced Technology for Digital Libraries*, Crete, September 1998.

7. Zamir, O. and Etzioni, O.: Grouper: A Dynamic Clustering Interface to Web Search Results. In Proceedings of WWW8, June 1999.

8. Guinan C. and Smeaton, A. F.: Information Retrieval from Hypertext Using Dynamically Planned Guided Tours. in: *Proceedings of ECHT'92 (European Conference on Hypertext),* Milan, Italy, pp.122-130, D. Lucarella *et al.* (Eds.), (1992).

9. Mani and Maybury M. T. (Eds).. Advances in Automatic Text Summarisation. MIT Press, 1999.

10. Lee, H. *et al.*: Implementation and Analysis of Several Keyframe-Based Browsing Interfaces to Digital Video. In *Proceedings of the Fourth European Conference on Digital Libraries (ECDL)*, Lisbon, Portugal, Springer-Verlag LNCS 1923, J. Borbinha and T. Baker (Eds), pp.206-218, September 2000.

11. Smeaton, A. F. *et al.*: The TREC-2001 Video Track Report. In *Proceedings of TREC-2001*, NIST Special Publication   (in press), E. M. Voorhees and D. K. Harman (Eds.), 2002.

12. Wactlar, H., Christel, M., Hauptmann, A. and Gong, Y. Informedia Experience-on Demand: Capturing, Integrating and Communicating Experiences across People, Time and Space. *ACM Computing Surveys*, Vol. 31, No. 9, June, 1999.

13. IBM CueVideo Project. http://www.almaden.ibm.com/cs/cuevideo/ Last visited 19 April 2002.

14. Smeaton, A. F.: Information Retrieval: Still Butting Heads with Natural Language Processing ? in *Information Extraction*, M. T. Pazienza (Ed), Springer LNCS, 1997.

15. Sadlier, D. A. *et al.*: MPEG Audio Bitstream Processing Towards the Automatic Generation of Sports Program Summaries.   Submitted to ICME Conference, 2002.

16. The VACE Project. http://www.informedia.cs.cmu.edu/arda-vace/ Last visited 19 April 2002.

17. Centre for Digital Video Processing, Dublin City University links to related projects. http://www.cdvp.dcu.ie/links.html Last visited 17 April 2002.

Spin Images and Neural Networks
for Efficient Content-Based Retrieval
in 3D Object Databases

Pedro A. de Alarcón, Alberto D. Pascual-Montano, and José M. Carazo

Biocomputing Unit. Centro Nacional de Biotecnología (CSIC)
Campus Universidad Autónoma de Madrid, 28049 Madrid, Spain
{pedro,pascual,carazo}@cnb.uam.es
http://www.biocomp.cnb.uam.es

Abstract. We describe a system for querying 3D model databases using
the *spin image* representation as a shape signature for objects depicted
as triangular meshes. The spin image representation facilitates the task
of aligning the query object with respect to matched models (coarse-
grain registration). The main contribution of this work is the
introduction of a three-level indexing schema based on artificial neural
networks. The indexing schema improves significantly the efficiency in
matching query spin images against those stored in the database. Our
results are suitable for content-based retrieval in 3D general object
databases. A particular application to molecular databases is also
presented.

1 Introduction

Retrieving objects by their content as opposed to keyword indexing or simple
browsing has become an important operation, and consequently, an active field of
research. After more than a decade of intensive research, *content-based image
retrieval* (CBIR) technology moved out of the laboratory and into the marketplace, in
the form of commercial products like QBIC [1] and Virage [2]. CBIR draws many of
its methods from the field of image processing, computer vision, pattern recognition,
and database technology.

As web-based repositories of multi-dimensional scientific data continue to grow,
so does the need for content-based retrieval of three-dimensional (3D) scalar data.
One of the most prevalent source of volumetric data is medical and biological
imaging. During the last decade bioscientists have witnessed an spectacular growth of
molecular databases. Molecular databases store in some cases (Protein Data Bank [3])
thousands of molecular complexes described as three dimensional datasets. Public
access to these repositories boosts research targeted at the discovery of new drugs and
medicines. Certainly, content-based retrieval would provide scientists with a valuable
tool that facilitates, for example, the task of finding molecules that are structurally
similar to a given one. To do so, different features are used to represent the content of

M. S. Lew, N. Sebe, and J. P. Eakins (Eds.): CIVR 2002, LNCS 2383, pp. 225-234, 2002.

a 3D object. Among them, shape is the most relevant attribute for the task of object recognition. There exist a number of CBIR systems that use different shape features in order to perform similarity retrieval in 3D general model databases. Most of them use the object's surface (e.g: VRML description) to derive a signature that represents its shape. Some examples are Nefertiti [4], the Ferret Project [5], or the 3D search engine depicted in [6]. Other systems are tailored to domain-specific databases of volumetric datasets. In [7] a 3D neuroradiologic image retrieval system is presented. It deals directly with multimodal 3D brain images. They use thematic data and image-based descriptors in order to better characterize pathological versus normal brains. This is a good example of how a CBIR system is perfectly adapted to a particular application domain. Nevertheless, it results hard to figure out how to extend their developments to other 3D model databases. Ankerst *et al.* [8] introduced a novel approach for similarity search in 3D protein databases. Shape histograms and chemical properties of proteins form the objects´ feature vectors. The shape histograms are built from partitioning the 3D space into concentric shells and sectors. They assume proteins to be given as sets of 3D points. However, we still believe that some difficulties may appear from using that shape signature since it is neither scale-invariant, nor translation invariant. Also, it does not provide a description of the protein's shape at the local level. This issue is quite important as many molecular interactions take place at very precise locations on the protein surface (active sites).

In the present manuscript, we describe a system for querying 3D model databases using the *spin image* representation [9]. The spin image representation provides a local description of the surface that is translation/rotation-invariant. A normalization step is used in order to make it invariant to scale changes. Nevertheless, the main contribution of this work is the introduction of a 3-level indexing schema based on artificial neural networks. The indexing schema improves significantly the efficiency in matching query spin images against those stored in the database. Our schema makes the spin image representation potentially suitable for content-based retrieval in large 3D model databases

The rest of the paper is organized as follows. Section 2 is devoted to briefly analyse the representation of 3D free-form objects and to describe the spin image representation. Then, the indexing mechanism is presented (section 3). Our system is tested with some small 3D object databases in section 4. The paper concludes with the main conclussions and some guidelines for future work (section 5).

2   Object Representation and Shape Features

Three-dimensional object recognition uses the true 3D shape of objects in its model representation. 3D data can come in the form of depth-maps, isolated 3D points and lines, or 3D intensity images, depending on the sensor and sensing algorithm. In this work, we will consider the shape as a geometric concept derived from the object's surface. Definitions of free-form surfaces and objects are often intuitive rather than formal. Besl [10] stated, "a free-form surface has a well defined surface normal that is continuous almost everywhere except at vertices, edges and cusps". Many other definitions exist though.

2.1   Object Representations in Computer Vision

Unfortunately, most 3D file formats (VRML, 3Dstudio, etc) have been designed for visualization. They contain only geometric and appearance attributes, and usually lack of semantic information that would facilitate object recognition. In a computer vision context [11], the object representation is designed for use in the specific vision application (such as navigation, recognition, or tracking), and visual fidelity may not be a criterion of interest. On the other hand, efficient execution, saliency and locality are desirable features. Several surface representations in the context of computer vision have been proposed in the literature [12]. Among those that adopt complete mathematical forms, parametric surfaces, algebraic implicit surfaces, superquadrics and generalized cylinders are the most popular ones. They all are global and compact representations but lack of some characteristics. For example, only parametric forms allow local control of the surface, but in contrast, they are difficult to fit to the actual object's contour. Polygonal meshes have become a popular representation for 3D objects. Meshes allow local control of the surface and can faithfully approximate complex free-form objects to any desired accuracy given sufficient space to store the representation. In the past this was their major limitation, but with the decreasing cost of computer memory and advances in hardware for computer graphics, even very dense meshes have become practical. In our system, we will assume 3D objects to be represented as polygonal meshes (also known as polygon soups). A polygonal mesh is defined by two components: a list of 3D vertices and an indexed list of polygons each specified as al list of vertex indices. We restrict polygons to be triangles because triangle meshes are easy to manipulate and efficiently rendered.

2.2   The Spin Image Representation

The problem of determining the similarity of two shapes is a fundamental task in shape-based recognition, retrieval, clustering, and classification [13,14]. In general, matching methods are grouped according to their representation of shape: 2D contours, 3D volumes, 3D surfaces, structural models and statistics. Unfortunately, most of the methods for 2D shape matching are not extensible to the 3D case. Nevertheless, a number of methodologies have been proposed for three-dimensional shape matching   [14]. For the present work we have chosen the spin image representation [9]. This model-based representation combines the descriptive nature of global signatures with the robustness to partial views and clutter of local features. The representation allows achieving both object recognition and coarse-grain registration. Also, as no assumptions about the topology or shape are made, arbitrarily shaped objects can be recognized.

Given a polygonal mesh, an oriented point O on the surface is specified by its 3D position p and surface normal n (fig. 1). We have chosen the centroids of triangles in the mesh as the location of oriented points and the corresponding triangle's normals as their associated direction. Thus, a given object has as many oriented points as the number of triangles that form the mesh. Each oriented point defines a local coordinate system using the tangent plane P through p oriented perpendicularly to n and the line L through p parallel to n. The two coordinates of the basis are α, the perpendicular distance to the line L, and β the signed perpendicular distance to the plane P. There is an unique function ("spin map") that maps each 3D point x on the surface into (α, β) coordinates with respect to the oriented point p. The spin map provides a mapping from points on the surface into a two-dimensional space where some of the 3D metric is preserved. A 2D histogram can be built from the spin map by accumulating 2D positions into discrete bins. Bilinear interpolation of the spin-map coordinates is used in order to reduce the effects of possible noise in the data. This histogram is called "spin image" and it is unique for each oriented point in the surface. For each oriented point on the surface of the object, its spin image is computed. The set of all spin images forms the "spin image stack" and it is a global description of the object's shape.

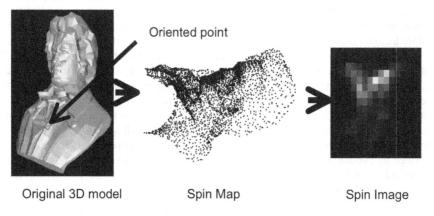

Original 3D model   Spin Map   Spin Image

Fig. 1. For each oriented point (triangle center) on the triangular mesh, a spin map can be computed by an unique mapping into a 2D space. The spin map is discretized into bins so that the spin image from that oriented point is obtained

A spin image is an object centered (i.e: pose-independent) encoding of the shape of an object because the spin-map coordinates of a point with respect to a particular oriented point basis are independent of translational and rotational transformations. However they are still scale-dependent since the α,β values depend on the scale of the original 3D space. To avoid this, we normalize each vector (spin image) to unit length. As a consequence, spin images are much more robust against resolution changes. The resolution of a mesh is commonly defined as the average length of its edges for uniformly distributed meshes. Let's suppose two meshes representing the same object but one surface has a richer sampling than the other. Without normalization, their spin images would be different due to the difference in their ranges. Normalization also makes spin images to be scale-invariant and less sensitive to the metric that will be used thereafter for comparison. In [9] a modified version of the linear correlation coefficient is used as metric distance between two images. The metric considers the amount of overlap in the comparison of two spin images, which is important for partial matching purposes. In our case, Euclidean distance produced excellent results quite similar to those obtained with the linear correlation coefficient.

3   Artificial Neural Networks for Efficient Indexing

At recognition time, spin images from points on the model are compared to spin images from points in the scene. When two spin images are similar enough, the object (or a partial view of it) is recognized and a point correspondence between model and scene is established (registration). Models are stored together with their spin images stacks in a database. The user presents a query object to the database and ask for a "top list" of similar objects as well as which surface regions of the query object resemble to those of the retrieved objects. Efficiency is a critical factor in this procedure. The worst case computational cost of the query is $O(N_{db}N_q)$, where N_{db} is the total number of spin images stored in the database and N_q the number of spin images that were selected from the query object. It is essential to provide some indexing mechanism in order to quickly match scene spin images to model ones. Moreover, in [15] the authors pointed out that spin images coming from the same surface can be correlated for two reasons: first, spin images generated from oriented point bases that are close to each other on the surface will be correlated. Second, surface symmetry and the inherit symmetry of spin image generation will cause two oriented point bases on equal but opposite sides of a plane of symmetry to be correlated. Even more, surfaces from different objects can be similar on the local scale. Therefore, it may exist a correlation between spin images of small support generated from different objects. We can benefit from this correlation to make spin image comparisons more efficient through mapping the total number of spin images into a reduced representative set of them. A different approach is proposed in [9]. Spin images are compressed to reduce the amount of correlation among them. Spin images can be considered as p-dimensional vectors. Principal component analysis (PCA) was used in order to map p-dimensional vectors into s-dimensional space where s<p.

Artificial neural networks (ANN) have been applied in similar contexts [16,17]. In this work we propose an improved indexing method based on ANN to efficiently access to spin images. Our method achieves both compression and indexing of the original set of spin images. Basically, a self-organized map (SOM) is built from the stack of spin images of a given object. This is a way of "summarizing" the whole stack into a set of representative spin images. Then, a clustering algorithm (kcmeans) is applied in order to group representative views in the SOM map into a reduced set of clusters (fig. 2). At query time, spin images will be first compared with the clusters centers resulting from the kcmeans method and subsequently with the SOM map if finer answer is requested.

3.1 Self-Organizing Feature Maps (SOM)

The Self-Organizing Map is a neural network that simulates the hypothesized self-organization process carried out in the human brain when some input data are presented [18]. The structure of this neural network is composed of two layers: an input layer formed by a set of units (on for each component of the input vector) and an output layer formed by units or neurons arranged in a low dimensional grid (usually two-dimensional). The algorithm maps a set of input vectors (spin images in our case) onto a set of output vectors (neurons), but unlike other mapping algorithms, the output response is ordered according to some characteristic feature of the input vectors. It can be interpreted as a nonlinear projection of the p-dimensional input onto an output array of nodes. Each neuron has a vector of coefficients associated with it ($v_i \in \Re^p$) usually known as "code vectors".

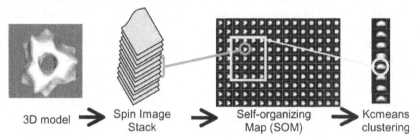

| 3D model | Spin Image Stack | Self-organizing Map (SOM) | Kcmeans clustering |

Fig. 2. Three-level indexing mechanism. A self-organized 2D map "summarizes" the information provided by spin images. Each output vector has an attached set of spin images. The third-level of indexing clusters the map into a reduced set of representative images (centers of clusters)

The functionality of the algorithm can be described as follows: when a spin image ($x_i \in \Re^p$) is presented to the net, the neurons in the output layer compete with each other and the winner (whose code vector has the minimum distance from the input spin image) as well as a predefined set of neighbor code vectors update themselves. This process is continued until some stopping criterion is met, usually, when code vectors "stabilize" or when a number of iterations are completed. Details of the algorithm can be found in [18]. The result of applying SOM to the spin image stack is a reduced set of spin images that represent the inherent clustering in the original set. At this level of indexing, comparison with a given object is reduced from thousands of spin images to less than two hundreds.

3.2   Kernel Probability Density Estimator Clustering Algorithm (Kernel C-means)

Kernel c-means clustering algorithm is a process of grouping similar objects into the same class by maximizing an objective function explicitly designed to estimate a set of the cluster's centers whose probability density function resembles as best as possible the probability density of the input data. The theoretical basis of these methods has been reported in detail elsewhere [19,20] and will only be briefly reviewed here. Given n data items of dimension p, $X_i \in \Re^{p \cdot 1}$ with $i=1 \ldots n$, the problem is to find c surrogate "representative" data items (centroids), $V_i \in \Re^{p \cdot 1}$ with $j=1 \ldots c$, such that the estimated probability density:

$$D(\mathbf{X}) = \frac{1}{c} \sum_{j=1}^{c} K(\mathbf{X} - \mathbf{V}_j; \alpha) . \qquad (1)$$

where K is a kernel function, with $\alpha > 0$ the kernel width parameter that controls the smoothness of the estimated density. Intuitively, the kernel estimator can be seen as a sum of "bumps" placed at the observations (data). The kernel function K determines the shape of the bumps while the parameter α determines the width. A commonly used kernel function is the well-known Gaussian kernel. In general, let $D(X;\theta)$ be the probability density for the random variable $\mathbf{X}$, where θ is some unknown parameter vector. Let $X_i \in \Re^{p\cdot1}$, $i=1\ldots n$, denote the data items then:

$$L = \prod_{i=1}^{n} D(\mathbf{X}_i;\theta) \qquad (2)$$

is the likelihood function, and the most common statistical estimator for θ is obtained my maximizing eq. 1. Note that in this case the parameter vector is composed by the (code vectors) $V_i \in \Re^{p\cdot1}$ and the Kernel width α, i.e: $\theta=\{\{Vj\},\alpha\}$. The log-likelihood is:

$$l = \sum_{i=1}^{n} \ln\left(D(\mathbf{X}_i)\right) = \sum_{i=1}^{n} \ln\left(\frac{1}{c}\sum_{j=1}^{c} K(\mathbf{X}_i - \mathbf{V}_j;\alpha)\right) \qquad (3)$$

The Kernel c-means algorithm consists of an iterative optimization of the above objective function (eq. 3). As a result, we add an additional level of indexing following the self-organizing map to further reduce the number of comparisons to be done at query time. The indexing will be defined by two parameters (d,c) where d is the number of nodes in the self-organizing map (second level) and c the number of clusters produced by the kcmeans algorithm (third level).

Efficiency in accessing to stored spin images is thus related to these parameters. As said above, if no indexing mechanism exists worst case computational complexity is $O(N_{db}N_q)$ where N_{db} is the total number of spin images in the database. More particularly:

$$N_{db} = \sum_{i=1}^{N_{obj}} N_{spin\_images}(obj_i) \qquad (4)$$

With our methodology, the parameter N_{db} is reformulated as $N_{db} = dN_{obj}$ at the second level, and $N_{db} = cN_{obj}$ at the third one. The reader should note that the number of "original" spin images per object equals to the number of triangles (which is quite often in the range of thousands). On the other hand, we have performed efficient and accurate retrieval with (d,c) parameters in a range of 150 and 10 feature vectors respectively.

4   Experiments

Our system was tested against three small databases (fig. 3). The aircraft and molecular database contains very similar objects. It was intentionally done in order to measure the accuracy in the recognition process. The database of macromolecules was built from real datasets depicted as volumetric density images. A method for obtaining a density-oriented mesh representation from volumetric datasets is detailed in [20].

For each query object the complete spin image stack is computed. The accuracy is defined by the number of query spin images that were correctly classified when compared to the third level of indexing. All the experiments were performed with (p,c)=(150,10). Note that the number of comparisons per object is reduced from several thousands to 10.

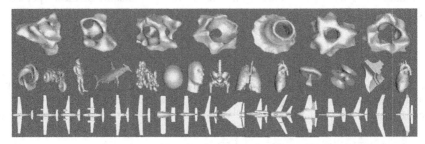

Fig. 3. Three test databases (one per row). Each object was presented to the database as a query input. The query object was retrieved in first place 100% of times. Similarity between the query object and target is measured (average recognition accuracy, a.r.a for short). (a) Molecular database (first row). 7 models, a.r.a: 70% (b) Mixture database (second row). 14 models, a.r.a: 92% (c) Aircraft database, 17 models a.r.a: 73%. Note that recognition accuracy is lower than that for molecular and aircraft databases since they contain very similar objects. Nevertheless, for all query examples the similarity value with the second best matched object is lower than 10%. No ambiguity in the answer is possible

Given an arbitrarily shaped query object, a number of spin images are calculated. Let be $T_{obj}=\{t\}$, the set of triangles t that compose the mesh of such object and $SP_{obj}=\{s\}$ the corresponding set of spin images where spin image s_i corresponds to triangle t_i. The query spin image set QSP_{obj} is a subset of SP_{obj}. This is important as the user might be interested in retrieving those objects in the database that have similarity to a particular surface patch of the query object. Thus, locality is well supported.

At query time, Euclidean distance of spin images in QSP_{obj} and those at the 3rd (or 2nd) level of indexing is computed. A vote is added to the database object with closest spin image. The average recognition accuracy is the number of spin images in QSP_{obj} that were correctly classified assumed that the query object was already stored in the database (fig. 3). Another interesting way of visualizing the query result is by the use of locality. Triangles associated to spin images in QSP_{obj} might be colored according to the color assigned to the three best-matched objects in the database (fig. 4).

5   Conclusions and Future Work

In this work we have presented a shape-based retrieval system that is suitable to both general and domain-specific 3D object databases. It assumes free-form 3D objects to be provided as triangular meshes. The spin image representation has been employed as a robust shape signature. Our main contribution is the use of a self-organizing map and a clustering algorithm as an indexing mechanism so that the overall number of comparisons is significantly reduced. Thus, our method achieves indexing and data compression as only "representative" spin images are chosen. Also, it increases the robustness against noise (non-representative spin images). The experimental results confirm a remarkable improvement in the efficiency of the comparison procedure so that interactive queries are possible. Even more, the recognition accuracy penalty is minimum. An important issue for further work is to incorporate metrics (better than euclidean distance) at the indexing level that would allow to retrieve objects when only partial views are provided. Also, we plan to provide the user not only with a top list of matching objects but with the rigid transformations that would align the query object with the target. Application to large molecular databases is underway.

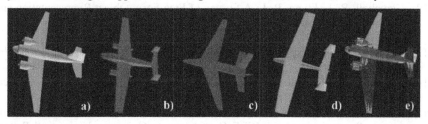

Fig. 4. An additional advantage of our system is that spin images support locality. A new plane is presented to the aircraft database (a). When compared with the third indexing level (kcmeans), the three most similar aircrafts are shown with a.r.a 27%(b) 13%(c) and 12%(d). Triangles in the query object were colored (d) according to the correspondence of their associated spin images with respect to matching items

References

1. Flickner, M. et al: Query by image and video content: the QBIC system. IEEE Computer Vol. 28(9) (1995) 23-32.
2. Gupta, A. et al.: The Virage image search engine: an open framework for image management. Storage and Retrieval for Image and Video Databases IV, Proc SPIE 2670 (1996) 76-87.
3. Sussman, Lin, Jiang, Manning, Prilusky, Ritter and Abola: Protein Data Bank (PDB): database of 3D structural information of biological macromolecules. Acta Crystallogr. Sect D 54 (1998) 1078-1084.
4. Paquet, E., Rioux, M.: Nefertiti: a tool for 3-D shape databases management, Proceedings of the SAE International Conference on Digital Human Modeling for Design and Engineering, The Hague, Netherlands. (1999)
5. Elvins, T., Jain, R.: Web-based volumetric data retrieval. Symposium on Virtual Reality Modeling Language, San Diego USA. (1995) 7-12.
6. Osada, R., Funkhouser, T., Chazelle, B., Dobkin, D.: Matching 3D Models with Shape Distributions. Shape Modeling International , Genova, Italy, May, 2001.
7. Liu, Y., Rothfus, W. E., Kanade, T.: Content-based 3D Neuroradiologic Image Retrieval: Preliminary Results. IEEE International Workshop on Content-based Access of Image and Video Databases. (1998) 91 - 100.
8. Ankerst M., Kastenmüller G., Kriegel H.-P., Seidl T.:Nearest Neighbor Classification in 3D Protein Databases. Proc. 7th Int. Conf. on Intelligent Systems for Molecular Biology (ISMB), Heidelberg, Germany. AAAI Press (1999) 34-43.

9. Johnson, A., Hebert, M.: Recognizing objects by matching oriented points. Proceedings of the IEEE Conference on Computer Vision and Pattern Recognition (CVPR '97) (1997) 684–689.
10. Besl, P. J.: The free-form surface matching problem. Machine Vision for Three-dimensional Scenes. Academic Press, San Diego (1990) 25-71.
11. Hebert, M., Ponce, J., Boult and Gross, A.: Object Representation in Computer Vision. Springer-Verlag Eds., Berlin 1995.
12. Campbell, R. J., Flynn, P. J.: A Survey of Free-Form Object Representation and Recognition Techniques. Computer Vision and Image Understanding, Vol 1 (2001) 166-210.
13. Loncaric, S.: A survey of shape analysis techniques. Pattern recognition, Vol 31(8) (1998) 983-1001.
14. Veltkamp, R. C., Hagedoorn, M.: State of the art in shape matching. Technical Report UU-CS-1999-27, Utretch University, the Netherlands, 1999.
15. Johnson, A., Hebert, M.: Efficient multiple model recognition in cluttered 3-D scenes. Proceedings of the IEEE Conference on Computer Vision and Pattern Recognition (CVPR '98) (1998) 671 - 677.
16. Böhm C., Kriegel H.-P., Seidl T.: Adaptable Similarity Search Using Vector Quantization. Proc. Int. Conf. on Data Warehousing and Knowledge Discovery (DaWaK 2001), Munich, Germany, (2001).
17. Bernard, S., Boujemaa, N., Vitale, D., Bricot, C.: Fingerprint Classification using Kohonen Topologic Map. The IEEE International Conference On Image Processing, ICIP (2001).
18. Kohonen T.: Self-Organizing maps, Second Edition, Springer-Verlag Eds (1997).
19. Pascual-Montano, A., Donate, L. E., Valle, M., Bárcena, M., Pascual-Marqui, R. D., Carazo, J. M.: A Novel Neural Network Technique for Analysis and Classification of EM Single Particle Images. J. of Struct. Biol. 133 (2/3), (2001) 233-245.
20. De-Alarcón, P. A., Pascual-Montano, A., Gupta, A., Carazo, J. M.:Modeling shape and topology of low-resolution density maps of biological macromolecules. (2002) (Biophysical Journal, in press).

Size Functions for Image Retrieval:
A Demonstrator
on Randomly Generated Curves

A. Brucale, M. d'Amico, M. Ferri, L. Gualandri, and A. Lovato

Dip. di Matematica, Università
Piazza di Porta S. Donato, 5, I-40126 Bologna Italy
ferri@dm.unibo.it

Abstract. Size functions, a class of topological-geometrical shape descriptors, are applied to the search of image datasets by a hand-drawn input: This is the core of a demonstrator accessible through the Internet. Two datasets are provided, of about 700 curves each; both are unstructured sets of randomly generated curves. One consists of piecewise smooth curves, the other of polygonals. The curve generator has been explicitly designed so that the curves have no semantic content; they don't have any kind of indexing or textual caption either. The characteristics of the demonstrator and some experimental results are presented.

1   Introduction

We present a demonstrator of an image retrieval engine, based on size functions, working on two unstructured datasets of randomly generated curves. The datasets consist of 700 curves each; curves are closed unions of smooth (but not rectilinear) arcs in one case, closed polygonals in the other. The query is performed by drawing an example by hand. The demonstrator is reachable through the Internet (http://vis.dm.unibo.it/EVAM/).

Size functions are a relatively new class of shape descriptors, which can be modularly adapted to specific recognition or comparison problems. From the mathematical viewpoint, they are based on geometric–topological theory of critical points, as will be specified in the next Section. So long, they have been applied to several classification problems, mostly on images of natural origin.

In the terminology of [14], our system belongs to the categories of "content based", "search by association", "narrow domain", "approximate query by image example", but the difference with the systems reported in that survey are: Semantic interpretation (as, e.g., in [12]) is impossible or at least highly subjective (Rorschach-like); no colors or textures are present to direct selection (as, e.g., in [5]); also salient geometric features (as used, e.g., in [13]) would be of little of no use while comparing a hand-drawn sketch with a piecewise smooth curve or a polygonal. No indexing or other structure (as, e.g., in [1]) is present in the image domain (it is a dataset, not a database).

M. S. Lew, N. Sebe, and J. P. Eakins (Eds.): CIVR 2002, LNCS 2383, pp. 235–244, 2002.
© Springer-Verlag Berlin Heidelberg 2002

The demonstrator is far from being a general purpose engine: Image processing problems have been by-passed, so that the user can concentrate on our main issue, i.e. the value of size functions in finding similarities and differences of "pure" shapes. We are well aware that a commercial (or at least a useful) product should integrate size functions with other techniques, but this is far from our present scope.

2   The Definition of Size Function

Size functions are a simple, but effective tool for shape comparison. Their application is particularly useful when no standard, geometric templates are available. Examples of applications are tree-leaves, hand-drawn sketches, monograms, hand-written characters, white blood cells and the sign alphabet.

Let us recall the definition of a size function. Consider a continuous real-valued function $\varphi : M \to \mathbb{R}$, defined on a subset M of a Euclidean space. The *size function* of the pair (M, φ) is a function $\ell_{(M,\varphi)} : \mathbb{R}^2 \to \mathbb{N} \cup \{\infty\}$. For each pair $(x, y) \in \mathbb{R}^2$, consider the set $M_x = \{P \in M : \varphi(P) \le x\}$. Two points in M_y are then considered to be equivalent if they either coincide or can be connected by a continuous path in M_y. The value $\ell_{(M,\varphi)}(x, y)$ is defined to be the number of the equivalence classes obtained by quotienting M_x with respect to the previous equivalence relation in M_y.

Figure 1 shows a simple example of size function. In this case the topological space M is a curve, while the measuring function φ is the distance from the centre of mass C.

The use of size functions in shape description and comparison is motivated by the following properties:

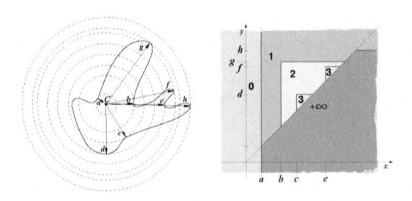

Fig. 1. A curve and its size function with respect to the distance from the centre of mass

Invariance. Size functions inherit the same invariance of the corresponding measuring functions.

Modularity. Changing the invariance group simply requires changing the measuring functions.

Multilevel analysis. Details are described close to the diagonal $\Delta = \{(x, y) \mid x = y\}$, while more general aspects of shape appear far from Δ.

Resistance to noise. Information is distributed all over the real plane, so that size functions can be used in presence of noise and occlusions.

Fast computation. Size functions are easily computable.

Standardization. The problem of comparing shapes is changed into a simple comparison of functions.

3 Representation and Comparison of Size Functions

As Figure 1 shows, size functions have a typical structure. In the halfplane $\{x < y\}$ they are linear combination of characteristic functions of triangular regions. In other words, if ℓ is a size function we have that $\ell(x, y) = \sum_{i \in I} \chi_i(x, y)$ (for $x < y$), where I is countable set and each χ_i is the characteristic function of a triangle T_i, having vertices $(a_i, a_i), (a_i, b_i), (b_i, b_i)$ (i.e., $\chi_i(x, y) = 1$ if $a_i \leq x \leq y \leq b_i$ and 0 otherwise). That means that each size function can be described by a formal linear combination $\sum_{h \in I} m_h C_h$ of points $C_h = (a_h, b_h)$ (the right angle vertex of T_h) with integer multiplicities m_h. If $b_h < \infty$ we call C_h a *cornerpoint*, otherwise we say that C_h represents a *cornerline* (in this case the corresponding triangle T_h degenerates to an "infinite" triangle). Each distance between formal series naturally produces a distance between size functions. A formal and detailed treatment of this subject can be found in [6].

Of the many available distances between formal series, the one we use in this paper is the matching distance (similar to the "bottleneck matching" of [15], but performed on the size functions, instead of the images). In plain words, if we wish to compute the distance between two formal series $\sum_{i \in I} m_i C_i, \sum_{j \in J} m'_j C'_j$

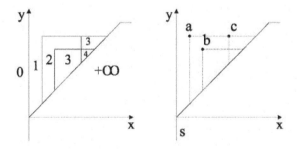

Fig. 2. A size function on the left and the corresponding cornerpoints a, b, c and cornerline s on the right. The resulting formal series is $s + a + b + c$

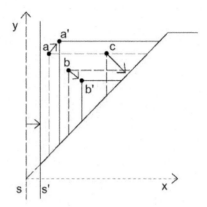

Fig. 3. A possible way to deform a size function, represented by dashed lines, into another one, represented by continuous lines. The cornerpoint c is destroyed

we consider any possible matching between the C_i's and the C''_j's, considering the multiplicities. Each of these matchings is weighted by a deformation cost (in this work: Euclidean displacement). The wanted distance is the infimum in the set of all possible costs of matchings. We also allow the "destruction" of cornerpoints by taking them onto the diagonal $\Delta = \{(x, y) | x = y\}$. The inverse process of cornerpoint creation is allowed, too.

   The computation of the matching distance is performed by means of a Max Flow algorithm. Only the first ten cornerpoints are considered, with respect to the decreasing distance from the diagonal Δ. A similar version of the matching distance (more along the lines of "elastic matching" [2]) can be found in [7].

4   Curve Generation

Two sets of 700 curves have been built for the demonstrator. The curves of the first set have a finite (possibly empty) set of singular points, out of which they are smooth and non-rectilinear (see Figure 4 for the first 70 of them). The second set consists of closed polygonals (see Figure 5 for the first 70 of them).

   The piecewise smooth curves are all generated by random parameter variations of the formula

$$\begin{cases} x(t) = \lambda\rho(t) \cos\theta(t) \\ y(t) = \frac{1}{\lambda}\rho(t) \sin\theta(t) \end{cases}$$

with

$$\theta(t) = s \left(2\pi t + \alpha_1 \sin(\ell_1 2\pi t) + \beta_1 \cos(\ell_2 2\pi t)\right) + \frac{k\pi}{2}$$
$$\rho(t) = r_0 + \alpha_2 \sin(h_1 2\pi t) + \beta_2 \cos(h_2 2\pi t)$$

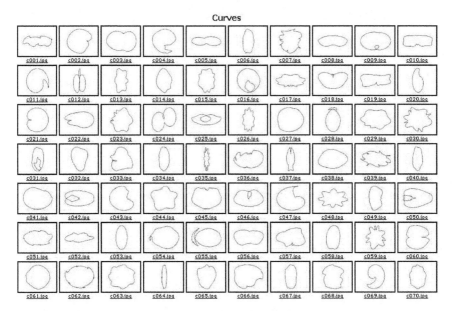

Fig. 4. A sample of our dataset of randomly generated smooth curves

where λ is randomised with normal distribution, with mean $m = 1$ and variance $\sigma^2 = 0.3$, and r_0 is also randomised with normal distribution, $m = 3$, $\sigma^2 = 1$. The values of the other parameters are uniformly distributed in their sets:

$$s \in \{-1, 1\}$$
$$k \in \{0, 1, 2, 3\}$$
$$\alpha_1 \in [0, 2]$$
$$\ell_1 \in \{0, 1\}$$
$$\beta_1 \in [0, 0.3]$$
$$\ell_2 \in \{0, 1, 2, 3, 4, 5, 6, 7\}$$
$$\alpha_2 \in [0, 3]$$
$$h_1 \in \{1, 2, 3\}$$
$$\beta_2 \in [0, 0.6]$$
$$h_2 \in \{1, 2, 3, 4, 5, 6, 7, 8, 9, 10\}.$$

Polygonals are generated by connecting random points in the prescribed frame with segments.

5   Experimental Results

The measuring functions used in the demonstrator are the distances from 16 fixed points forming a regular grid around the center of mass of the image; they are

Polygons

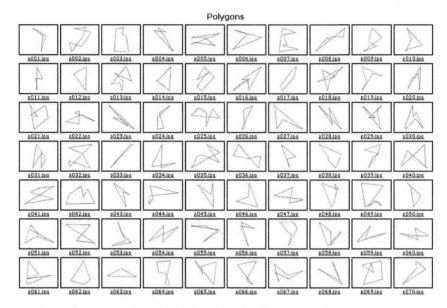

Fig. 5. A sample of our dataset of randomly generated polygonal curves

disposed in four horizontal rows, each of length equal to 2.8 times the average distance of the image pixels from the center of mass. The invariance granted by this type of measuring functions is under translations and scale changes. We remind that invariance under other transformation groups (rotations, affine transformations etc.) could be easily imposed by a different choice of measuring functions.

Two types of queries were performed: by submitting an image extracted from the dataset, and by submitting a hand–drawn curve. The latter is of course the more interesting experiment, and it is the one on which we are reporting. This was done separately for the smooth curves and for the polygonals.

It should be noted that it is not possible to establish a hit/miss ratio or an analogous mark, since shape similarity is subjective. Moreover, the curves of our dataset do not represent any concrete object, so they cannot be classified even according to a semantic criterion. Next, we present some of the queries (Figures 6 and 7), with the first five output curves.

Some remarks about the depicted results:

- Some queries were aimed to specific curves present in the datasets, and some were totally generic.
- Given an input drawing, it is unlikely that there are several resembling shapes in the dataset, so even among the first five output curves there are some scarcely related.
- Finally, it should be noted that in most cases even the most resembling output curves are not superimposable to the input example. Moreover, even

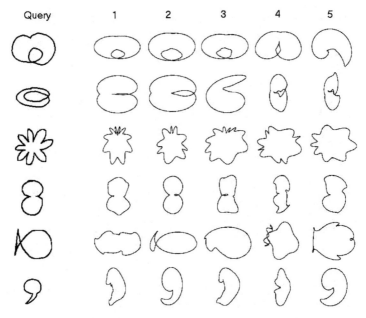

Fig. 6. The results of some sample queries in our dataset of smooth curves

salient points or any commonly used geometric features would not have helped to retrieve those curves (actually, they might even be misleading!). The reason why the system succeeds in retrieving them all the same, resides in the highly qualitative kind of shape description and comparison due to size functions.

The demonstrator runs on a Athlon 900 MHz based PC, under Linux OS. Response time for a query is, for the moment, between 40 and 60 seconds.

6   Comparisons

Of course, there are several shape descriptors and shape matching methods (see, e.g., [15]). As Carlsson [3] points out, there are two major categories: recognition by components (requiring some *a priori* knowledge) and recognition on similarity of views (based on geometrical features). Our system is rather in the second class, although in a generalised sense. In fact, the similarities which are caught by size functions may not be so apparent: they depend on the choice of the measuring function. That no semantics is involved, is clear in the present experiment. But also geometric features, in the usual sense, are not there to favour a matching.

Take the classical Fourier descriptors, for instance [17,8]. A coincidence in the first few coefficients would grant that two shapes are roughly superimposable, differences being limited to the higher frequencies. Size functions (with distance

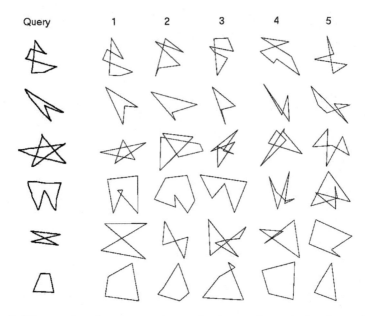

Fig. 7. The results of some sample queries in our dataset of polygonal curves

from centre of mass as a measuring function) recognise the presence and size of comparable bumps even if they are differently disposed in the two images to compare; this is a simple case of similarity with no (even rough) superimposition. Such cases are easy to find already in the few examples of Figures 6 and 7. We remind that, anyway, also size functions enjoy completeness theorems like Fourier descriptors, in the case of the present experiment [4].

Other descriptors, like order structure [3], turning function [16], chain code histogram [11], need a sort of local superimposition and are mainly limited to silhouettes. A much closer relative to size functions is the Reeb Graph [9], which has been used so long — as far as we know — only for 3D objects and for heights and distances as the only involved functions.

As for the type of experiment, we are aware only of [11,10], where irregular objects with no perceivable semantics are treated, but again working on silhouettes and with a smaller dataset.

Anyway, we do not claim that our method be better than the usual ones, but at least independent and possibly complementary.

Following the indications of the referees, whom we thank, we shall extend the experiment to a database of drawings — which we are preparing — and to a standard trademark database. Moreover, we shall compare the ranking of our system with the one independently produced by human subjects.

7   Conclusions

An on–line demonstrator of a qualitative type of image retrieval runs on two sets of randomly generated curves. Size functions build the mathematical engine. The output shapes are remarkably similar to the input example, but they could not be retrieved by superimposition nor by comparison of currently used feature vectors. Similarity is of an intuitive kind, not relying on geometrical features nor on semantic content. This appears to be a good evidence that size functions are an efficient tool for reducing the "semantic gap" — in the words of [14] — between data and its interpretation by the user. The next steps will be the experiment with real databases and the comparison with human selectors.

Acknowledgements

Work performed within the activity of ARCES "E. De Castro", under the auspices of INdAM-GNSAGA and of the University of Bologna, funds for selected research topics.

References

1. Brown, L., Gruenwald, L.: Tree-based indexes for image data. J. Visual Comm. and Image Representation **9** (1998) 300–313. 235
2. Burr, D. J.: Elastic matching of line drawings. IEEE Trans. on PAMI, 3 (1981) 708–713. 238
3. Carlsson, S.: Order structure, correspondence, and shape based categories. in: D. A. Forsyth et al. (Eds.): Shape, Contour and Grouping in Computer Vision, LNCS 1681 (1999) 58–71. 241, 242
4. Ferri, M., Frosini, P.: Range size functions. Proc. SPIE Conf. on Vision Geometry III, Boston, 1994 Nov. 2-3 (1995) 243-251. 242
5. Forsyth, D. A., Fleck, M. M.: Automatic detection of human nudes. Int. J. Computer Vision **32** (1999) 63–77. 235
6. Frosini, P., Landi, C.: Size functions and formal series. Applicable Algebra in Engineering Communication and Computing, 12 (2001) 327–349 237
7. Donatini, P., Frosini, P., Landi, C.: Deformation energy for size functions. In: E. R. Hancock, M. Pelillo (eds.) Energy Minimization Methods in Computer Vision and Pattern Recognition,, LNCS 1654 (1999) 44–53. 238
8. Granlund, G. H.: Fourier Preprocessing for hand print character recognition. IEEE Trans. Computers, C-21 (1972) 195–201. 241
9. Hilaga, M., Shinagawa, Y., Kohmura, T., Kunii, T. L.: Topology matching for fully automatic similarity estimation of 3D shapes. SIGGRAPH 2001, Computer Graphics Proc., Annual Conference Series (2001) 203–212. 242
10. Iivarinen, J., Peura, M., Särelä, J., Visa, A.: Comparison of combined shape descriptors for irregular objects. In: A. F. Clark (ed.) Proc. 8th British Machine Vision Conference, BMVC'97, Essex (1997) 430–439. 242
11. Iivarinen, J., Visa, A.: Shape recognition of irregular objects. In: D. P. Casasent (ed.) Intelligent Robots and Computer Vision XV: Algorithms, Techniques, Active Vision, and Materials Handling, Proc. SPIE 2904 (1996) 25–32. 242

12. Joyce, D. W., Lewis, P. H., Tansley, R,H., Dobie, M. R., Hall, W.: Semiotics and agents for integrating and navigating through multimedia representations. Proc. Storage and Retrieval for Media Databases, vol. 3972 (2000) 120–131. 235
13. Mokhtarian, F.: Silhouette–based isolated object recognition through curvature scale–space. IEEE Trans. PAMI **17** (1995) 539–544. 235
14. Smeulders, A. W. M., Worring, M., Santini, S., Gupta, A., Jain, R.: Content–based image retrieval at the end of the early years. IEEE Trans. PAMI **22** (2000) 1349–1380. 235, 243
15. Veltcamp, R. C., Hagedoorn, M.: State–of–the–art in shape matching. Technical Report UU-CS-1999-27, Utrecht University (1999). 237, 241
16. Vleugels, J., Veltkamp, R. C.: Efficient image retrieval through vantage objects. in: D. P. Huijsmans and A. W. M. Smeulders (Eds.) Visual Information and Information Systems — Proc. 3rd Int'l Conf. VISUAL'99, Amsterdam, LNCS 1614 (1999) 575–584. 242
17. Zahn, C. T., Roskies, R. Z.: Fourier descriptors for plane close curves. IEEE Trans. Computers, C-21 (1972) 269–281. 241

An Efficient Coding of Three Dimensional Colour Distributions for Image Retrieval

Jeff Berens and Graham D. Finlayson

School of Information Systems, University of East Anglia
Norwich NR4 7TJ, UK
{jeff,graham}@sys.uea.ac.uk

Abstract. The distribution of colours in an image provides a useful cue for image indexing and object recognition [1,3,2,4]. Previously, we have shown how chromaticity distributions can be coded using a hybrid compression technique: histograms are coded with a Discrete Cosine Transform and then Principal Component Analysis is applied to a reduced set of the DCT coefficients, resulting in excellent indexing results, using just the first eight Principal Components [5,6]. We have investigated compression on colour distributions independent of colour intensity, however, colour is generally represented by a 3-D model, (two chromaticity channels and one intensity channel). One difficulty with 3-D chromaticity distribution histograms is their sparseness - many bins contain no or few image pixels. This becomes a problem when attempting to derive PCA statistics: it becomes necessary to analyse an unrealistically large number of histograms. We show that applying the Discrete Fourier Transform to colour distribution histograms leads to a dimensionality reduction that makes PCA possible. We also demonstrate the general case that 3-D and n-D distributions, particularly sparse ones, can be significantly reduced in dimension.

1 Introduction

It is now well established that the distribution of colours in an image can be a highly useful cue for image indexing and object recognition. Much research, including some by ourselves, has taken place into image retrieval and object recognition using image colour content. Our own work has been concerned with how much colour information is contained in an image, and how that information can be significantly and efficiently compressed. We have shown how chromaticity distribution histograms can be coded using a hybrid compression technique: the histograms are first coded using the Discrete Cosine Transform (DCT) and then a further coding is carried out using Principal Component Analysis (PCA). The importance of this type of coding of an image colour distribution for the purpose of indexing, is that a colour distribution can be represented by just the first eight principal components. Indeed we have presented results demonstrating successful indexing into small and moderately sized databases. However, our initial studies investigated compression on colour distributions independent of colour intensity.

M. S. Lew, N. Sebe, and J. P. Eakins (Eds.): CIVR 2002, LNCS 2383, pp. 245–252, 2002.

One reason for this is that the distribution of image chromaticities is a 2-D histogram; in effect it is a 2-D matrix. We interpreted the matrix as an image and then applied standard signal processing methods for image compression.

However, colour is generally represented by a three-dimensional model, comprising two chromaticity channels and one intensity channel. Commonly used three-dimensional colour spaces such as Opponent-colour, HSV (Hue, Saturation, Value) and HSI (Hue, Saturation, Intensity) [7] are all examples of this type of 3-D colour model. When considering the properties of 3-D distribution histograms, it is clear that the fundamental difference from 2D is that the pixels are clustered together in much smaller numbers - in 2D they are projected, generally as a single coherent cluster, onto a single plane, whereas in 3-D, the clusters are unevenly split between a set of planes. Figure 1 demonstrates this: a typical colour image has been converted from its RGB's to opponent-colour, quantised to eight levels per colour axis, and a distribution histogram formed. Histogram (a) represents the distribution of the 2-D chromaticities, Red-Green (RG) against Blue-Yellow (BY). A zero bin count is shown as black, and the brighter the bin, the higher the pixel count in that bin. The 3-D histogram represents the 2-D chromaticities in relation to their colour intensity value. The sequence of eight histograms (b) represents the RG,BY chromaticity distribution at each of the eight quantised levels of the Black-White axis (BW), the colour intensity, with the left-hand histogram containing the RG,BY counts at the lowest BW intensity level, and the right-hand one the highest intensity level. This figure clearly demonstrates that many of the intensity levels contain very few pixels.

Figure 1 also demonstrates another important factor which must be considered when coding colour distribution histograms: for a typical colour image, most histogram bins formed from the colour distribution will be empty, many bins will contain just a few pixels, and only a small number of bins will contain any significant pixel quantities [1]. This characteristic has implications when obtaining a statistical model using PCA - in order to derive an accurate representation from such sparse structures, it is necessary to analyse a very large number of histograms (i.e. images), in fact, far more images are needed than are typically found in an image database.

We have previously employed a first-stage coding, prior to carrying out PCA, as a means of both smoothing the data and reducing the dimensionality of the 2-D distribution histogram. That is, project 2-D histograms onto a smooth basis and then code the smoothed histogram. Since this smooth representation is, by definition, parameterised by a small number coefficients, PCA can be used to reliably extract useful statistics for image indexing. Simply, we would like to apply the same workflow that we previously used for 2-D in 3-D (and in the future, n-D).

A feature of our previous work has been to employ, wherever possible, standard image processing tools, Our 2-D histograms were coded using the DCT, and in the second stage, the DCT coefficients were coded with respect to the optimal PCA bases. The DCT transform was used because in chromaticity space,

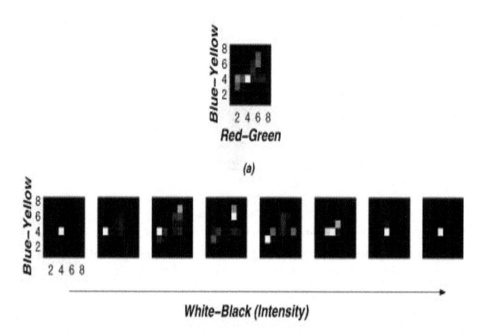

Fig. 1. Opponent-Colour distribution histograms (8 bins per axis) for a typical colour image: (a) shows the 2-D chromaticity histogram, (b) shows the 3-D histogram, displayed for ease of visualisation as eight 2-D chromaticity histograms, one for each of the intensity levels, increasing from left to right. A zero pixel count is shown as black, and the brighter the bin colour, the higher the count in that bin. Only a small proportion of the 2-D bins, and even fewer of the 3-d bins, contain any significant pixel count, clearly demonstrating the sparsity of 3-D histograms

colour distributions are often non-zero at the edges, and so we need to use flexible boundary conditions (the DCT assumes neither periodicity nor zero-boundary conditions). However, in the 3-D case, highly saturated colours, that is, histogram bins near the boundary are rare (if indeed they are even possible) and so the distributions naturally go to zero near the edges. Figure 2 shows a cumulative opponent-colour distribution formed from chromaticities of all the images in a medium sized database (as used in later experiments) - few of the image chromaticities fall near the boundary. Under these conditions, the use of Fourier transform encoding becomes more appropriate than the DCT (since we have zeros at the boundary of the cube, we have, from the viewpoint of Fourier encoding, a natural periodicity).

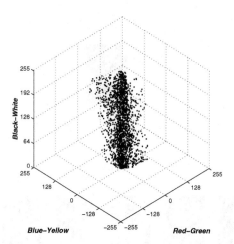

Fig. 2. A cumulative opponent-colour distribution comprising 240 diverse colour images - note how few of the pixels occur near the boundaries

In producing this work, the authors were aware of the literature regarding image coding, specifically, the technique *Vector Quantisation*. In vector quantisation, the number of colours is quantised, typically from 24 bit (16.7 million colours) to 8-bit (256 colours) or 16 bit (65536 colours). The quantisation can be carried out by reassigning every pixel RGB value to the most similar colour in a predefined set of reduced colours, or by clustering the pixel RGB's to produce a variable set containing the most commonly occurring colours in the image. Vector quantisation is the basis of the GIF image compression standard [11]. A drawback of vector quantisation is that information between quantisation control points is lost in the quantisation process. Initial investigations indicate to us that the shape of the colour distribution between control points is important.

This work presents the results of using the DFT as a means of providing a first-stage coding of both 2- and 3-D histograms. We show in this paper that successful indexing on small and medium sized databases can be carried out on 3-D colour distributions represented by the first eight principal components, derived from the low-order Fourier components. Furthermore, we believe that this method provides a general and attractive feature analysis: by frequency coding the 3-D histogram we effectively reduce the dimensionality enabling accurate PCA modelling. Our approach and use of the DFT is in fact quite general: DFT extends to n-dimensions, making this procedure suitable for higher-dimensional data reduction. One further advantage of the DFT is that it is very fast - the 2-D DFT operates at $O(\frac{N^2}{2}log_2N^2)$, and the 3-D DFT at $0(\frac{N^3}{2}log_2N^3)$, where N is the dimension of the histogram axis (assuming equal dimensions per histogram axis).

In section 2 of this paper we show how image distributions are compared using the first eight principal components, derived from the low-order DFT coefficients. In section 3, we present results showing indexing performance for a range of 3-D colour spaces.

2 Distribution Comparison Using Low-Order DFT Statistics

Let us firstly consider how histogram distributions and their derived statistics can be compared: If H denotes a colour histogram then H_i denotes the number of RGB's (or any other colour representation) falling in the ith region of colour space. The distance between two images I_1 and I_2 is defined as the distance between their respective histograms:

$$||I_1 - I_2|| \equiv ||H_1 - H_2|| \tag{1}$$

We use the notation $\underline{H}$ to denote the one dimensional vector counterpart of H (H is stretched out as a vector), and so the distance between histograms H_1 and H_2 is defined to be:

$$||H_1 - H_2|| = ||\underline{H}_1 - \underline{H}_2|| = \sqrt{\sum_i (\underline{H}_{1,i} - \underline{H}_{2,i})^2} \tag{2}$$

Using the result that $||\underline{v}||^2 = \underline{v}.\underline{v}$ we rewrite the distance as:

$$||H_1 - H_2||^2 = (\underline{H}_1 - \underline{H}_2).(\underline{H}_1 - \underline{H}_2) \tag{3}$$

The vector dot-product can be written as $\underline{v}.\underline{v} = \underline{v}^t\underline{v}$ (where ^{t} denotes transpose) and so,

$$||H_1 - H_2||^2 = (\underline{H}_1 - \underline{H}_2)^t(\underline{H}_1 - \underline{H}_2) \tag{4}$$

Assuming that each histogram is an n dimensional vector of DFT coefficients then $\mathcal{B}$ is an $n \times k$ matrix (where $k \leq n$):

$$\underline{H} = \sum_i \mathcal{B}_i b_i \tag{5}$$

where $\mathcal{B}_i$ denotes the ith column of $\mathcal{B}$ and b_i is a scalar weight. We chose the columns of $\mathcal{B}$ to be the first n low-order Fourier functions. In so doing, we are enforcing a bandlimit on histogram data. If we group these weights into the k-vector $\underline{b}$ then:

$$\underline{H} = \mathcal{B}\underline{b} \tag{6}$$

$\mathcal{B}$ is an orthogonal basis, therefore $\mathcal{B}^t\mathcal{B} = \mathcal{I}$ (the $k \times k$ identity matrix). Substituting $\underline{H}_1 = \mathcal{B}\underline{b}_1$ and $\underline{H}_2 = \mathcal{B}\underline{b}_2$ into (5b):

$$||H_1 - H_2||^2 = (\underline{b}_1 - \underline{b}_2)^t(\underline{b}_1 - \underline{b}_2) \tag{7}$$

That is,

$$||H_1 - H_2||^2 = ||\underline{b}_1 - \underline{b}_2||^2 \qquad (8)$$

Equation (8) is important since it informs us that we can compute the distance between two n dimensional histograms by calculating the difference of corresponding k dimensional basis weight vectors. If $k < n$ then the cost of histogram comparison is reduced. More importantly, if $k << n$ then this tells us that only a small part of histogram space is important.

So, to derive the best k-dimensional subspace of histogram space, and given a set of m histograms $M = \{\underline{H}_1, \underline{H}_2, \cdots, \underline{H}_m\}$, $\mathrm{PCA}(M, k)$ returns the k-dimensional basis $\mathcal{B}$ that minimises:

$$\mathrm{PCA}(M, k) = \mathcal{B} \ , \ \min_{\mathcal{B}} \sum_{i=1}^{m} ||\underline{H}_i - \mathcal{B}\underline{b}_i||^2 \qquad (9)$$

and where it is understood that the weight vectors b_i are defined relative to the basis $\mathcal{B}$, it is quite straightforward to show that $b = \mathcal{B}^t H$. [10].

3 Experiments

To test our 3-D coding technique, we created feature vectors for each image, and their related test images, from two image databases. We generated 16x16x16 bin colour distribution histograms for each image, applied the first stage coding, and then derived the PCA's from the transform coefficients, selecting the first eight PCA's to produce a feature vector for indexing. The first database consisted of 240 diverse photographic images, including those used by Swain in his initial work on Colour Indexing [1]. We created a test set of 100 images, containing database images viewed from different positions and varying degrees of occlusion, (this set included the original Swain test images). We calculated the distance between the feature vector for each test image and all database image feature vectors. The smallest distance overall was used to define the identity of the test object. We had previously achieved our best performance on this database using 8 PCA feature vectors derived from DCT coding of 16x16 bin distribution histograms in opponent-colour space, with an Average Match Percentile (AMP) of 99.86% on this database. This level of performance is indicative of correct recognition much of the time. When we repeated this with the DFT coding of 3-D opponent-colour space, we achieved an AMP of 99.13%, which, although lower than our original results, is still highly satisfactory. We repeated the experiment in another 3-D space, this time using the image RGB's, and achieved similar (given the small size of the database) results - 99.3% AMP. We tested the performance using different quantities of DFT coefficients, as shown in Table 1, and found that the optimum number was 125. Using less coefficients produced lower AMP's, and using more made little difference. As a comparison, the results for PCA applied to the full 16 x 16 x 16 bin histograms is shown. As expected, smoothed histograms deliver better results.

Table 1. Table of results for small database, showing AMP% for number of DFT coefficients used

#DFT COEFFICIENTS USED	Hist only	8	27	64	125	216	343	512
OP-COL	97.66	98.13	98.41	99.03	99.12	99.13	99.14	99.09
RGB	98.03	98.57	99.01	99.21	99.32	99.30	99.30	99.18

In order to test the scalability of the technique, we then repeated the experiments on a much larger database containing 4100 Corel photostock images. We created a test set by creating two test images from image, the first by cropping it to a panoramic format by removing the top and bottom 15% from the original image, and the second by removing 15% from each side of the original image to produce a portrait format. Thus 8200 test images were created. By adopting this approach we are modelling typical camera output formats used in the APS (Advanced photo system). The results, shown in Table 2, were highly satisfactory, as almost perfect indexing was achieved.

Table 2. Table of results for Corel database, showing AMP% for number of DFT coefficients used

#DFT COEFFICIENTS USED	Hist only	8	27	64	125	216	343	512
OP-COL	97.35	98.23	99.40	99.71	99.79	99.82	99.82	99.81
RGB	98.01	98.71	99.52	99.82	99.84	99.84	99.84	99.82

4   Conclusions

We have shown in this paper how 3-D chromaticity distribution histograms can be successfully coded in such as way as to compress all the salient information into a very small feature vector, and that it can be useful for image indexing. In order to minimise information loss, we choose to apply the DFT as a first stage coding that models *every* data point in the (histogrammed) distribution, rather than modelling just a reduced set of control points. Future work will compare the performance of the DFT against vector quantisation as a first-stage coding.

Interestingly, we found that RGB provided better indexing performance than opponent-colour, unlike our results in 2-D, where consistently opponent-colour out-performed all other colour spaces tried. This is to be investigated further. More importantly, this work has shown that the 3-D DFT is both effective and simple to apply, and that extending its use to n-D provides no difficulties, making it a very powerful pre-coding step prior to deriving higher-order statistics, such as PCA. We are currently investigating spatial and colour information contained in 3-D distribution histograms in order to further improve indexing efficiency. Initial results show great promise.

Acknowledgements

This work has been funded by EPSRC Award GR/L60852. The authors would like to thank the Corel Corporation for the ease of access to their royalty free photostock.

References

1. M. J. Swain and D. H. Ballard. Color indexing. *International Journal of Computer Vision*, 7(11):11–32, 1991. 245, 246, 250
2. M. A. Stricker and M. Orengo. Similarity of color images. In *Storage and Retrieval for Image and Video Databases III*, volume 2420 of *SPIE Proceedings Series*, pages 381–392. Feb. 1995. 245
3. M. Flickner, H. Sawhney, W. Niblack, J. Ashley, Q. Huang, B. Dom, M. Gorkani, J. Hafner, D. Lee, D. Petkovic, D. Steel and P. Yanker. "Query by Image and Video Content" IEEE Computer Magazine, Sept. 1995, pp 23-31. 245
4. B. V. Funt and G. D. Finlayson. Color constant color indexing. *IEEE transactions on Pattern analysis and Machine Intelligence*, 17:522-529, May 1995. 245
5. G. D. Finlayson, J. Berens and G. Qiu. "A Statistical Image of Colour Space". In *IEE 7th International Conference on Image Processing and its Applications*, 1999. 245
6. G. D. Finlayson, J. Berens and G. Qiu. "Image Indexing Using Compressed Colour Histograms". *IEE Proceedings, Vision and Image Signal Processing*, Vol 146, Aug. 2000. 245
7. B. Wandell. "Foundations of Vision". Sinauer Associates, Inc. 1995. 246
8. R. C. Gonzalez and R. E. Woods *Digital Image Processing* Adisson-Wesley, 1993.
9. C. Faloutsis *Searching Multimedia Databases by Content* Klewer Academic Publishers, 1996.
10. G. H. Golub and C. F. van Loan. *Matrix Computations* John Hopkins U. Press, 1983. 250
11. M. Sonka, V. Hlavac and R. Boyle. *Image Processing, Analysis, and Machine Vision, 2nd Edition.* PWS Publishing, 1999. 248

Content-Based Retrieval
of Historical Watermark Images: I-tracings

K. Jonathan Riley and John P. Eakins

Institute for Image Data Research, University of Northumbria
Newcastle NE1 8ST, UK
{jon.riley,john.eakins}@unn.ac.uk

Abstract. Providing content-based access to archives of historical watermarks involves a number of problems. This paper describes the development and evaluation of SHREW, a shape-matching system for watermark images based on techniques developed for the ARTISAN trademark retrieval system. Encouraging retrieval results have been obtained using tracings of watermark images, whether whole-image features or component-based measures are used for matching. Further development of the system is continuing.

1 Introduction

1.1 Background

Watermarks are permanent pictorial devices introduced into paper during sheet formation. The device is formed from copper-bronze wire sewn to the surface of the perforated papermakers' mould. Each sheet of paper taken from the same mould is identical. The combination of watermark and perforations provides important indicators of the origins and authenticity of specific papers. Watermark analysis is used by historians, art historians, criminologists and others in order to identify and date all kinds of paper samples.

Capturing the watermark images from the original papers can be problematic since many can be obscured by various types of media on top of the paper or secondary supports underneath it. A number of reproduction methods are in use, including tracings, transmitted light photography, and radiographic techniques. Tracings are cheap and easy to produce, but are subjective and often lack accuracy. Transmitted light photography can produce a clear image, but details of the watermark are often obscured by the media layer. Radiographic techniques can overcome many of these problems, although they risk introducing additional noise into the recorded images.

1.2 Watermark Matching and Identification

Currently, matching an unknown watermark to known ones is a difficult and time consuming task. Researchers must work manually and painstakingly through thousands of drawings of marks and their supporting text. A number of manual classification codes have been devised to assist scholars in this task, such as the

M. S. Lew, N. Sebe, and J. P. Eakins (Eds.): CIVR 2002, LNCS 2383, pp. 253-261, 2002.

Briquet classification [1] and the more recent IPH (International Association of Paper Historians) code [2].

So far, only one instance of the use of content-based image retrieval (CBIR) techniques for historical watermark images has been reported in the literature: the SWIC (Search Watermark Images by Content) system of Rauber et al [3]. This system can retrieve images using either manually-assigned codes such as Briquet's, or automatically extracted features such as the number, shape and relative locations of image regions. Rauber et al report promising results on a test database of 3000 images, though it is not clear how they judge retrieval success.

1.3 Techniques for Shape Matching

Many techniques for shape similarity matching have been proposed over the last twenty years. Some, such as boundary deformation [4], are based on direct matching of shape boundaries. However, these tend to be computationally very expensive, so most techniques are based on the comparison of shape features extracted from the image under consideration. These include simple features such as aspect ratio, circularity, and transparency [5], "natural" measures of triangularity, ellipticity and rectangularity [6], Fourier descriptors [7], moment invariants [8], and Zernike moments [9]. Two types of shape feature are of particular significance because of their incorporation in the new MPEG-7 standard [10] - the angular radial transform (ART), and curvature scale-space coefficients [11].

A recent paper comparing the effectiveness of a range of such features for trademark image retrieval [12] indicated that acceptable retrieval performance could be achieved with a wide variety of image feature combinations. Although a combination of simple, "natural" and Fourier descriptors proved most effective in these experiments, differences between these and many other combinations proved only marginal. The majority of retrieval failures observed resulted from failure of image segmentation to reflect human perceptual judgements [13].

1.4 Aims of Current Investigation

The principal aim of our investigation is to investigate the applicability of techniques developed for our ARTISAN trademark image retrieval system [14] to the similarity retrieval of historical watermarks. A subsidiary aim is to compare the effectiveness of whole-image and component-based matching for this application. This work forms part of the Northumbria Watermarks Archive project [15], which aims to create a collection of digitised watermark images - reproduced by a variety of methods, catalogued in accordance with IPH standard, and searchable by shape similarity - to be made available to researchers and scholars via the World-Wide Web

2   Methodology

2.1 Watermark Image Preprocessing

The initial step in our investigation of the applicability of the techniques developed within our ARTISAN project was the conversion of raw watermark images into a form suitable for feature extraction. Two test collections were available to us at the start of this project - a set of 806 tracings of watermark images from the Churchill collection [16], digitised by the Conservation Unit at the University of Northumbria, and 2687 digitised electron-radiographs kindly donated by the Koninklijke Bibliotheek in Holland. Examples of these images are shown in Fig 1.

Fig. 1. Examples of tracings (left) and electron-radiographs (right) of historical watermarks

The principal differences between these images and the majority of the modern trademarks for which ARTISAN was developed are:

- they are grey-scale rather than bilevel (black and white) images;
- the majority are line-based rather than consisting of large solid or textured regions;
- they are often more ornate, containing significant amounts of fine detail;
- electron-radiograph images in particular have large amounts of associated noise, an inevitable side-effect of the extraction process.

This paper describes our experiments with watermark tracings; our investigation of techniques for retrieval of electron-radiographs is still continuing, and will be reported elsewhere.

All images were first normalized to fit within a 512×512 bounding box, to facilitate the use of segmentation algorithms developed for ARTISAN. For different experiments they were then subjected to one or more of the following procedures:

(a) None (feature extraction was performed directly on grey-scale images).
(b) Thresholding to convert the grey-scale to bilevel images, by replacing the intensity $I(x,y)$ of each pixel with the value 255 where $I(x,y) > T_I$, and 0 otherwise. The threshold $T_I = \mu_I + \kappa\sigma_I$, where μ_I is the mean pixel intensity over the entire image, σ_I its standard deviation, and κ an empirically-determined constant. Best results were achieved in our experiments with $\kappa = 0.67$.
(c) Noise reduction. Several different filters were applied so that their effectiveness in smoothing out the noise inherent in the digitisation process could be investigated. These comprised two linear filters, a 5×5 mean filter and a 5×5 Gaussian ($\sigma = 1$), and two non-linear filters, a 5×5 median and a 5×5 Kuwahara. The latter replaces the centre pixel with the mean value of the quadrant of the convolution kernel with smallest intensity variance; it is claimed to be particularly effective at preserving edges during the smoothing process [17].

(d) Morphological operations. The morphological *close* operator ($\mathbf{X} = \mathbf{X} \bullet \mathbf{B}$, where **B** is a structuring element - for our experiments, a discrete circular kernel of radius r pixels was used) was applied in some experiments to strengthen the thin line elements present in many of the tracings, in an attempt to prevent these lines from being lost in subsequent processing.

2.2 Feature Extraction

Two separate sets of features were extracted for our experiments:

-   Three sets of **whole-image features** were computed, both directly from raw images and from images thresholded as indicated in (b) above. The features extracted were ART as defined in the MPEG-7 standard [10], the 7 normal moment invariants defined by Hu [8], and the 4 affine moment invariants defined by Flusser and Suk [18].

-   Three sets of **component-based features** were computed, from images thresholded as in (b), from images smoothed as in (c), from images closed as in (d), and from images processed by selected combinations of these operations (see below). Following the selected combination of these steps, images were processed using the technique described in [12], generating a set of regions defined by closed boundaries for each image. From the boundary of each of these regions, the following shape descriptors were then extracted:

    (i)   aspect ratio, circularity, and convexity as defined in [12];
    (ii)   triangularity, ellipticity and rectangularity as defined by Rosin [6]
    (iii) 8 normalized Fourier descriptors as defined in [12].

2.3 Shape Matching and Evaluation of Effectiveness

For each combination of image preprocessing and feature extraction techniques selected for study, a test database was built from the 806 tracing images at our disposal. A prototype search system known as SHREW (**SH**ape **RE**trieval of **W**atermarks) was built to allow input and matching of query images, and display of search results as a ranked list of thumbnail images. Several alternative means of computing overall similarity scores between query and stored images were provided, as with our earlier ARTISAN system [12]. For matching based on whole-image features such as ART, image similarities were computed as city-block distances $L_1 = |v_q - v_s|$ between vectors comprising selected features from each image. For matching on the basis of *component* features, component similarities were first computed as city-block distances between feature vectors based on the simple, "natural" and Fourier descriptors defined above. These were then combined into image similarity measures using the *asymmetric simple* method defined in [14]. Although SHREW has the facility to select the feature set used for matching at run time, the entire feature set was used in the experiments reported here. Fig 2 shows the results of a typical search.

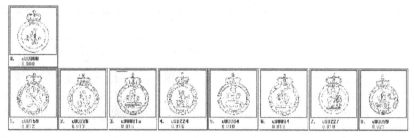

Fig. 2. Typical retrieval results from SHREW, showing the query image (top line) and the first eight retrieved images in order of similarity

The overall effectiveness of each combination of image enhancement and shape matching technique was evaluated, using a set of 15 test query images and sets of expected search results, assembled by colleagues in the Conservation Unit at Northumbria University. As in previous trials, the effectiveness of each combination of techniques evaluated was determined by averaging normalized precision P_n and normalized recall R_n scores over each of the 15 queries. P_n and R_n were defined by Salton [19] as:

$$P_n = 1 - \frac{\sum_{i=1}^{n}(\log R_i) - \sum_{i=1}^{n}(\log i)}{\log\left(\frac{N!}{(N-n)!n!}\right)}$$

$$R_n = 1 - \frac{\sum_{i=1}^{n}R_i - \sum_{i=1}^{n}i}{n(N-n)}$$

where N is the total number of items in the collection, n the number judged relevant to the current query, and R the rank at which relevant item i is actually retrieved. Each of these measures gives an estimate of retrieval effectiveness in the range 0-1, but emphasizes different aspects of system performance. Broadly, P_n gives an overall measure of system performance at all retrieval ranks, while R_n emphasizes good performance at high retrieval ranks.

3   Retrieval Results

The first set of experiments investigated the effectiveness of matching on features computed from whole images. Two databases were built, one deriving features from the original grey-scale images, the other from images thresholded at a pixel intensity of $\mu_I + 0.67\sigma_I$. Each of the 15 queries was then run against each database using each of the three whole-image features defined above. Table 1 shows the results of these experiments:

Table 1. Effectiveness of whole-image features for retrieval. All results are expressed as mean ± standard error over 15 test queries

Feature used for matching	Grey-level images		Thresholded images	
	P_n	R_n	P_n	R_n
Angular Radial Transform	0.53 ± 0.06	0.81 ± 0.04	0.67 ± 0.05	0.89 ± 0.02
7 Normal moment invariants	0.44 ± 0.03	0.79 ± 0.03	0.37 ± 0.04	0.77 ± 0.03
4 Affine moment invariants	0.28 ± 0.01	0.68 ± 0.03	0.34 ± 0.03	0.72 ± 0.04

Overall, ART is significantly more effective than either the normal or affine moment invariants, whether P_n or R_n measures are used ($P < 0.005$ or better, Wilcoxon signed-rank matched-pairs test). Its effectiveness is also significantly enhanced by image thresholding ($P < 0.001$).

The second set of experiments was designed to measure the effectiveness of component-based matching. Five databases were built, all using thresholded images (the shape extraction routines derived from ARTISAN require bilevel images as input), using no filter, mean, median, Gaussian and Kuwahara filters respectively. The 15 queries were run against each of these databases, using asymmetric simple matching with simple, "natural" and Fourier descriptors, the combination that proved most effective in the evaluation of ARTISAN [12]. Results are shown in Table 2.

Table 2. Effectiveness of component-based features for retrieval, and the effect of filtering to remove noise, using thresholded images ($T_1 = \mu_1$)

Noise filter used	P_n	R_n
(none)	0.33 ± 0.05	0.60 ± 0.05
5×5 median filter	0.25 ± 0.04	0.54 ± 0.05
5×5 Gaussian filter	0.51 ± 0.06	0.74 ± 0.05
5×5 Kuwahara filter	0.52 ± 0.06	0.73 ± 0.05
5×5 mean filter	0.57 ± 0.06	0.81 ± 0.05

Use of component-based matching on these images clearly brings no compelling advantages - when used on images on which no noise removal had been performed, it gave poorer results than any of the whole-image based methods listed in Table 1. Noise removal using mean, Gaussian or Kuwahara filtering improved performance significantly ($P < 0.01$ for Kuwahara or Gaussian filters, $P < 0.001$ for the mean), though the differences between mean, Gaussian and Kuwahara filters were not significant. The poor performance of the median filter in this context was noteworthy.

A further set of experiments was performed, to investigate the effects of varying the threshold for segmentation, and of applying the morphological *closure* operator after thresholding. All images were thresholded at a level of either μ_1 or $\mu_1 + 0.67\sigma_1$, subjected to noise removal with either the mean or Kuwahara filter, and subjected to a single closure operation with structuring element radius $r = 2$, 4 or 6 pixels. The results of these experiments are shown in Table 3.

Table 3. Effectiveness of component-based matching using different thresholds, and the effects of applying the *closure* operator together with either mean or Kuwahara filtering

Filter used	Radius of structuring kernel	$T_I = \mu_I$		$T_I = \mu_I + 0.67\sigma_I$	
		P_n	R_n	P_n	R_n
Mean	0 (no closure)	0.57 ± 0.06	0.81 ± 0.05	0.37 ± 0.05	0.65 ± 0.04
	2 pixels	0.53 ± 0.05	0.83 ± 0.03	0.40 ± 0.05	0.70 ± 0.04
	4 pixels	0.54 ± 0.05	0.81 ± 0.04	0.48 ± 0.04	0.79 ± 0.03
	6 pixels	0.55 ± 0.06	0.79 ± 0.05	0.47 ± 0.05	0.78 ± 0.03
Kuwahara	0 (no closure)	0.52 ± 0.06	0.73 ± 0.05	0.37 ± 0.05	0.64 ± 0.04
	2 pixels	0.58 ± 0.06	0.83 ± 0.05	0.55 ± 0.07	0.79 ± 0.04
	4 pixels	0.50 ± 0.06	0.79 ± 0.04	0.61 ± 0.04	0.86 ± 0.02
	6 pixels	0.50 ± 0.06	0.77 ± 0.05	0.61 ± 0.05	0.88 ± 0.03

From these results it can be seen that while closure makes no significant difference to retrieval effectiveness when either the lower threshold is chosen or mean filtering is used, it does have a significant effect at the higher threshold level when Kuwahara filtering is used ($P < 0.01$ for both R_n and P_n using closure with $r = 6$). For the images in our database, the combination of a high threshold level, edge-preserving filtering and closure clearly pays off in terms of improved retrieval performance. However, retrieval effectiveness scores using the best combination of parameters for component-level matching are still marginally inferior to those for whole-image matching based on ART (Table 1), though the difference is significant ($P < 0.01$) only for the R_n measure.

4   Conclusions and Further Work

It is clearly possible to base an effective retrieval system for historical watermarks on techniques developed for trademark image retrieval. Features based on whole images and components can offer broadly comparable levels of retrieval effectiveness, provided a modest amount of image preprocessing is performed to reduce noise and strengthen lines from the original image. It is however worth noting that the best retrieval effectiveness scores obtained in the current study were still markedly lower than those achieved by ARTISAN (P_n 0.70 ± 0.04 and R_n 0.94 ± 0.02).

Given the extra computational cost of component-based matching, it clearly does not seem worthwhile to use this for watermark retrieval in preference to matching on the basis of whole-image features such as ART, at least for whole-image queries. (Matching on whole-image features is clearly of little use for part-image matching, where a user wishes to query a database to identify all watermarks which match a specified fragment of an image). The effectiveness of matching by a combination of whole-image and component features remains to be explored. Preliminary analysis of results from individual queries suggests that whole-image features may sometimes be effective when component-based features are not, and *vice versa*, suggesting that there is scope for synergy between the two types of measure.

Future work on this project will include a detailed analysis of individual cases of retrieval failure (a technique which proved highly beneficial in the development of ARTISAN), in order to inform the task of improving image preprocessing and feature extraction techniques. We are also currently extending our methods to handle electron-radiographs, where more sophisticated noise reduction methods will be needed, and plan to develop more powerful techniques for partial shape matching in a future phase of the project.

Acknowledgements

We would like to express our sincere appreciation to Richard Mulholland of the Conservation Unit, University of Northumbria, for digitizing the tracings from the Churchill collection and providing queries and ground truth for the evaluation experiments, and to Jonathan Edwards, School of Computing, University of Northumbria, for advice and assistance throughout the project. Thanks are also due to Hes and de Graaf, Netherlands for permission to reproduce the tracings from the Churchill collection shown in Figs 1 & 2, and Koninklijke Bibliotheek, Netherlands for permission to reproduce the radiograph image in Fig. 1. The financial support of the UK Arts and Humanities Research Board is also gratefully acknowledged.

References

1. Briquet, C. M.: Les Filigranes. Karl Hiesermann, Leipzig (1907)
2. International Standard for the Registration of Paper with or without Watermarks, International Association of Paper Historians (IPH), English Version 2.0, 1997
3. Rauber, C., Pun, T. and Tschudin, P.: Retrieval of images from a library of watermarks for ancient paper identification. In: Proceedings of EVA 97, Elektronische Bildverarbeitung und Kunst, Kultur, Historie, Berlin (1997)
4. Jain, A. K. et al: Object matching using deformable templates. IEEE Transactions on Pattern Analysis and Machine Intelligence 18(3) (1996) 267-277
5. Levine, M. D.: Vision in man and machine, ch 10. McGraw-Hill, N Y (1985)
6. Rosin, P. L.: Measuring shape: ellipticity, rectangularity and triangularity. In: Proceedings of 15th International Conference on Pattern Recognition, Barcelona (2000) vol 1, 952-955
7. Zahn, C. T. & Roskies, C. Z.: Fourier descriptor for plane closed curves. IEEE Transactions on Computers C-21 (1972) 269-281
8. Hu, M. K.: Visual pattern recognition by moment invariants. IRE Transactions on Information Theory IT-8 (1962) 179-187
9. Teh, C. H. and Chin, R. T.: Image analysis by methods of moments. IEEE Transactions on Pattern Analysis and Machine Intelligence 10(4) (1988) 496-513
10. Available on the Web at http://www.cselt.it/mpeg/public/mpeg-7_visual_xm.zip
11. Mokhtarian, F. S. et al: Efficient and Robust Retrieval by Shape Content through Curvature Scale Space. In: Proceedings of International Workshop on Image DataBases and MultiMedia Search, Amsterdam (1996) 35-42

12. Eakins, J. P. et al: A comparison of the effectiveness of alternative feature sets in shape retrieval of multi-component images. In: Storage and Retrieval for Media Databases 2001, Proc SPIE 4315 (2001) 196-207

13. Ren, M., Eakins, J. P. and Briggs, P.: Human perception of trademark images: implications for retrieval system design. Journal of Electronic Imaging 9(4) (2000) 564-575

14. Eakins, J. P., Boardman, J. M. and Graham, M. E.: Similarity retrieval of trademark images. IEEE Multimedia 5(2) (1998) 53-63

15. Brown, J. E. et al: When images work faster than words - the integration of CBIR with the Northumbria Watermarks Archive. To be presented at ICOM-CC 13th triennial meeting, Rio de Janeiro, Sept 2002

16. Churchill, W. A. Watermarks in Paper in Holland, England, France etc. in the XVII and XVIII Centuries and their Interconnection, Menno Hertzberger & Co. Amsterdam, 1935

17. Kuwahara, M. et al: Digital Processing of Biomedical Images. Plenum Press, New York (1976) 187-203

18. Flusser, J. and Suk, T.: Pattern recognition by affine moment invariants. Pattern Recognition 26(1) (1993) 167-174

19. Salton, G.: The SMART retrieval system - experiments in automatic document processing. Prentice-Hall, Englewood Cliffs, New Jersey (1971)

Object Recognition for Video Retrieval

Rene Visser, Nicu Sebe, and Erwin Bakker

LIACS Media Lab, Leiden University
Niels Bohrweg 1, 2300 CA Leiden, The Netherlands

Abstract. Recognition of objects in video can offer significant benefits to video retrieval including automatic annotation and content based queries based on the object characteristics. This paper describes our preliminary work toward recognizing objects in video sequences and gives a brief survey of the relevant research in the literature. We use the Kalman filter to obtain segmented blobs from the video, classify the blobs using the probability ratio test, and apply several different temporal filtering methods, which results in sequential classification methods over the video sequence containing the blob. Results from real video sequences are shown.

1 Introduction

Detecting particular objects in video is an important stop toward semantic understanding of visual imagery. For example, in content based retrieval, the ability to detect people and automobiles gives the option of advanced queries such as "Find a video clip which contains a crowded area or a fast moving car."

In this paper, we give a summary of the current work in video segmentation for moving objects followed by the implemented tracking and classification algorithm. The novel aspect of our work is described in Sections 4 and 5 where we use temporal filtering for classifying the moving objects and show some early results. Conclusions are given Section 6.

2 Background

The segmentation of moving objects is an important problem in image sequence analysis and in the problem of video retrieval (i.e. [2, 5, 13]). There is significant related research [1-18]. Some approaches toward visual matching apply stochastic parsing [14]; integrate learning [1, 5, 8]; and adaptive background mixtures [7]. A real-time system for tracking people, called Pfinder ("Person finder"), is proposed by Wren, et al. [16]. First, a model of the scene is built by observing the scene when no person is present. For each pixel, the mean color value and the covariance of the associated distribution is determined. Then when a person enters the scene, the system

M. S. Lew, N. Sebe, and J. P. Eakins (Eds.): CIVR 2002, LNCS 2383, pp. 262-270, 2002.

begins to build up a model of that person. This is done by first detecting a large change in the scene, and then building up a multi-blob model of the person over time. The model building process is driven by the distribution of color on the person's body with additional blobs added for other colored regions.

In the work by Gu, et al., [3], a method for spatio-temporal segmentation of long image sequences of scenes that include multiple independently moving objects is presented. This method is based on the Minimum Description Length (MDL) principle.

A similar application with different techniques has been presented in [6]. In this paper, two approaches for motion based representations are described. The first approach demonstrates that dominant 2D and 3D motion techniques are useful for computing video mosaics through the computation of dominant scene motion and/or structure. However, this may not be adequate if object level indexing and manipulation is to be accomplished efficiently. The second approach addresses this issue through simultaneous estimation of an adequate number of simple 2D motion models. A unified view of the two approaches naturally follows from the multiple model approach: the dominant motion method becomes a particular case of the multiple method if the number of models is fixed to be one and only the robust EM algorithm without the MDL stage is employed.

Another application in the area of videos is presented in [19]. It addresses the problem of automatic video browsing as it describes methods that use edges and motion for detecting production effects, e.g. cuts, fades, dissolves, wipes and captions, and computing motion segmentation. This segmentation involves the computation of the primary and the secondary motion. This is achieved by finding isolated peaks in the error surface formed by the similarity of two images, a scalar function of the displacement that makes the two images most similar. These motions are then used to classify the individual pixels. This involves the individual pixel error scores at the primary and secondary displacements.

The problem of detecting semantic events in video can be solved by a three-level approach as proposed in [4]. At the first level the input video sequence is decomposed into shots using a simple color histogram based technique, global motion is estimated, 76 low-level color and texture features are extracted, and motion blobs are detected. At the second level a multi-layer perceptron neural network is used to classify the detected motion blobs as moving object regions based on the extracted color and texture features. This level also summarizes each shot in terms of intermediate-level descriptors. At the third level the generated shot summaries are analyzed and the presence of the events of interest are detected based on an event inference model which incorporates domain-specific knowledge.

In this approach, the difference between the current and the motion compensated previous frame is used to detect motion blobs using a robust statistical estimation based algorithm [20]. Based on the frame difference result, the algorithm constructs two 1D histograms by projecting the frame difference map along its x and y direction, respectively. The histograms, therefore, represent the spatial distributions of the motion pixels along the corresponding axes. The instantaneous center position and size of a moving object in the video can be estimated based on statistical measurements derived from the two 1D projection histograms.

In [17], ASSET-2 (A Scene Segmenter Establishing Tracking), a motion segmentation and shape tracking system has been presented. The paper describes a way image sequences taken by a moving video camera may be processed to detect and track moving objects against a moving background in real-time.

In this approach, each frame is initially processed to find two dimensional features and edges. The two dimensional features are found using the SUSAN corner detector or the Harris corner detector. The SUSAN edge detector is used to find the edges. The two dimensional feature list is passed to a feature tracker which uses a two dimensional motion model to match and track features over as many frames as possible. A two dimensional vector field can then be created by taking either feature velocities or displacements over a fixed number of frames. The resulting vector field is passed to a flow segmenter which splits the list of flow vectors into clusters which have similar flow within them and are different to each other. Next, this cluster list is compared with a temporally filtered list of clusters, and the filtered list is updated using the newly found clusters. This list of clusters is finally used to provide information about the motions of objects.

In [20], an object matching system is described. It is able to extract objects of interest from outdoor scenes and match them. The application involves measuring the average travel time in a road network. Two cameras are placed about a mile apart from each other. Image analysis is performed on both the road image sequences to extract the objects of interest. An integration of multiple cues, including a motion segmentation mask of the moving areas in the image sequence, homogeneous color regions, edges, and model information is used to identify the moving objects. Two objects extracted from images captured by the two independent cameras at different times are then matched to evaluate their similarity.

The method to extract moving objects from an image sequence is based on the fusion of a motion segmentation mask obtained by image subtraction and a set of homogeneous color regions obtained by the integration of a split-and-merge algorithm and edges resulting from the Canny edge detector. Motion segmentation based on image subtraction is used to obtain an approximate location of the moving objects in the image sequence.

3   Video Object Tracking

In order to find the moving objects we maintained a weighted average of the scene to be used as the background reference frame. Subtracting the current frame from the background reference frame results in blobs of moving objects. We described the blobs using an eigenvector/value decomposition of the region within the blob. A blob is thus represented by a vector of n values and n parameters; and the (x,y) origin parameters.

We applied an approach similar to [1] which used Kalman filtering (Appendix) to maintain the object identity over the video sequence and to optimize the tracking process of each blob. Figure 1 displays an example of blob segmentation where the left image is the original frame and the right image contains the segmented person.

The tracking method is important in that it is the first stage of segmentation. Any errors which appear in the tracking stage will also propagate to the classification stage described next.

Fig. 1. Frame from sequence (left); Segmented blob using background subtraction and Kalman Filtering (right)

4   Classification

Understanding video sequences is a well researched topic (see [1], [5], [16], and [8]). The blobs are represented using their eigenvector decomposition [15]. We selected N eigenvectors based on maximizing the information content [5, 13, 22]. For the classification, we turned to the statistical literature for the probability ratio (PRT) test, which we use to classify the blobs as people, automobiles, or unknown.

1. Video segmentation and tracking similar to [1].
2. Blob region representation using eigenvector selection [5, 13, 15, 16, 22].
3. Initial blob region classification into 3 classes: people, automobiles, or unknown using PRT.

Median Classification and Standard Rate (SR)

The misdetection rate for the single frame classification method is defined as the standard rate (SR). As a certain object is tracked during a particular period of time, a certain history is built up. For example, if an object is tracked during eleven successive image frames, eleven classifications have been made, eleven classification values have been assigned. These classification values can be used to determine a global classification value for the whole sequence. The number of 0 classifications and the number of 1 classifications is determined. If the number of 0 classifications is greater than the number of 1 classifications, the global classification will be 0 ("no moving person"). If the number of 1 classifications is greater than the number of 0 classifications, the global classification will be 1 ("moving person"). If the number of 1 classifications equals the number of 0 classifications, the global classification will be 0.5 ("undecided").

Instead of counting the different classifications, the average classification value of the sequence is computed. This is a faster way to determine the global classification value. If the average equals 0.5, then the classification will be "undecided". If the average is less than 0.5, then the classification is "no moving person". If the average is greater than 0.5, the classification will be "moving person".

In order to reduce noise, to reduce the number of misclassifications in the SR sequence, a median filter can be used to determine new classification values. In this way, a smoothing of the classification sequence can be achieved.

Five methods, all based on median filtering, are developed. In the following sections, these methods are presented and evaluated. In these methods, the median filtering as presented is actually achieved by the average classification approach, as described above.

Inside Filtering (IF)

A median filter can be seen as a rectangular window moving from the left to the right over the sequence of classification values. A median filter of size five for example will first take the first five values. The values are sorted and the middle value is taken to adjust the third value of the classification sequence as shown in Figure 2.

Then the filter is moved one place to the right and the process starts again. Five values are taken. The median of these five values will be assigned to the fourth value of the classification sequence. In this way, when using a filter of size five, the filter runs from the third element from the beginning of the sequence to the third element from the end of the sequence. We can define an inside and an outside. The outside elements are in this example the first two and the last two elements of the sequence

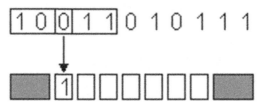

Fig. 2. A median filter: Filter Size = 5

As the method works on the inside of the sequence, it is called Inside Median Filtering (IF). The outside values remain the same.

Outside Filtering (OF)

In this method, the filter used only works on the outside elements. The process is as follows. Imagine two virtual values before the first element and two virtual values after the last element of the classification sequence. The filter of size five can be used on the two virtual values and the first three values of the sequence. As we have only three values available, the median of these three values will be determined (see Figure 3). The second value will be based on the first four values.

Inside & Outside (IOF)

This method is a combination of the Inside and the Outside methods. These methods are applied at the same time. The filter starts assigning a value to the first element of the sequence. In our example, the Outside method will be used for the first two elements. After that, the Inside method is used seven times. Finally, the Outside method is used again to assign values to the last two elements of the sequence.

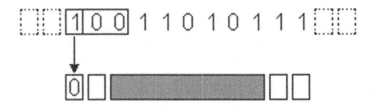

Fig. 3. Outside Median Filtering: Filter Size = 5

InsideOutside (IFOF)

A different combination is used here. First, the Inside method is used. This provides a new sequence of classification values. Then the Outside method is used on the new values.

OutsideInside (OFIF)

As in the previous method, only now the Outside method is used first.

5   Results

It is notable that our system uses off-the-shelf components which can be found in a typical computer/electronics store. We expect our system to be challenged by the following sources of noise:

- low resolution/detail images
- color distortion
- lens distortion
- loss of brightness and contrast from the video capture process.
- artifacts from the video tracking/segmentation process.
- block compression artifacts inherent in MPEG-1

Fig. 4. Example of detecting automobiles and people

In Figure 4, we display an example of the detection of automobiles and people in a busy street scene. For our tests, we used 6 video sequences of city street intersections. Each sequence was 5 minutes (at 25 fps) in length and captured using a PAL camcorder. The video was extracted to 1.5 Mbps MPEG-1 digital format using an ATI All-In-Wonder Rage 128 card. The frame resolution was half PAL resolution.

Moreover, the system is expected to function in situations where the size of the moving object is quite small (less than 20 pixels wide).

On a PIII-800 Mhz computer, our system was able to capture, track, segment, and classify all of the blobs at a rate of 19 fps.

Figure 5 shows the results of using different filter sizes to improve the initial classification rate (SR).

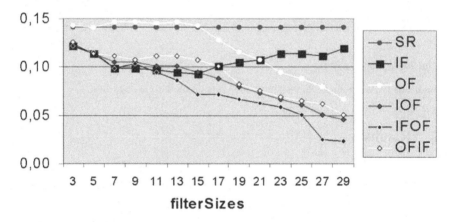

Fig. 5. Misdetection (y-axis) vs. filter size (x-axis)

6   Discussion and Conclusions

In the previous sections we have described a system for recognizing two categories, people and automobiles from video sequences. The novel aspect of this paper is in applying several temporal averaging methods toward classifying the blobs. These methods exploit the sequential nature of video to improve the classification results.

In particular, each frame in the video containing the blob adds additional evidence toward the classification process. As more frames are processed, it can be shown that the error probabilities decrease.

From the tests, the IFOF method had the best misdetection rate. However, all of the median filter based methods improved the initial misdetection rate when sufficient frames were used.

Regarding future work, we think that significant improvement needs to be achieved in increasing the number of object categories and more abstract concepts. For example, it would be beneficial to classify abstract concepts such as a Christmas scene, or the emotional mood of the scene. Learning algorithms such as [5] and [8] appear to have the most potential. Therefore, methods based on techniques found in

[2], [5], [7], [8], [15], and [22] are intended for exploration as well as matching in the compressed space [10]. Furthermore, the theoretical generalization of the PRT to sequential data, SPRT, will be investigated in later work.

References

1.   A. M. Baumberg, "Learning Deformable Models for Tracking Human Motion," PhD thesis, The University of Leeds, School of Computer Studies, UK, October 1995.
2.   S.-F. Chang, W. Chen, H. J. Meng, H. Sundaram, D. Zhong, "VideoQ: An Automated Content Based Video Search System Using Visual Cues," *ACM Multimedia '97 Proceedings*, pp. 313-324, (Seattle, Washington, USA), November, 9-13, 1997.
3.   H. Gu, Y. Shirai, M. Asada, "MDL-Based Segmentation and Motion Modeling in a Long Image Sequence of Scene with Multiple Independently Moving Objects," *IEEE Transactions on Pattern Analysis and Machine Intelligence*, Vol. 18, No. 1, pp. 58-64, January 1996.
4.   M. Irani, P. Anandan, "A Unified Approach to Moving Object Detection in 2D and 3D Scenes," *IEEE Transactions on Pattern Analysis and Machine Intelligence*, Vol. 20, No. 6, pp. 577-589, June 1998.
5.   M. Lew, "Next Generation Web Searches for Visual Content," *IEEE Computer*, November, pp. 46-53, 2000.
6.   H. S. Sawhney, S. Ayer, "Compact Representations of Videos Through Dominant and Multiple Motion Estimation," *IEEE Transactions on Pattern Analysis and Machine Intelligence*, Vol. 18, No. 8, pp. 814-830, August 1996.
7.   C. Stauffer, W. E. L. Grimson, "Adaptive background mixture models for real-time tracking," *Proc. IEEE Computer Vision and Pattern Recognition*, (Fort Collins, Colorado), June 23-25, 1999.
8.   M. Lew, T. Huang, K. Wong, "Learning and Feature Selection in Stereo Matching," *IEEE Transactions on Pattern Analysis and Machine Intelligence*, September, pp. 869-881, 1994.
9.   S. A. Niyogi, E. H. Adelson, "Analyzing and Recognizing Walking Figures in XYT," *Proc. IEEE Computer Vision and Pattern Recognition*, pp. 469-474, (Seattle, WA), June 21-23, 1994.
10. M. Lew, T. Huang, "Image Compression and Matching," *Proc. IEEE International Conference on Image Processing*, pp. 720-724, 1994.
11. M. Betke and N. Makris, "Fast Object Recognition in Noisy Images Using Simulated Annealing," *Proc. International Conference on Computer Vision*, pp. 523-530, June 1995.
12. M. Betke, E. Haritaoglu and L. S. Davis, "Real-Time Multiple Vehicle Detection and Tracking from a Moving Vehicle." *Machine Vision and Applications*, July 2000.
13. M. Lew, N. Sebe, "Visual Websearching Using Iconic Queries," *Proc. IEEE Computer Vision and Pattern Recognition*, pp 788-789, 2000.

14. Y. Ivanov and A. Bobick, "Recognition of Visual Activities and Interactions by Stochastic Parsing," *IEEE Transactions on Pattern Analysis and Machine Intelligence*, Vol. 22, No. 8, pp. 852-872, August 2000.
15. A. Pentland, B. Moghaddam and T. Starner, "View-Based and Modular Eigenspaces for Face Recognition," Proc. *IEEE Computer Vision and Pattern Recognition*, 1994.
16. C. R. Wren, A. Azarbayejani, T. Darrell, A. P. Pentland, "Pfinder: Real-Time Tracking of the Human Body," *IEEE Transactions on Pattern Analysis and Machine Intelligence*, Vol. 19, No. 7, pp. 780-785, July 1997.
17. S. M. Smith, J. M. Brady, "ASSET-2: Real-Time Motion Segmentation and Shape Tracking," *IEEE Transactions on Pattern Analysis and Machine Intelligence*, Vol. 17, No. 8, pp. 814-820, August 1995.
18. I. Haritaoglu, D. Harwood and L. Davis, "W4: Real-Time Surveillance of People and Their Activities," *IEEE Transactions on Pattern Analysis and Machine Intelligence*, Vol. 22, No. 8, pp. 809-830, August 2000.
19. R. Zabih, J. Miller, K. Mai, "Video Browsing Using Edges and Motion," *Proc. IEEE Computer Vision and Pattern Recognition*, pp. 439-446, (San Francisco, California), June 18-20, 1996.
20. R. J. Qian, M. I. Sezan, K. E. Matthews, "Face Tracking Using Robust Statistical Estimation," *Proc. IEEE International Conference on Image Processing*, , vol. 1, pp. 131-135, 1998.
21. M.-P. Dubuisson, A. K. Jain, "2D Matching of 3D Moving Objects in Color Outdoor Scenes," *Proc. IEEE Computer Vision and Pattern Recognition*, pp. 887-891, (Seattle, WA), June 21-23, 1994.
22. M. Lew, D. Denteneer, "Fisher Keys for Content Based Retrieval," *Journal of Image & Vision Computing*, vol. 19, pp. 561-566, 2001.

Appendix

Kalman Filtering (see [1])

The Kalman equations used for the tracking were:

Time update equations:

$$\hat{x}_{k+1}(-) = A_k \hat{x}_k(+) \tag{A.1}$$

$$P_{k+1}(-) = A_k P_k(+) A_k^T + Q_k \tag{A.2}$$

Measurement update equations:

$$P_k^{-1}(+) = P_k^{-1}(-) + H_k^T R_k^{-1} H_k \tag{A.3}$$

$$K_k = P_k(+) H_k^T R_k^{-1} \tag{A.4}$$

$$\hat{x}_k(+) = \hat{x}_k(-) + K_k\left(z_k - H_k \hat{x}_k(-)\right) \tag{A.5}$$

Semantic Video Retrieval Using Audio Analysis

Erwin M. Bakker and Michael S. Lew

LIACS Media Lab, Leiden University
Niels Bohrweg 1, 2300 CA Leiden, The Netherlands
{erwin,mlew}@liacs.nl

Abstract. Semantic understanding of video is an important frontier in content based retrieval. In the research literature, significant attention has been given to the visual aspect of video, however, relatively little work directly uses audio content for video retrieval. Our paper gives an overview of our current research directions in semantic video retrieval using audio content. We discuss the effectiveness of classifying audio into semantic categories by combining both global and local audio features based in the frequency spectrum. Furthermore, we introduce two novel features called Frequency Spectrum Differentials (FSD), and Differential Swap Rate (DSR), that both model the shape of the spectrum.

1   Introduction

In most video retrieval systems, the semantic aspect of the retrieval process is focused on the imagery within the video, largely ignoring the audio content. We investigate the possibility of automatic categorization of video into semantic categories using the audio content. Our intention is to be able to identify parts of the video which contain categories such as speech, music, automobiles, and explosions using the audio information. Specifically, if one poses a query such as "find video shots containing explosions," then our system should be able to locate the explosions using the audio content and return to the user the video shots corresponding to the explosions.

Another motivation for this research is that in many cases, the visual portion of video simply does not contain sufficient information for semantic categorization. An example of this would be finding video segments which contain human speech. In some video segments, we hear the speech as the camera is pointed at presentation slides. In this case, the visual portion of the video does not contain the information to accurately determine whether speech is present.

2   Background

This research topic is related to audio retrieval. There have been several important works in the past. An excellent overview is given by Foote [1]. The vast majority of research has focused on high level feature approaches toward content based audio recognition.

M. S. Lew, N. Sebe, and J. P. Eakins (Eds.): CIVR 2002, LNCS 2383, pp. 271–277, 2002.
© Springer-Verlag Berlin Heidelberg 2002

In the algorithm used by the MuscleFish System (Wold [8]), the authors attempt to use high level features such as loudness, pitch, brightness, bandwidth, and harmonicity in computing similarity between audio samples.

In the CueVideo system (Srinivasan [6]), the authors suggest the zero crossing rate (ZCR) as a feature in audio recognition. They found that the ZCR is moderately effective in discriminating between the categories of silence, speech, and music, however, ZCR related features appear to have a high percentage of false positives, except when the false negatives are very high.

Melih, et al. [4] introduce the possibility of combining audio compression with retrieval. They experimented with the MDCT and the Hartley transforms and chose the Hartley transform because it is purely real and it is possible to explicitly derive the phase information from it. The retrieval aspects were not directly benchmarked.

Clustering audio content is clearly necessary to semantic retrieval. One representative work is by Matichuk, et al. [3]. They perform clustering based on the frequency and harmonic characteristics. In addition they suggest some new metrics for comparing the similarity of sound segments.

Lu, et al. [2] examine the possibility of using KNN and LSP VQ based classification to categorize 1 second audio windows into the categories of speech, music, environmental sound, and silence. They used features based on the zero crossing rate, frequency spectrum, and LSP distance.

One work which used low level features is given by Subramanya, et al. [7]. They investigated the possibility of directly using features from the DCT, DFT, Haar, and Hadamard transforms for classifying audio into the categories of speech, music, and other simple sounds. The paper showed that the DCT coefficients could be effective for indexing, but typically only reached approximately 0.53 retrieval precision.

3   Feature Extraction and Algorithm

Our algorithm for automatic categorization of video into semantic categories using the audio content is based on the extraction of both global and localized audio features. For the extraction of global features, we implemented a well-known recursive band-pass filter $H_{a,b}$. It was used to gather the global frequency spectrum information. Consequently, the different spectrum segments were used in the Extended Feature Extraction Algorithm to obtain Frequency Spectrum Differentials (FSD). These formed the additional coefficients that described the important localized features of the audio signal.

The different spectrum segments were also used in the ZCR-Algorithm and the DSR-Algorithm to calculate the Zero Cross Rate (ZCR), and the Differential Swap Rate (DSR) for each segment, respectively. The algorithm described by Srinivasan [6] used only the original signal to compute the ZCR, and added band-passed energy features to obtain a richer feature set. Our ZCR-Algorithm uses all the spectrum segments, and hence is a new application of the ZCR-feature. The DSR-Algorithm is a novel algorithm that calculates the swap rate of the differentials in the band-passed segments.

The band-pass filter $H_{a,b}$ that we used is a recursive filter, i.e., the output signal y_i at a given time is calculated using both the current and the previous M input signals (x_i, and $x_{i-1,...}$) and the previous N output signals ($y_{i-1,...}$). In general recursive filters allow effective and computationally efficient implementations. For clarity reasons only, we will give a well-known derivation here.

For a sequence (x_i) of inputs and a sequence (y_i) of outputs our linear recursive band-pass filter $H_{a,b}$ can be described by the following formula:

$$y_n = \sum_{k=0}^{M} c_k x_{n-k} + \sum_{j=1}^{N} d_j y_{n-j} , \tag{1}$$

where c_k, and d_j are given constant coefficients.

We can now go to the frequency domain and formulate a corresponding frequency response function H for frequency f:

$$H(f) = \frac{\sum_{k=0}^{M} c_k e^{-2\pi i k(f\Delta)}}{1 - \sum_{j=1}^{N} d_j e^{-2\pi i j(f\Delta)}} , \tag{2}$$

where Δ is the sampling interval. Now we can re-parameterize the frequency f into a new variable ω as follows:

$$\omega = \tan(\pi(f\Delta)) = i\left[\frac{1 - e^{2\pi i(f\Delta)}}{1 + e^{2\pi i(f\Delta)}}\right] = i\left[\frac{1-z}{1+z}\right] \tag{3}$$

Taking the inverse of this equation gives:

$$z = e^{2\pi i(f\Delta)} = \frac{1 + i\omega}{1 - i\omega} \tag{4}$$

Now we can formulate the design of the band-pass filter $H_{a,b}$ that we used in our algorithm. The filter's lower cutoff frequency corresponds to a value $\omega = a$, and the filter's upper cutoff frequency corresponds to a value $\omega = b$. The corresponding frequency response function $H(f)$ is now given by the following rational function:

$$|H(f)|^2 = \left(\frac{\omega^2}{\omega^2 + a^2}\right)\left(\frac{b^2}{\omega^2 + b^2}\right) , \tag{5}$$

which can be rewritten as:

$$H(f) = \frac{-\dfrac{b}{(1+a)(1+b)} + \dfrac{b}{(1+a)(1+b)} z^{-2}}{1 - \dfrac{(1+a)(1-b) + (1-a)(1+b)}{(1+a)(1+b)} z^{-1} + \dfrac{(1-a)(1-b)}{(1+a)(1+b)} z^{-2}} \tag{6}$$

This defines the coefficients for our band-pass filter $y_i = H_{a,b}(x_i)$ ($i = 1,...n$, x_i the input signal, and y_i the output signal). The filter $y_i = H_{a,b}(x_i)$ can thus be described by the following recursive function:

Coefficients:

$$c_0 = -\frac{b}{(1+a)(1+b)}$$

$$c_1 = 0$$

$$c_2 = \frac{b}{(1+a)(1+b)}$$ (7)

$$d_1 = \frac{(1+a)(1-b)+(1-a)(1+b)}{(1+a)(1+b)}$$

$$d_2 = -\frac{(1-a)(1-b)}{(1+a)(1+b)}$$

Recursion:

$$y_0 = 0$$

$$y_1 = 0$$ (8)

$$y_i = (c_0 x_i + c_1 x_{i-1} + c_2 x_{i-2}) + (d_1 y_{i-1} + d_2 y_{i-2}), \text{ for } i = 2,...,n-1$$

We used the band-pass filter H to obtain the global energies of several frequency bands. The global energies of the frequency bands constituted the global features that were used to describe the audio samples in our audio database. In the following we describe the (Global-) Feature Extraction Algorithm for frequency bands of 1000Hz:

Feature Extraction Algorithm to Obtain the Global Features (GF-Algorithm):

1. get the data $X = (x_i)_{i=0,...,n-1}$ of the audio sample, (x_i are positive 8-bit values)
2. for $f = 1000$Hz, $S=$ sampling rate, and each k, $k=0,...,10$, determine the energy:

$$E_k = \sum_{i=0}^{n-1} H_{a_k,b_k}(x_i), where \; a_k = kf/S, and \; b_k = (k+1)f/S$$

3. determine the total energy of the sample $E_k = \sum_{i, (i=0,...,n-1)} x_I$

the sample X is now described by the normalized feature vector $F = (F_k) = (E_k/E)$, $k=0,...,10$.

Extended Feature Extraction Algorithm to Obtain the Global and Local Features (GLF-Algorithm):

In the GLF-Algorithm we extend the global features with additional features derived from Frequency Spectral Differentials (FSD). From the band-passed signal

$$Y^k = H_{a_k,b_k}(X),$$ (9)

the differential

$$diff(Y^k) =_{def} (Y_i^k - Y_{i+1}^k),$$ (10)

and the second order frequency spectral differential $\text{diff}^2(Y_k)$ is determined. For each band-passed signal the energies of the resulting 1st and 2nd order DFS signals are added as extra features to the global-feature set of the given audio signal.

The Zero Cross Rate Algorithm (ZCR-Algorithm):

In the ZCR-Algorithm we calculate the zero cross rate for each band-passed signal Y^k. That is, we calculate the number of times the signal swaps its sign and divide this by the number of samples in the signal.

The Differential Swap Rate (DSR-Algorithm):

In the DSR-Algorithm we calculate for each band-passed signal Y^k the number of times that the $diff(Y^k)$ signal swaps its sign, and divide this number by the number of samples in the signal.

For the classification of a given audio sample we used the method of vector quantization. Vector quantization is implemented with a given metric d. For our experiments we implemented different metrics d: L_2, L_1, and $L_{0.5}$.

Classification Algorithm Using Vector Quantization :

1. Calculate the feature vector F_X for the given audio sample X.
2. Determine the audio sample C that is an element of our audio database such that its feature vector F_C minimizes $d(F_X,F_C)$ over all the available audio samples in our audio database, where d is a given metric (i.e., either L_2, L_1, or $L_{0.5}$).

Finally, the audio sample X is classified with the class to which C belongs.

4 Experimental Results

The goal of our classification algorithm was to classify the sound samples into one of the classes Speech, Music, Piano, Organ, Guitar, Automobile, Explosion, or Silence. In our tests, we used a database composed of sound samples from the movies Star Trek: The Final Frontier, Tomb Raider, the TV show - Friends, home video, and some performances of classical-piano, -organ, and -guitar pieces.

The global features were chosen to be the normalized energies of the frequency spectrum quantized into 1000Hz segments. In our case this led to 12 features per sample. In Table 1, we see the misdetection rates for the global features across the different classifications, using the L_1 norm.

For the quantification of the localized features, we additionally computed the energy of the first and second order DFS's of the original audio signal, and the ZCR and DSR of the band-passed audio signal. The distances in the feature space were measured with different metrics: L_2, L_1, and $L_{0.5}$. The L_1 metric outperformed the L_2 metric in all classification experiments using our vector quantization algorithm. This in accordance to the findings of Sebe [5]. The misdetection rates using both global and localized features using the L_1 metric are also shown in Table 1. Furthermore, we found that the $L_{0.5}$ metric in almost all cases gave slightly better results than the L_1 metric. Our best classification result is obtained using the $L_{0.5}$ metric in a feature space of normalized DSR+FSD vectors.

Using global features and a quantization of the frequency spectrum into 100Hz segments showed better results for the automobile class only. This at the cost of a ten-fold increase in the number of features. When using both global and local features the quantization into smaller segments did not lead to significant improvements over the experiments with a quantization into 1000Hz segments. Therefore, only the results with quantization into 1000Hz segments are shown in Tables 1 and 2.

Table 1. Probability Of Misdetection (GF = Global Features, GLF = Global and Local Features, ZCR = Zero Cross Rate, DSR = Differential Swap Rate)

Class	GF L_1	GLF L_1	GLF $L_{0.5}$	ZCR $L_1 , L_{0.5}$	DSR $L_1, L_{0.5}$	DSR+GLF $L_{0.5}$
Speech	0.09	0.03	0.03	0.03	0.00	0.00
Music	0.03	0.03	0.07	0.14	0.07	0.03
Piano	0.22	0.11	0.00	0.00	0.00	0.00
Automobile	0.40	0.10	0.00	0.10	0.00	0.00
Explosion	0.00	0.00	0.00	0.00	0.00	0.00
Silence	0.00	0.00	0.00	0.00	0.00	0.00

Speech, music, and explosions were classified well, however, the system using purely global features had difficulties with automobile and piano sounds. Whereas the ZCR-Algorithm showed moderate problems with the classification of automobile and music. By adding local features, we were able to significantly improve the misdetection rates for the categories of speech and automobile. The DSR-Algorithm showed the relative best results. By using the $L_{0.5}$ metric and the DSR plus DFS (localized) features we were able to obtain our best results, but they were only slightly better than the DSR results.

Table 2. Probability Of Misdetection (GF = Global Features, GLF = Global and Local Features, ZCR = Zero Cross Rate, DSR = Diff Swap Rate)

Class	GF $L_{0.5}$	GLF $L_{0.5}$	ZCR $L_{0.5}$	DSR $L_{0.5}$	DSR+GLF $L_{0.5}$
Speech	0.06	0.03	0.09	0.00	0.00
Music	0.03	0.07	0.14	0.07	0.03
Piano	0.78	0.44	0.78	0.00	0.11
Organ	0.00	0.00	0.00	0.20	0.00
Guitar	0.50	0.30	0.50	0.30	0.20
Automobile	0.20	0.00	0.10	0.10	0.00
Explosion	0.00	0.00	0.00	0.00	0.00
Silence	0.00	0.00	0.00	0.00	0.00

After the initial experiments we added the two extra classes guitar and organ to our audio database. Especially the guitar class was expected to lead to classification difficulties as the classical guitar in some cases closely resembled the classical piano. In Table 2 the results of the different algorithms are shown. We only used the $L_{0.5}$ metric. The GF-, GLF-, and ZCR-Algorithms failed on correctly classifying the guitar and the piano. The DSR-Algorithm had some difficulties with the guitar and the organ, but outperformed the other algorithms in all cases but the organ. The best results were obtained by the combination of the DSR and GLF features.

5   Conclusions

Semantic video retrieval has the potential of bringing intuitive searching for video to the general public. Regarding the World Wide Web, we can bring video search to anyone with a Web browser, if these visual learning technologies mature. We are currently creating a large library of training databases and detectors. In this paper, we gave an overview of those video retrieval methods that are using audio content. We showed that the use of global and/or local features of the audio can be used in effective classification algorithms. Our preliminary results show that combining global and local features gives the best classification accuracy. Regarding future work, we will be investigating other perceptual audio features and multi modal based video retrieval.

Acknowledgments

The authors would like to thank the LIACS Media Lab for supporting this research.

References

1. Foote, J.: An Overview of Audio Information Retrieval. ACM Multimedia Systems Journal (1999)
2. Lu, L., Jiang, H., Zhang, H.: A Robust Audio Classification and Segmentation Method. Proceedings of the ninth ACM International Conf. on Multimedia (2001) 203-211
3. Matichuk, B., Zaiane, O.: Unsupervised Classification of Sound for Multimedia Indexing. MDM/ ACM SIGKDD'2000 (2000)
4. Melih, K., Gonzalez, R.: Structured Coding for Content Based Interactive Audio. Proc. IEEE International Conf. On Multimedia Computing and Systems '99 (1999)
5. Sebe, N.: Improving Visual Matching. Similarity Noise Distribution and Optimal Metrics. Ph.D. Thesis (2001)
6. Srinivasan, S., Petkovic, D., Ponceleon, D.: Towards Robust Features for Classifying Audio in the CueVideo System. Proceedings of the seventh ACM International Conf. on Multimedia (1999) 393-400
7. Subramanya, S. R., Youssef, A., Narahari, B., Simha, R.: Use of Transforms for Indexing in Audio Databases. ICIMA (1998)
8. Wold, E., Blum, T., Keislar, D., Wheaton, J.: Content-Based Classification, Search, and Retrieval of Audio. IEEE Multimedia (Fall) (1996) 27-36

Extracting Semantic Information
from Basketball Video Based on Audio-Visual Features

Kyungsu Kim[1], Junho Choi[3], Namjung Kim[2], and Pankoo Kim[3]

[1] Key Technology and Research Center
Agency for Defense Development
arieskim@empal.com
[2] Dept. of Computer Science
Chosun college of Science & Technology
njkim@mail.chosun-c.ac.kr
[3] Dept. of Computer Science and Engineering
Chosun University, Gwangju 501-759 Korea
{pkkim,spica}@mina.chosun.ac.kr

Abstract. In this paper, we propose a mechanism for extracting semantic information from basketball video sequence using audio and video features. After we divide the input video into shots by a simple cut detection algorithm using visual information, we analyze audio signal data to predict the location of an important event from which a cheering sound happens to start using the combination of MFCC features and the LPC entropy. Finally, we extract semantics about class of shot by computer vision techniques such as basketball tracking and related objects detection. Experimental results show that the proposed scheme can concretely extract semantics from basketball video data as compared to the existing methods.

1   Introduction

Recently, many researchers have tried to extract semantic information from video data automatically. Specially, the analysis of sports video with rich sound sources has been an interesting topic in video indexing and retrieval. Babaguchi, et al. proposed an event-based video indexing method where they use inter-modal collaboration to combine multi-modal information streams, which include visual, auditory and text [1]. Rui, et al. developed a system for extracting highlights from TV baseball programs [11]. They employed speech endpoint detection using phoneme-level features, and used reporter's exciting speech to infer important events. They employ a template matching method to detect specific events. Zhou, et al. examined basketball video structure and categorized accordingly, for example, left fast-break, right fast-break left dunk, right dunk, close-up shots and so on, classified by distinct visual and motional characteristics features by a rule based classifier [13]. Nepal, et al. proposed temporal models based on a number of automatic techniques for feature detection used in com-

M. S. Lew, N. Sebe, and J. P. Eakins (Eds.): CIVR 2002, LNCS 2383, pp. 278-288, 2002.
© Springer-Verlag Berlin Heidelberg 2002

bination with heuristic rules determined through manual observations of sports foot-age. In their work, they find 'Goal' segments in Basketball video automatically [9].

In this paper, we propose a mechanism to extract high-level semantics from video data using audio and visual feature. In general, we can use visual feature, such as color histogram and motion, or audio information, or both to extract semantic information from video data. We apply the proposed scheme to basketball games. Since the bas-ketball game has richer sound source than outdoor games, such as football or tennis, we firstly examine audio information to detect important events and then analyze visual information for further processing in our scheme. We have defined some se-mantics about scoring shots in our final step. Figure 1 shows the overall procedure for our scheme.

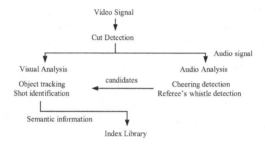

Fig. 1. Overall procedure for extracting semantic information

2   Primary Video Data Processing

We describe a simple cut detection algorithm using histogram difference between frames, since we don't require exact algorithm in our case. We extract DC image only for Y channel in the MPEG compressed domain and get the frame-to-frame histogram difference of Y-component. The used color histogram comparison metric is defined as follows.

$$D(H_i, H_j) = \frac{1}{256} \sum_{1}^{256} \left| H_i(k) - H_j(k) \right| \tag{1}$$

where H_i and H_j represent the histogram for frames i and j, respectively, and k is one of the 256 possible gray levels. To detect shot changes in the next step, we select thresholds adaptively to extract key frames from each segment according to the com-plexity of each flying window based on adaptive key frame extraction methods [14]. In experiments, the flying window containing 900 frames is used.

We can consider camera motion to help to extract more exact semantic information. For each incoming B or P frame, we extract motion vectors from them, then the cam-era movements are detected by the following two quantities, namely, the magnitude of the average motion vectors (denoted as Mag-avg) and the average magnitude of mo-tion vector (denoted as Avg-mag). Based on the observation of optical flow of differ-ent kinds of camera motions including camera panning, tilting, zoom-in and zoom-out, we can detect a global camera motion based on the combinations of the following two features. One feature is that during camera panning and tilting, most of motion vectors are parallel to the modal vector, thus two values of Mag-avg and Avg-mag should be of the same order. The other is that during camera zoom-in and zoom-out operation, the motion vectors are all pointing outwards or inwards, thus the quantity Mag-avg should be much smaller than Avg-mag.

3   Essential Sound Extraction from Audio Signal

After we segment a video signal into shots, we analyze the audio signal to detect cheering sound that give us important information about events. It is based on the following observations and proper domain knowledge of the subject.

Cheering is the important sound event in a basketball game analysis, because it occurs after or before an exciting event like a dunk shot. It has its several distinct characteristics. Firstly, cheering sound usually lasts quite a long time and makes the spectrum stationary in the time domain. Secondly, cheering sound is usually louder than other sounds, which means sound energy can be used to detect cheering. Lastly, there is no phoneme structure in the spectrum. Figure 2 represents the procedure for audio processing.

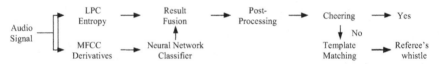

Fig. 2. Procedure for classifying the cheering sound

Based on the above observations, we employ MFCC (Mel-Frequency Cepstrum Coefficient) features to characterize spectral properties [10], and the entropy of LPC (Linear Prediction Coefficient) in a large time window for temporal characteristics. MFCC features have been successfully used in speech recognition. MFCC computes the cepstral on the warped mel-scale spectrum and is different from the traditional cepstral coefficients. Specially, it has finer resolution at the low frequency while coarser resolution at high frequency. The mel-scale spectrum zooms into the phoneme area of the spectrum of a segment of speech. Various features defined on the mel-scale spectrum have been developed for speech detection and recognition. Basically, MFCC is one of these features. We extract 12 MFCC features and only use 10 features of them at 50 frames per second. Five more MFCC derivatives are also extracted which captures the spectrum relationship between the adjacent frames.

Figure 3 shows original sound in the time domain, spectrogram and MFCC features for an audio clip with the cheering sound mostly. We can find high amplitude for quite a long time in the low frequency area of the spectrogram. Figure 4 shows original sound in the time domain, spectrogram and MFCC features for an audio clip with the speech sound mostly. We can also find a format representing a phoneme structure.

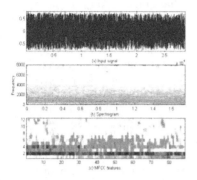
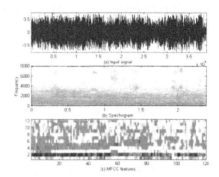

Fig. 3. Spectrogram and MFCC features of a cheering sound

Fig. 4. Spectrogram and MFCC features of a speech sound

We apply a neural network based classifier to classify MFCC windows initially. The neural network is a kind of non-linear classifier, which performs high in classification in high dimensional and nonlinear feature space. The node number of input, hidden and output layer is 15, 5, and 3, respectively. In our work, MFCC windows are classified into three classes of cheering sound, speech sound and other sound. Figure 5 shows the spectrogram and classification results of a scoring shot audio clip for each processing window. Most of the cheering sound is actually located on the left side in the figure. In the bottom graph of figure 5, the numbers 1, 2, and 3 represent cheering sound, speech, and other sound, respectively. Although speech is confused with other sounds in the classification results, we can see that cheering sound rarely confuses.

The MFCC features only take advantage of the spectrum feature in a small window. However, we also define the signal stationary measure in time domain to consider the large windows of sound track, since the spectrum of cheering sound is almost constant. The property does not factor speech sound and other sounds.

We use an entropy feature to measure the spectral variations in the time domain. Entropy can be used as a measure of stationary signal. LPC coefficients are a kind of polynomial approximation for the signal spectrum envelop. Therefore, it can effectively eliminate the effect of noise in the audio signal. We define the signal stationary measure in the time domain as the average entropy of each LPC component:

$$\text{Entropy} = -\frac{1}{D}\sum_{d=1}^{D}\sum_{n=1}^{w} P_{dn}\log P_{dn} \qquad (2)$$

where P_{dn} is $\dfrac{|a(n,d)|^2}{\sum_{n=1}^{w}|a(n,d)|^2}$, a is the LPC coefficients, w is the time window size.

We use LPC coefficient frames in a window, where each LPC coefficient is extracted in 512 audio samples. D is the LPC order. The equation obtains its maximum value when each LPC component is invariant on the large time window. We can see that the entropy measure of cheering is approximate to the maximum value. Speech has lower value because of the variation of LPC coefficients.

In the final step, we need to combine these two outputs, which is called information fusion. All signal are binarized into true or false signal, and these binary signals are combined. However, post-processing is also needed for this scheme because of the noise caused by neural network classification and thresholding. For our system, the fusion output is 1 if the sound window is cheering, otherwise 0. However, in order to eliminate the noise, we use hysteresis thresholding, which employs two thresholds to binarize the signal, the first threshold is used to get the initial '1' windows, and the second threshold is used to extend the '1' windows to their neighborhood area until the signal value falls below the second threshold. We use a morphological filter for post-processing to fill the gap by narrowing the '0' window. Figure 6 shows LPC entropy(first graph), final classification result(second graph), after result fusion and filtration of the results of audio samples.

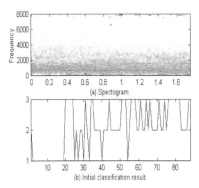

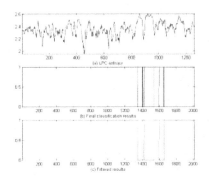

Fig. 5. Classification by a neural network classifier

Fig. 6. Classified result of audio clip(40 sec)

In addition to cheering sound, the referee's whistle is also very important in the basketball game. Therefore, we detected referee's whistle through template matching in final audio processing step. When a referee gives a whistle, we examined how the state of a game is changed. Firstly, if the whistle sound is detected in the rear of a processing shot, the next shot will be a close up shot and the state of the game will be stopped. So, both of the current processing shot and the next shot need not be processed, since they don't include any important events. However, if whistle sound is detected in the front or middle of the current processing shot and the shot is much longer than the average shot length, there will be a possibility that another referee's whistle is included and the attacking direction can be changed more than once. Using the observed domain knowledge, we apply template matching to the rest area except for the area including cheering sound. The results are listed in the Table 1. False detection ratio means the ratio that the system recognizes a non-cheering sound window as cheering and a whistle sound as one of other sounds. False rejection ratio is vice versa. Each ratio represents the ratio about the total processing window.

Table 1. Experimental results for sound detection

Metrics Sound	Total sound segments (#)	Total relevant sound (#)	System detected relevantly	System detected irrelevantly	False detection ratio	False rejection ratio
Cheering	11000	950	720	310	$230/950 \cong$ 24 %	$310/10050 \cong$ 3.1 %
Referee's whistle	1369	519	400	76	$119/519\cong$ 23 %	$76/850 \cong 9$ %

4   Recognizing the Object Movements by Visual Information

Our visual analysis is started from the candidates detected by the audio analysis. We employ visual analysis to candidate regions. In our work, the aim is to extract some semantic information. And also, we focus on tracking the basketball in this paper. Ball-tracking method is based on the head-tracking algorithm [4]. The original elliptical head-tracking algorithm is modified slightly to track basketball with a circular projected image in the two dimensional space. In the original algorithm, a human operator should initialize the first two frames by clicking, to the best of his ability, around basketballs in each frame. However, we locate the region of basketball by template matching automatically, therefore, it takes time to process the ball in the first frame of the candidate region.

The ball is modeled as circle with the following state.

$$\text{Ball's state or location: } S = (x, y, r) \qquad (3)$$

where (x, y) is the center of the ball, r is the length of the radius. Moreover, a Kalman filter is utilized to allow behavior of our implemented algorithm to be influenced by a dynamic model. The dynamic model simply assumes that the basketball moves in straight lines in the plane of the computer monitor, which shall be denoted as the x-y plane. The z coordinate is reflected in the state space of the Kalman filter indirectly through a parameter that describes the apparent radius of the basketball. Along the axis the model predicts the ball is stationary. In other words, the dynamics expect the radius of the ball to remain constant between frames. The prediction point (x^P, y^P) in frame t arises from the assumption of constant velocity, using the positions found in the previous two frames:

$$x_t^P = 2x_{t-1} - x_{t-2}, \quad y_t^P = 2y_{t-1} - y_{t-2}, \quad r_t^P = r_{t-1} \qquad (4)$$

In our experiment, we can know that the occlusion problem was not solved perfectly, despite the use of the Kalman filter, since basketballs simply do not move in straight lines. In special, when the ball passes behind a player, it's likely to move in an unpredictable fashion. However, varying the search window can alter the performance of the tracker to some extent. Varying the search window can enhance the performance of the tracker, no matter which algorithm for measurement is used for the matching. When the field is too narrow, the ball quickly migrates out of the search window. On the other hand, using a large search window is not an optimal solution because it greatly increases the computational time of the measurement.

The original algorithm performs well when tracking human faces. However, because a basketball moves much more rapidly than a head, we have a more difficult tracking problem in our work. Moreover, color segmentation algorithms can be often confused with the background clutter or other objects in the image. In particular, the color of the ball is sometimes similar to players skin colors or the uniform color in certain lighting conditions. This would cause the algorithm not to be able to locate the ball correctly.

Fig. 7. Frame sequences tracking a ball among 54 total frames

Figure 7 shows the result of basketball tracking in dunk shot sequence. In the first sequence with black background, the ball in all frames is correctly detected perfectly. However, in the second sequence, our ball-tracking algorithm starts to fail from the 20th frame. Our algorithm mistake any part of the player's exposed body for the basketball, since the color distribution of the player's leg is nearly the same as that of the ball in the searching area.

5   Semantic Information Extraction from Video Objects

5.1   Dominant Colors and Edge Extraction

We use color features and edge features to locate the backboard automatically. To do this, key frames extracted adaptively in the longest segment are used. Specially, we assume that the dominant colors in the long distance camera shot of basketball games consists of the court. Firstly, we used the median-cut algorithm to reduce the color map to about 256 colors, colors in the image were then mapped to their closest match in the new-color map. Thus the colors of the original image were clustered. Then we transform the color space from YCbCr to RGB to HSV. Finally, all pixels were mapped into homogeneous regions with one of the first 5 maximum dominant hues. Thus, we can regard the color of the court as the average of the dominant colors of each key frame.

We also used a gradient edge operator to detect edges that consist of the backboards and baskets. However, edge extraction process is time consuming. In our work, we apply edge detection algorithm to only the rest area excepting the dominant area. Figure 8 shows the results for dominant color extraction represented by hue. The hue value of the first dominant color is between 45 degree and 50 degree.

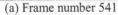

| (a) Frame number 541 | (b) Frame number 4328 |

Fig. 8. Example of the frames extracted by dominant color

5.2   Shot Identification

We extract some semantics from a basketball shot using the results of the previous processing. Firstly, we have to know the location of the backboard to get a relation between the basketball and the backboard. The processing to locate the backboard is actually processed with basketball tracking in parallel. However, when we locate the backboard, we just apply from frames near the cheering sound. We know that backboard mainly consists of vertical and horizontal lines and the baskets have the shape of an ellipse from our observations about basketball. Therefore, we used the Hough transform to detect lines and ellipse from images in candidate region.

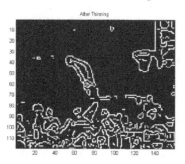 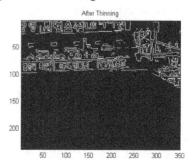

Fig. 9. Edge extraction for identification of backboard and basket

Figure 9 shows the results of edge extraction applied to the rest area except for the court area representing the dominant color. Edge extraction process is applied to an entire image, since it does not include dominant colors.

Finally, we consider the relationship between the ball and the location of the backboard and the basket. In order to decide whether a shot is a dunk shot or a long distance shot, when the ball is nearing the backboard, we also consider whether a player is holding the ball or not using the edge and color information of neighboring pixels of the ball. In the work, a bounding box of the ball is used.

6   Conclusions

We have proposed a scheme for semantic indexing for video data using audio and visual information. We select TV basketball programs as input video. In this paper, we processed audio signals to detect important events in sports video rather than visual analysis. We assume that an important event means a scoring shot or a dunk shot. After we segment the input video into shots, we detect the starting point of cheering sound by analyzing the audio signal by combining MFCC features and LPC entropy. We also processed noise elimination by a post-processing. The experimental results show that the classifier is very effective for identifying cheering sound from other sounds such as speech or basketball specific sound.

After detecting cheering sound, we analyze visual information to extract semantic information. In this paper, we originally hoped that visual analysis would perfectly help to enhance the performance of the event detection by audio signal analysis. However, we introduce only a basketball tracking method in this paper. We modified the original algorithm for face tracking to apply to the basketball tracking. Though it did not perfectly work in a dynamic environment such as a basketball game, the experimental results show that it can use the method with consideration about lighting condition is given. In our final step, we used a dominant color detection, edge detection and Hough transform to extract semantic information, that is to decide whether a shot is a dunk shot or a long distance shoot. Our algorithm comparatively works well. However, some other techniques need to be added to extract more accurately.

Acknowledgements

This work supported by grant no. R05-2001-000-00979-0 from the Basic Research Program of the KOSEF. And Dr. Pankoo Kim was a corresponding author.

References

1. Babaguchi, N., Kawai, Y., Kitahashi, T.: Event Based Video Indexing by Intermodal Collaboration, Proc. of the First International Workshop on Multimedia Intelligent Storage and Retrieval Management (MISRM'99) in conjunction with ACM Multimedia Conference, (1999) 1-9
2. Benitez, A. B. and Smith, J. R.: New Frontiers for intelligent Content-Based Retrieval, Proc. of the SPIE 2001 Conference on Storage and Retrieval for Media Databases (IS&T/SPIE-2001), Vol. 4315, Jan. (2001)
3. Benitez, A. B., Smith, J.R., Chang, S. F.: MediaNet: A Multimedia Information Network for Knowledge Representation, Proc. of the SPIE 2000 Conference on Internet Multimedia Management Systems (IS&T/SPIE-2000), Vol. 4210, Nov. (2000)
4. Birchfield, S.: Elliptical Head Tracking Using Intensity Gradients and Color Histogram, IEEE Conf. On Computer Vision and Pattern Recognition, Jun. (1998)
5. Chang, S.F. and Messerschmitt, D.G.: Manipulation and compositing of mc-dct compressed video, IEEE Journal on Selected Areas in Communications, vol.13, (1995) 1:1-11
6. Chang, Y., Zeng, W., Kamel, I., Aonso, R.: Integrated Image and Speech Analysis for Content-Based Video Indexing, Proc. of the Third IEEE International Conference on Multimedia Computing and Systems, (1996) 306-313
7. Kittler, J, Messer, K., Christmas, W. J., Obadia, B.L., Koubarroulis, D.: Generation of Semantic Cues for Sports Video Annotation, Proc. of the ICIP2001, Oct. (2001)
8. Naphade, M. R. and Huang, T. S.: Semantic Filtering of Video Content, Proc. of the SPIE Conference on Storage and Retrieval for Media Databases, Jan. (2001)
9. Nepal, S., Srinivasan, U., Reynolds, G.: Automatic Detection of 'Goal' Segments in Basketball Videos, Acm Multimedia Sept. (2001)
10. Rabiner, L. and Schafer, R.: Digital processing of speech signals, Prentice Hall, 1973.
11. Rui, Y. and Gupta, A., Acero, A.: Automatically Extracting Highlights for TV Baseball Programs, Proc. of the ACM Multimedia, Oct. (2000) 105-115

12. Zhong, D. and Chang, S.F.: Structure Analysis of Sports video Using Domain Models, IEEE Conference on Multimedia and Exhibition, Aug. (2001)
13. Zhou, W, Vellaikal, A., Kuo, C.-C. J.: Rule-based Video Classification System for Basketball Video Indexing, ACM Multimedia Oct. (2000)
14. Zhuang, Y., Rui, Y., Huang, T.S., Mehrotra, S.:Adaptive Key Frame Extraction Using Unsupervised Clustering, Proc. of IEEE International Conference on Image Processing, Oct. (1998) 866-870
15. Zhuang, Y., Rui, Y., Huang, T.S., Mehrotra, S.: Applying Semantic Association to Support Content-Based Video Retrieval, Proc. of IEEE VLBV98 Workshop, (1998) 45-48
16. Xu, P., Xie, L., Chang, S.F.: Algorithms and Systems for Segmentation and Structure Analysis in Soccer Video, IEEE International Conference on Multimedia and Expo, Aug. (2001)

Video Indexing and Retrieval
for Archeological Digital Library, CLIOH

Jeffrey Huang, Deepa Umamaheswaran, and Mathew Palakal

Indiana University Purdue University at Indianapolis (IUPUI)
Department of Computer Science, Indianapolis, IN 46202
{huang,dumamahe,mpalakal}@cs.iupui.edu

Abstract: With the ever-increasing amount of digitally archived libraries that are being collected, new techniques are needed to organize and search these collections, retrieve the most relevant selections, and effectively reuse them. This helps a user find contents of interest in faster and more precise fashion than searching a single track. This paper introduced a video indexing and retrieval system for an archeological database, CLIOH (Cultural Digital Library Indexing our Heritage), using wavelet best basis and self-organizing neural networks. Texture similarity matching provides the functionality of video retrieval by comparing the Euclidean distance of encoded wavelet quadrature tree structures generated from probe texture icon and gallery texture icons. Experimental result using video sequences drawn from the CLIOH database proves the feasibility of our approach.

1 Introduction

A large number of historical sites that encase our heritage are in danger of being lost of destroyed. Steady deterioration, largely due to the pollution, war and natural disasters, is wiping away the history and heritage that has stood for so long. There has been a constant effort to collect as much information (in the form of video) as possible from many of these sites and to save them for the future generations. Now the most impending question is how to manage the huge amounts of video libraries and make the best use of them. Through digital video library systems, video data are now free to wander the same path as any other digital data. Advances in data compression, storage, and telecommunications have enabled the rapid technological growth of these systems. In the scenario of these kinds of libraries, video might have been captured at different sites each from different angles and video is then transmitted and archived in a centralized storage device, such as video server. Therefore, data might be overwhelming at most times. New techniques are needed to organize and search these vast data collection, retrieve the most relevant selections, and effectively reuse them.

M. S. Lew, N. Sebe, and J. P. Eakins (Eds.): CIVR 2002, LNCS 2383, pp. 289-298, 2002.
© Springer-Verlag Berlin Heidelberg 2002

The CLIOH[1] (Cultural Digital Library Indexing our Heritage) project at Indiana University aims at creating a digital archive of ancient treasures. The present video sequences in the database are mainly shot in Chichen Itza and Uxmal in Mexico, which documented Mayan ruins and ancient structures. Like most other documentary videos these sequences contain video with slow pan, zoom and tilt motions. The common approaches for video indexing concentrate on extracting the global spatial features such as color histograms, texture, shape, etc. However, using spatial information for feature analysis is not sufficient enough to describe local contents of image and the performance is easily degraded under variety of image resolution. Although this might work on regular movie sequences [1], it fails to work with CLIOH like video sequences.

The wavelet basis functions [2] provide one possible tessellation of the conjoint spatial/spectral signal domain. Its representation also provides for multi-resolution analysis (MRA) through the orthogonal decomposition of a function along basis functions. Wavelet networks ('wavenets') using stochastic gradient descent akin to back propagation (BP) [3] are an example using wavelet decomposition coupling with feed forward neural networks for classification purpose. Huang and Wechsler [4] introduced an approach for the eye detection task using optimal wavelet packets for eye representation and radial basis functions (RBFs) for classification ('labeling') of facial areas as eye vs. non-eye regions. Our approach uses the features generated by wavelet best basis decomposition [5] for keyframe clustering and texture classification. The video indexing and retrieval architecture for the CLIOH database consists of two major components - (i) video indexing module with wavelet best basis and Self-Organize Map (SOM) neural network [6] (ii) video retrieval from user inscribed texture icons using wavelet quadrature tree structure.

2   System Architecture

The video indexing and retrieval architecture for CLIOH database consists of two major components, video indexing module and video retrieval module using either key frames similarity match or user subscribed texture icons. In this paper, we only focus on texture searching. The Video index module includes the process of: (i) video break detection and extraction of key frames using conventional features such as histogram, edge, texture, and motion, (ii) wavelet best basis decomposition and extraction of features (coefficient matrix) from these key frames, and (iii) clustering of key frames using self-organizing [6] neural networks with the features generated by wavelets. Fig.1 shows the overall process followed by our system, while the sect. 4 and 5 describe the indexing module in more details.

Video retrieval module based on user-inscribed texture includes the process of (i) selection of unique textures to be a part of the gallery (ii) wavelet best basis decomposition and encoding for representing as a quadrature tree structure [5] (iii) user enquiry for a texture followed by a similarity match and retrieval of key frame and corresponding video shot. Fig 2. explains the architecture of the video retrieval module.

---

[1] CLIOH – A digital Cultural Library Indexing Our Heritage project is supported by the Institute of Museum and Library Services (IMLS) and funded by National Leadership Grant

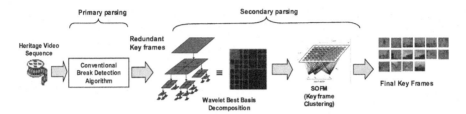

Fig. 1. Video indexing module consisting of the wavelet best basis decomposition followed by the SOM clustering for key frame extraction

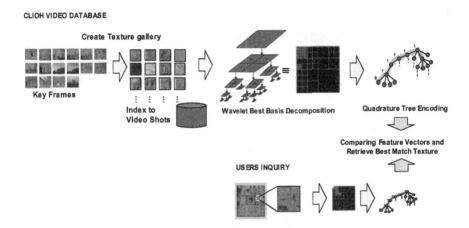

Fig. 2. Video texture retrieval module involving the texture comparison between the users inquiry and the texture gallery present in the CLIOH database

3   Finding Video Breaks and Key Frame Extraction

Cuts or camera/video breaks are perceived as instantaneous changes from one shot to another [7]. So, in other words, a break is declared when the video sequence has a change of scene. The global spatial features such as histogram, texture and the edge detection from each video frame are first extracted as feature vector. A comparison between feature vectors gives an indication whether difference between two adjacent images is significant enough to find a break and declare a video shot. The graphs in Fig. 3 show the fluctuation in intensity values for gray scale images, hue, saturation values and the histogram differences of them in a sample video sequence. The peaks show evidently that there have been video breaks.

Once the shot has been located we use the criteria that the frame in a shot displaying the least motion $M(t)$ is found to be sufficiently good enough to be declared the key frame, where $M(t)=\Sigma_i\Sigma_j(|O_i(i,j,t)|+|O_j(i,j,t)|)$ and $O\ (i,j,t)$ is the optical flow [8] of pixel i, j in frame t. Optical flow is the distribution of apparent velocities of movement of brightness patterns in an image and is able to easily detect the change in motion in two successive frames in a video sequence.

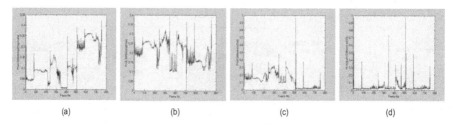

(a)   (b)   (c)   (d)

Fig. 3. Plots of intensity variation of consecutive video frames using (a) grayscale (b) hue (c) saturation information, and (d) gray-scale histogram to identify video breaks

I=1:F=81   I=82: F=245   I=246:F=345   I=346:F=415

I=416: F=468   I=469:F=512   I=513: F=588   I=589:F=752   Frame # 626   Frame # 678

4(a)   4(b)

Fig. 4. Key frames extracted using global features showing the (a) Primary key frames extracted (b) Missing frames

The CLIOH database contains documentary/archeological video sequences that have characteristics such as slow camera panning/ zooming motion and long continuous shot. The difference between two successive frames t and $t+1$ might not be significant, while after n frame duration the difference between frame t and $t+n$ could be quite large. The conventional key frame finder by comparing only the consecutive video frames using window sliding method easily fails to detect some important frames within a video shot. Fig 4(a) shows the found primary key frames, while two key frames shown in Fig 4(b) are missing. To avoid this mistake, we re-compare all frames against the first frame in a video shot detected by previous process. When the difference is larger than a threshold, we launch a new shot and mark the current frame as a beginning frame to continue comparison.

4   Wavelet Best Basis and SOM for Keyframe Extraction

To remove redundant key frames, a clustering process is needed for parsing key frame. In general the spatial features for clustering is not sufficient enough to describe local contents of image. The wavelet basis functions [2], a self-similar and spatially localized code, are spatial frequency/orientation tuned kernels, whose representation provides for multi-resolution analysis (MRA) with the characteristic of retaining scale invariance. Key frame clustering uses the features generated by wavelet best basis decomposition.

4.1   Wavelet Best Basis Representation

The wavelet hierarchical (pyramid) is obtained as the result of orientation-tuned decompositions at a dyadic (powers of two) sequence of scales. The discrete wavelet transform (DWT) is:

$$\phi_{m,n}(x) = 2^{-m/2}\phi(2^{-m}x - n) \tag{1}$$

$$\psi_{m,n}(x) = 2^{-m/2}\psi(2^{-m}x - n) \tag{2}$$

where m, n are integer numbers, and $\phi_{a,b}$, and $\Psi_{a,b}$ correspond to the scaling and mother wavelet functions dilated by 2. The dilation equation, relating the mother wavelet to the scaling function is:

$$\psi(x) = \sqrt{2}\sum_{k} h_1(k)\phi(2x - k), \quad \text{where } h_1(k) = (-1)^k h_0(1-k) \tag{3}$$

Daubechies [9] has shown how one can derive the corresponding low (h_0) and high (h_1) pass filters for designing appropriate families of scaling and mother wavelet functions. Using the sequences h_0 and h_1, one computes the Discrete Wavelet Transform (DWT) and can decompose a signal into downsampled lowpass and highpass components. Fig.5 shows two level decomposition of a signal and its subsequent reconstruction using upsampling and interpolation. Mallat [2] has showed that for any orthonormal wavelet basis, the sequences of two-channel half-band filters, h and g, can be utilized to compute the DWT with perfect reconstruction. The design of Quadrature Mirror Filters (QMF) is a useful way to implement PR filter banks for multirate signal analysis and directly links to multi-resolution analysis (MRA) supported by wavelet theory.

The concept of optimal sampling can be expanded using different fitness criteria. As an example, Wilson [10] mentions the requirement for more complex approaches to signal representation, whose common feature is *adaptivity*. It should be possible to automatically adjust the resolution of the representation, i.e. its reconstruction ability, in order to provide the best 'fit' to a given data set, rather than using a fixed representation, whose resolution is only bound to be a compromise between space / time and frequency. Examples of such an approach include wavelet packets [11], where the wavelet dictionary is drawn using maximal energy concentration and/or least Shannon entropy (μ) which is defined as

$$\mu(v) = -\Sigma \|v_i\|^2 \ln\|v_i\|^2 \tag{4}$$

where $v = \{v_i\}$ is the corresponding set of wavelet coefficients. The Shannon entropy measure is then used as a cost function for finding the best subset of wavelet coefficients.   Note that minimum entropy corresponds to less randomness ('dispersion') and it thus leads to clustering. If one generates the complete wavelet representations (called wavelet packets) as a quadature tree, the selection of the best coefficients is done by comparing the entropy of wavelet packets corresponding to successive tree levels. One compares the entropy of each adjacent pair of nodes to the entropy of their union and the sub tree is expanded further only if it results in lesser entropy. The difference of tree structure can be observed, encoded, and used for discriminating image contents [5].

Fig. 5. Two channel DWT (two-level) decomposition and reconstruction of a signal using ideal filter bank

The process begins with the video key frames being fed into the best basis wavelet decomposition module and produces a quadrature tree structure for each frame. We limit the decomposition to take place down to six levels. Fig. 6 shows the quadrature tree structure derived by wavelet best basis decomposition from key frames found in different video shots. One can see that key frame A, B, and C contain different wavelet basis and hence their quadrature tree structures are evidently different.

(A)   (B)   (C)

Fig. 6. Results of Wavelet Best Basis Decomposition on three different images from the CLIOH database

4.2   Self-Organizing Map Clustering for Redundant Frames Removal

Clustering process was then carried out using Kohonen self-organizing map (SOM) network, which is an unsupervised neural network that has the abilities of self-organization. Through unsupervised learning (competitive learning) process, the neural network nodes become self-organized and specifically tuned to various clusters based on input features, which finally create topological mappings between the input data and sheet-like map units. This topology map is called the self-organizing map. The locations of the responses tend to become ordered as if some meaningful coordinate system for different input features were being created over the network. The spatial location or coordinates of a cell in the network match up to a particular domain of input signal patterns. It is this feature that is of particular interest since we need to figure out some pattern with which video shots occur and which cluster they may belong to. The wavelet coefficient matrix obtained after wavelet best basis decomposition is used as the input features for training the SOM. So, when the features of the image frames are passed in as the input they are rearranged on the network map and all the frames belonging to one cell on the network are inferred to belong to the same cluster. Hence we can deduce which frames are redundant in the set and only show one frame from every cell as the key frame for the video sequence. The frame with least quantization error was chosen as the best representation of a unique key frame within each cluster after network training.

5   Retrieval

Detected key frames were used for database video shot indexing available for users'
access. Video retrieval process consists of two steps: i) selection of textures from the
already acquired key frames, ii) clustering of textures to derive unique textures, and
iii) texture classification (similarity matching). Our system allows the user to submit
an image and select a texture in an image frame by cropping it and querying the
database for a clip containing similar texture. Index icons existing in the database are
sampled from key frame images internally linked to the corresponding shots by
default. Fig. 7 illustrates some sample textures in the gallery. As an example, user
can query the texture such as weaving brick contained in the image of the Governor's
palace in Uxmal. With similarity matching to texture 001 in gallery, the
corresponding video clip/shot is retrieved.

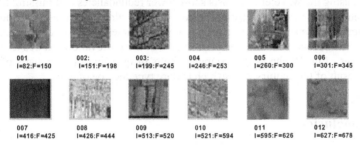

Fig. 7. Sample Texture Icons as stored in the texture gallery used to compare against in the
retrieval module

The system uses wavelet best basis decomposition and quadrature tree structure
representation [5] instead of the wavelet coefficients for texture similarity matching.
The quadrature trees were encoded using the binary system where a 1 represents a
node that is further decomposed and 0 indicates a leaf node. The tree was then
summed along each of its paths going to the deepest level (level 6). Fig 8. explains
the quadrature tree encoding scheme for a sample image. Since we only decompose
down to the sixth level and each edge along this path can be only either a 0 or a 1, we
have 1024 of such paths in the tree whose sum is bounded between 0 and 6.
Similarity matching measured the Euclidean distances by comparing pairwise feature
vectors of the user's selected texture against gallery textures. The shortest distance
after ranking indicated the closest match.

6   Experimental Results

The CLIOH project aims at establishing a new medium web based digital library of
the world's cultural treasures. The database contains video sequences shot in various
historical Mayan sites located at Uxmal, Mexico. Users of this system are mostly
intended to be professionals, academicians, and students. The system was tested to be
used to retrieve data that are of particular interest to the user including make search on

a particular texture that might be of interest to an archeologist, for example. The test-bed consisted of five video sequences drawn from the CLIOH database. Each video sequence has the length about four to five minutes. Video breaks and key frames were identified by the methods described earlier in Sec. 3. For redundant key frame removal, key frame images were then normalized to the size of 256x256 before performing wavelet best basis for feature extraction. After some trials we found that, for image size of 256x256, six-level wavelet decomposition was sufficient enough to generate distinguishable coefficients and train the self-organized neural networks. Fig. 9 shows the key frames extracted from two sample video sequences using (a) the wavelet best basis coefficients, versus (b) the use of global features only. Table 1 shows a comparison of the key frames extracted out of the sequences using the above two techniques. It also allows us to see a comparison between how efficiently the key frames are extracted in terms of redundant frames and missing frames. The results obtained using the wavelet best basis show fewer redundancies and missing frames as compared to those extracted using the conventional global features. It demonstrates our system performs robustly to reduce redundant frames and can detect essential key frames. The second part of the experiments is in the retrieval of textures as enquired by the user. Fig. 10 shows the example results of querying random parts of images and the result of the enquiry as to which texture type they belong to.

7   Conclusion

This paper has introduced an approach for the problem of video indexing and retrieval in general utilizing wavelet best basis tree structure coupling with Kohonen SOM neural network for archeological/ documentary type of videos in particular. We have described a framework using the optimal features generated through wavelet best basis decomposition for self-organizing neural networks to classify video sequences into shots and thereafter represent it with a key frame with minimal redundancy. Texture similarity matching provides the functionality of video retrieval by comparing the Euclidean distance of encoded wavelet quadrature tree structures generated from probe texture icon and gallery texture icons. Experimental result using video sequences drawn from the CLIOH database proves the feasibility of our approach. Wavelet best basis feature representation appears to provide robust image clustering and texture-matching schemes one should consider their use for the integration of multi-media library.

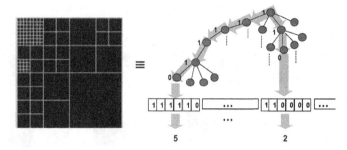

Fig. 8. Quadrature Tree Structure encoding method carried out after the wavelet best basis decomposition

Table 1. A comparison of the key frames extracted from five sample video sequences using (a) Wavelets & SOM vs. (b) conventional global features in terms of the number of key frames in total (T), the number of redundant frames (R), and the number of missing frames (M), while the last column (K) represents the number of video breaks which is the ideal number of key frames

Video seq. no.	(a)			(b)			K
	T	R	M	T	R	M	
Seq. 1	16	5	1	33	23	2	12
Seq. 2	15	4	0	9	6	8	11
Seq. 3	19	2	0	40	25	2	17
Seq. 4	18	2	0	32	20	4	16
Seq. 5	20	5	0	12	5	8	15

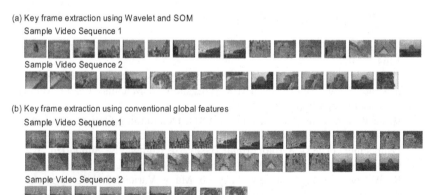

Fig. 9. Key frames extracted from two Sample video sequences using the method of (a) wavelet best bases coupling with SOM, versus (b) conventional key frame detection based on conventional global features

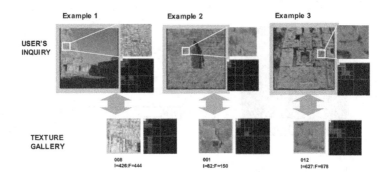

Fig. 10. Samples of the user inquiry and the resultant best texture match found from the gallery

References

1. W. Xiong, J. C. M. Lee, and R.H. Ma (1996), Automatic Video Data Structuring through Shot Partitioning and Key Frame Selection, Technical Report HKUST-CS96-13
2. S. Mallat (1989), A Theory for Multiresolution Signal Decomposition: the Wavelet Representation, IEEE Trans. on Pattern Analysis and Machine Intelligence, 11 (7), pp. 674-693
3. Q. Zhang and A. Benveniste (1992), Wavelet networks, IEEE Trans. on Neural Networks, 3(6), pp.889-89
4. J. Huang and H. Wechsler (1999), Eye Detection Using Optimal Wavelet Packets and RBFs, Int. Journal of Pattern Recognition and Artificial Intelligence, 13(6), pp.1009-1026
5. T. Chang and C. J. Kuo (1993), Texture Analysis and Classification with Tree-Structured Wavelet Transform, IEEE Trans. Image Proc., Vol. 2, No. 4, pp.429—441
6. T. Kohonen (1990), The Self-Organizing Maps, Proceedings of the IEEE, 78, 1464—1480
7. C. M. Lee and M. C. Ip (1994), A Robust Approach for Camera Break Detection in Color Video Sequences, IAPR Workshop on Machine Vision Application, (Kawasaki, Japan), pp.502—505.
8. B. K. P Horn and B. G. Schunck (1981), Determining Optical Flow, Artificial Intelligence, 17, pp.185-204
9. Daubechies (1988), Orthonormal Bases of Compactly Supported Wavelets, Comun. on Pure and Appl. Math., 41, pp. 909-996
10. R. Wilson (1995), Wavelets: Why so Many Varieties?, UK Symposium on Applications of Time-Frequency and Time-Scale Methods, University of Warwick, Coventry, UK
11. R. R. Coifman, & M. V. Wickerhauser, (1992), Entropy-based algorithms for best-basis selection, IEEE Transactions on Information Theory 38, pp.713-718

Fast *k*-NN Image Search with Self-Organizing Maps

Kun Seok Oh[1], Aghbari Zaher[2], and Pan Koo Kim[3]

[1] Department of Digital Media Division of Information Technology
Kwangju Health College
683-3 Shinchang-dong Kwangsan-ku Gwangju 506-701, Korea
okseok@www.kjhc.ac.kr
[2] Department of Intelligent Systems
Graduate School of Information Science and Electrical Engineering
Kyushu University
6-10-1 Hakozaki, Higashi-Ku Fukuoka, 812-8581, Japan
zaher@is.kyushu-u.ac.jp
[3] Department of Computer Engineering
Chosun University
375 Seosuk-dong Dong-ku Gwangju , 501-759, Korea
pkkim@mina.chosun.ac.kr

Abstract. Feature-based similarity retrieval became an important research issue in image database systems. The features of image data are
useful in image discrimination. In this paper, we propose a fast
k-Nearest Neighbor (*k*-NN) search algorithm for images clustered by the
Self-Organizing Maps algorithm. Self-Organizing Maps (SOM) algorithm maps feature vectors from high dimensional feature space onto a
two-dimensional space. The mapping preserves the topology (similarity)
of the feature vectors by clustering mutually similar feature vectors in
neighboring nodes (clusters). Our *k*-NN search algorithm utilizes the
characteristics of these clusters to reduce the search space and thus
speed up the search for exact *k*-NN answer images to a given query image. We conducted several experiments to evaluate the performance of
the proposed algorithm using color feature vectors and obtained promising results.

1 Introduction

Recently the accumulation of tremendous image data has been used in many fields
with a rapid development of multimedia technology, which has activated researches
on image databases. To make use of images stored in an image database, an image
search engine equipped with an efficient retrieval function is required. Image retrieval
methods are largely classified by the image attributes on which retrieval is based:
retrieval by text annotations, retrieval by visual features (color, texture, shape, etc.),
and/or retrieval by a combination of text annotations and visual features.

M. S. Lew, N. Sebe, and J. P. Eakins (Eds.): CIVR 2002, LNCS 2383, pp. 299-308, 2002.

Content-based image retrieval (CBIR) methods, which were discussed in many works [5,6,8], are based on the visual features of images. In image databases, an essential process to similarity retrieval is to search for k-nearest neighbor images to a given query image based on a distance measure provided by a system [10], where k equals one or more images. Therefore, similarity retrieval of images in such large-scale image databases should be finished in a practical time. The CBIR approach using the above technique has been proposed in many systems [6,14,15].

One way to improve performance of a search is to first cluster the data and then only the similar clusters to the given query are searched for k-NN answer images. In this paper, Kohonen Self-Organizing Map (SOM) algorithm [2] is used for the realization of automatic clustering of images based on their feature similarity. SOM generates a two-dimensional similarity-preserving topological map of nodes, where each node (cluster) represents a subset of images in the database. Each node N_i is represented by a weight vector that is also representative of the subset of images mapped into N_i. The weight vectors have the same dimensionality as that of the feature vectors of images. The weight vector is generally called a *codebook vector*. Neighboring nodes on a topological map usually have similar codebook vectors. With SOM clustering, two or more similar images are mapped into one node.

In this paper, we propose a k-NN search algorithm to efficiently search for the k-NN answer images to a given query image in the topological map generated by SOM. The k-NN search algorithm starts by finding a node whose codebook vector is the most similar to the query image. Then, the k-NN search algorithm, progressively, searches the nodes around the winner node. Based on some pruning strategy (discussed in Section 4.3), the unqualifying nodes are pruned; thus reducing the total number of visited nodes required to find the k-NN answer images. The k-NN search algorithm guarantees exact k-NN answer images in a practical time. Our experiments verify the efficiency of k-NN search method proposed in this paper.

The rest of this paper is organized as follows: Section 2 surveys some of the related works. Section 3 discusses the SOM clustering of images. The k-NN search algorithm is explained in Section 4. In Section 5, we discuss the conducted experiments. Finally, we conclude the paper in Section 6.

2   Related Work

SOM has been applied in visual information retrieval [1], especially in the application of similar image retrieval. There are two types of similar image retrieval methods. One type is the *retrieval-by-visualization* of the topological map [7], which is generated by SOM. In this method, information (i.e. image) retrieval is performed by navigating the topological map. During the retrieval process, relevance feedback[4] is suitable to perform interactive information retrieval. However, the main disadvantage of retrieval-by-visualization is that a user cannot retrieve similar images by inputting an arbitrary image as a query image.

The other type is a *query-by-example* (QBE) [17,18]. Netra [17] expresses the similar visual patterns by dividing the feature space into different parts produced by combining unsupervised learning and supervised learning of a neural network. In

[18], we utilized the wavelet transformation and SOM to extract image features and cluster the images based on the extracted features, respectively. The retrieval technique in [18] supports both query by image example and query by partial image example. The retrieval starts by finding the winner node (a node whose codebook vector is the most similar to the query), then the images associated with the winner node are returned to the user as the result of the query. The problem with this technique is that not all similar images (exact k-NN answer images) can be found in one node due to the limitations of the SOM as imposed by its input parameters. A naive approach to solve the above problem is to search all the nodes of map. This approach, however, is not practical because it takes so much time to retrieve all the similar images.

In this paper, we propose a k-NN search algorithm with some pruning strategy to perform efficient image retrieval based on SOM clustering of images in order to solve the above problem.

3   Clustering of Similar Images by SOM

We extracted color features of each image in the database using Haar wavelet transformation[12]. These color vectors represent the color contents of images and thus used for similarity retrieval of images [18]. The color elements of a vector are the computed wavelet coefficients, which are obtained by performing a 5-leveled Haar wavelet transformation for each of the three channels of the YIQ color space [16]. Consequently, we extracted the 4×4 coefficients of the lowest frequency sub-band as the color elements representing average color of the whole image. Thus, by combining the elements of the three-color channels, the color vector becomes a 48-dimensional color vector.

To cluster the computed color vectors, which represent the images in the database, we use a SOM algorithm. The nodes in the 2-dimensional array are initialized by a random weight value before the learning phase is performed. Thus, each node has an initial weight vector. The learning phase of the SOM is performed by inputting iteratively each color feature vector F_{col} obtained from wavelet transformation into the SOM. The process of clustering similar images consists of two steps: generation of a topological feature map and generation of best matching image lists.

3.1   Generation of a Topological Feature Map

To generate a topological feature map, some parameters, such as learning rate, neighborhood radius, size of map layer, neighborhood function, learning iteration, etc., are given to the SOM algorithm by the user before the learning phase is performed. Also, we use all the computed color vectors as a learning set of images in the learning phase of the SOM. During the learning phase, the topological feature map is generated as follows: First, a node with the most similar weight vector to an input color vector is selected from map layer. Next, the weight vectors of the selected node and the nodes within its neighborhood are updated to match the input color vector. This process is repeated as many times as specified by the learning iteration parameter. At the end of the learn phase of a SOM, a codebook vector(CBV) is associated with every node i, and we define CBV as:

$$CBV_i = [cv_{i1},\ cv_{i2},\ ...,\ cv_{ij},\ ...,\ cv_{im}]^T$$

Here, $i(1 \leq i \leq n)$ represents the node number in the map layer, m is the number of nodes in the input layer, and n is the number of nodes in the map layer. In the obtained topological feature map, the topological relationships (similarities) between color vectors are preserved in the 2-dimensional topological feature map in the sense that neighboring nodes contain similar images.

3.2　Generation of Best Matching Image List

In the second step, we generated the best matching image lists, which are the clusters that contain similar images. The distance between each color vector in the database and each CBV of the topological feature map is computed. Based on these computed distances, a color vector is associated with the node (cluster) whose CBV gives the smallest distance. The list of images associate with a node is called the *best matching node* (BMN). The distance between a color vector and CBV is computed by the Euclidean distance. Formally, the BMN_i is computed by:

$$BMN_i = \min_{1 \leq i \leq n} \{ \| F_{col} - CBV_i \| \}$$

The relationship between the feature vectors and weight vectors have many-to-one relationship based on the Euclidean distances between them. Thus, Based on the many-to-one relationship, some nodes (clusters) may be associated with more than one color vector and other nodes may be empty (with no color vectors).

4　Fast SOM-Based *k*-Nearest Neighbors Search

Based on the *k*-NN basic idea [10,11,13], efficient *k*-NN retrieval is realized for topological feature map. Simply, the *k*-NN search starts by finding the position of a query point in the map (that is, finding a winner node), and then the search considers the images associated with the winner node as *k*-NN candidates. The neighboring search continues by checking the neighboring nodes to the winner node, as shown in Fig. 1(left). First, we check the nodes on the 1-neighbor, which the nearest diamond around the winner node, and then check the nodes on 2-neighbor, and so on. To speed up the search, our k-NN search algorithm utilizes a pruning strategy to remove unqualifying nodes; thus reducing the search space. The pruning strategy (discussed in Section 4.3) requires the construction of a minimum bounding rectangle (MBR) of each node.

4.1　Searching the Neighborhood of the Winner Node

Initially, the position of the query point (winner node) is determined for *k*-NN search. As a property of the SOM topological feature map, there are similarities between winner node and neighboring nodes. Therefore, it is more effective that the search proceed progressively from the winner node to the neighboring nodes as shown in Fig.

1(left), because most of the *k*-NN answer image are contained in the first few neighboring areas of the winner node. By utilizing such a structure of the feature map, we can produce very fast approximate *k*-NN answer images if we stop the search after a checking only a few neighboring areas. However, the main goal of this paper is to find the exact *k*-NN answer images, thus the search stops after checking all qualifying nodes in the map.

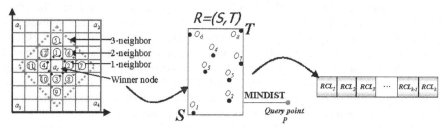

Fig. 1. k-NN search: (Left) The order of checked nodes by K-NN search algorithm in the SOM space, (Middle) The MBR, represented by R, inclosing objects of WN and (Right) Retrieved candidate list, RCL

4.2   MBR Definition of Each Node

All objects (i.e. images) associated with a node, where an object $O = \{o_{ij} \mid 1 \leq i \leq N,\ 1 \leq j \leq n\}$, are used to define the *MBR R* of the corresponding node as shown in Fig. 1(middle). Here, N and n represent the number of objects associated with a node and the number of elements in a color feature vector, respectively. Thus, *MBR R* is defined as:

$$R = (S, T)$$

Here, $S = \{s_j \mid 1 \leq j \leq n\}$ and $T = \{t_j \mid 1 \leq j \leq n\}$ represent the minimum and maximum points in *MBR R,* respectively.

Computation of *MBR R* is performed for all the nodes in the map layer except for the empty nodes. Next, we define the minimum distance *MINDIST* between a given query point *P* and *MBR R* as follows:

$$MINDIST(P, R) = (\sum_{i=1}^{n}(|p_i - r_i|)^{1/2}$$

$$r_i = \begin{cases} s_i & \text{if } p_i < s_i \\ t_i & \text{if } p_i > t_i \\ p_i & \text{otherwise} \end{cases}$$

That is, *MINDIST* is a point of minimum distance between *P* and all the points on boundary of the *MBR* R. The distance computation of *MINDIST*, which is distance measure between feature vectors, uses the Euclidean distance of L_2 space. This minimum distance *MINDIST(P,R)*   is used in deciding whether the current node is qualifying (should be visited) or unqualifying (should be skipped).

4.3   Pruning Strategy

Fast *k*-NN search is essential to retrieve the *k*-NN answer images to given queries within a practical time, especially, in a large image database environment. Therefore, during the search, our *k*-NN search algorithm employs a pruning strategy to disqualify some of the nodes, which leads to a reduction in the number of checked nodes, and thus speeds up the search. A node is considered unqualifying and should be skipped if one ,or both, of the following conditions hold: *(i)* it is an empty node (no images associated with it), *(ii)* its minimum distance *MINDIST(P,R)*, where *P* is the query point and *R* is the *MBR* of the current nodes, is larger than maximum distance value in the current *k*-NN candidate list. The application of the above pruning strategy greatly reduces the search space (number of nodes to be searched) leading to a fast *k*-NN search time.

4.4   *k*-NN Search Algorithm in SOM

The *k*-NN search algorithm consists of two phases as shown in Fig. 2 and Fig. 3. In first phase (Fig. 2), a winner node that is the nearest node to a given query image is computed from topological feature map(Step 4-11). The *k* images (where, *k*>0) associated with the winner node are used to update the *k*-NN candidate list, which is a list of the current *k*-NN answer images(Step 14). In this list, the *k*-th candidate is the current farthest image from the query, because the *k*-NN list is sorted in an ascending order with the closest image to the image query at the top of the list.

In second Phase (Fig. 3), the search proceeds outwards to check neighboring nodes of the WN in an orderly fashion (see Fig. 1(left)). At the same time, the two conditions of our pruning strategy are applied to remove the unqualifying nodes, as discussed above(Step 8-18). The remaining nodes (qualifying nodes) are used to update the k-NN_List *RCL*. That is, the distance *D(P, I)* between a query point *P* and every image *I* associated with the remaining nodes is compared against the distance *D(k-NN)* of the k-th candidate in the current k-NN candidate list, if *D(P, I)* <= *D(k-NN)* the k-NN list is updated with image *I* (Step 19); otherwise, *I* is discarded.

5   Experimental Results

The *k*-NN algorithm was implemented and several experiments were conducted to verify the effectiveness and efficiency of the *k*-NN search based on SOM clustering. The experiments were conducted on an Ultra5 (CPU:UltraSPARC-□, OS:Solaris8, Memory:512MB) SUN Microsystems workstation and we implemented using C++ program language. Image data used in the experiments was composed of 40000 images collected via the internet from two public sources: Stanford Univ. and H^{2}soft Company. To test the effect of database size on the performance of our k-NN search algorithm, we generated 6 data sets with different sizes. The size of each image was fixed at 128×128 pixels. Then, each image is represented by a 48-dimensional color vector as discussed in Section 3.

Phase 1: **ComputeWinnerNode**
Input : q_Object Output : WinnerNode, *k-NN_List*
1: *Nearest* = ∞ ; 2: q_N: feature vector of q_Object 3: *i* : node number on the Map 4: forall Node on the Map do begin 5:　　$Dist_i = \left\| cbv_i - q_N \right\|$; 6:　　if(*Nearest* > *Dist_i*) then do begin 7:　　　　*Nearest* = *Dist_i*; 8:　　　　*WinnerNode - i*; 9:　　end 10:　　$i = i + 1$; 11:end 12:/* initialize k-retrieved Image List */ 13:/* from WinnerNode */ 14:update **RCL** with *WinnerNode*

Phase 2: **SearchNearestNeighbor**
Input : γ, q_N, *k_NN* Output : *RCL*
1: γ: current NN radius 2: q_N: feature vector of q_Object 3: *R = (S,T)*;/* MBR */ 4: *i*: node number of the Map 5: get *NNodeList_i*; 6: if(*NNodeList* is equal to 0) return 7: for(*NNodeList_i* ∈ γ) do begin 8:　　/* prune empty Node */ 9:　　if(*NNodeList_i* is empty Node)then 10:　　　do begin 11:　　　　discard *NNodeList_i*; 12:　　end 13:　　/* prune visiting unnecessary Node */ 14:　　*Dist = MINDIST(q_N, R)*; 15:　　if(*Dist* > *RCL[k_NN].Dist*) then 16:　　　do begin 17:　　　　discard visiting unnecessary Node 18:　　end 19:　　update **RCL** with *NNodeList_i*; 20:　　*i=i+1*; 21: end 22: γ=γ+1; 23: call **SearchNearestNeighbor**(γ); 24:end

Fig. 2. Phase-1: Find Winner Node and initialize retrieval candidates

Fig. 3. Phase-2: Similarity retrieval within γ

Learning of a SOM was performed using color feature vectors of images as an input to the learning phase. The size of a map layer (an input parameter to the SOM algorithm) was initially established almost equal to size of the data set (number of images). First, the CBV of each node in the map layer is randomly initialized with a 48-dimensional vector, which has the same dimensionality as that of the color feature vectors of images. Thus, during clustering a data set of images, generally, many images, which are similar, are mapped into one node of the map layer. Consequently, empty nodes, which are associated with no images, are generated since the number of nodes in map layer is initially established almost equally to the number of images in the data set. The rate of empty nodes of map layer obtained at each of the 6 data sets. We notice that the rate of empty nodes is between 42% ~ 60%.

The performance of the *k*-NN search was evaluated with 20 query images selected randomly from each data set, and the results shown below (Fig. 4 and Fig. 5) are the average result of the 20 queries. As far as the effectiveness of the results, we compared the resulting images to those of the sequential scanning method and we found that our method produces exactly the same result as that of the sequential scanning method, but at a faster search time. The number of returned images is varied as follows: k=1,5,10,20,30,40,50,100. Through the experiment, we investigated the performance of the k-NN search algorithm as we increase the size of k and the size of the data set. We also examined the effect of *MINIDIST*. Fig. 4 shows the number of pruned nodes (nodes that need not to be searched) at each value of k for every data set. Notice that the number of pruned nodes do not include the number of empty. That is because the objective is to show the effectiveness of the pruning condition *MINDIST*. On the other hand Fig. 5 shows the number of visited nodes at each k for every data set. We notice, refer to Fig. 5, that the number of visited nodes is very small as compared to the number of non-empty nodes.

Compared Fig. 5 with Fig. 4, we notice that an increase of retrieval candidate k brought about an increase in the number of nodes to be visited (searched), but the number of pruning nodes was confined between 38% and 57% of overall nodes. This means that our pruning strategy represented by the metric MINDIST is very effective. We also found from Fig. 5 that the number of visited nodes to retrieve 1-NN (k=1) of the data set 40000 is one node (only the winner node). And, for data set 40000 for k=100, 1084 nodes are visited (2.7% of the non-empty nodes).

The efficiency of the k-NN search is evaluated as shown in Fig. 6 and Fig.7. Fig. 6 measures the overall retrieval time of the search, which includes time to find the winner node, and the time to search for k-NN images. Fig. 7 measures the k-NN search time. The results in Fig. 6 and Fig. 7 are the average of 20 queries. Here, we notice that search time, especially the overall time, is not affected by an increase in the value of k. The reason is that when evaluating a query all the qualifying nodes (non-pruned nodes) are visited regardless of the value of k.

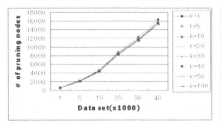

Fig. 4. Number of pruned nodes versus size of data sets for different k-NN

Fig. 5. Number of visited nodes versus size of data sets for different k-NN

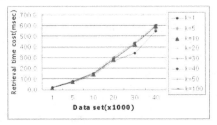
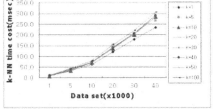
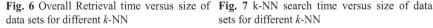

Fig. 6 Overall Retrieval time versus size of data sets for different k-NN

Fig. 7 k-NN search time versus size of data sets for different k-NN

6   Conclusions

In this paper, *k*-NN search algorithm is proposed to perform fast SOM-based similarity retrieval of images. The images in the database were clustered using a SOM algorithm, and MBR is computed for every cluster, and a minimum distance function *MINDIST* between a given query point and an MBR was defined. We implemented the proposed *k*-NN search algorithm and evaluated its effectiveness through several experiments using actual image data. The realization of fast similarity image retrieval was verified by these experiments due to the use of *MINDIST* metric which reduced the search space. As a result of comparing our method with the sequential scanning method, our method returns the same *k*-NN answer images, however at a faster time. Thus, retrieval of similar images in practical time became possible, even by using query-by-example technique.

Acknowledgement

This research was supported in part by the MIC in Korea through the University Research Program grant. Dr. Pankoo Kim was a corresponding author.

References

1.   Deboeck, G. and Kohonen, T.:Visual Explorations in Finance with Self-organizing Maps, Springer-Verlag, London(1998);Japanese ed., Springer-Verlag, Tokyo (1999).
2.   Kohonen, T.: Self-Organizing Maps, Series in Information Sciences, vol.30, Springer-Verlag, third edition, Berlin (2001).
3.   Baba, N., Kojima F., and Ozawa M.: Basis and Application of Neural Network, Kyoritu Shuppan, Tokyo, Japan (1994).
4.   Rui, Y., Huang, T. S., Ortega, M.,and Mehrotra, S.:Relevance Feedback: A Power Tool in Interactive Content-Based Image Retrieval,' IEEE Trans. on Circuits and Systems for Video Technology, Special Issue on Segmentation, Description, and Retrieval of Video Content, Vol.8, No.5(1998)644-655.
5.   Gudivada, V. N. and Raghavan, V. V. eds.: Content-based Image Retrieval System, IEEE Computer, Vol.28, No.9(1995)18-22.
6.   Flickner, M. et al.: Query by Image and Video Content: The QBIC System, IEEE Computer, Vol.28, No.9(1995)23-32.
7.   Laaksonen, J., Koskela, M., and Oja, E.:PicSOM:Self-organizing Maps for Content-Based Image Retrieval, Proceedings of IJCNN'99, Washington D.C.(July 1999).
8.   Eakins, J. P. and Graham, M. E.: Content-based image retrieval. Report to the JISC Technology Applications Programme (1999).
9.   Rui, Y., Huang, T. and Chang S-F.:Image Retrieval:Current Techniques, Promising Directions, and Open Issues, J. Visual Communication and Image Representation(JVCIR), Vol.10, No.1(1999)39-62.

10. Roussopoulos, N., Kelley, S., and Vincent, S.:Nearest neighbor queries, In Proceedings of the ACM SIGMOD Conference, San Jose, CA, May (1995)71-79.
11. Beyer, K., Goldstein, J., Ramakrishnan, R., and Shaft. U.:When is 'nearest neighbor' meaningful?, In Proc. 7th International Conference on Database Theory (1999)217-235.
12. Mallat, S. G.: Multifrequency channel decompositions of images and wavelet models, IEEE. Trans., Acoust.,Speech and Signal Proc., Vol.37, No.12 (1989) 2091-2110.
13. Ferhatosmanoglu, H., Tuncel, E., Agrawal, D., and El Abbadi, A.:Approximate nearest neighbor searching in multimedia databases, Technical Report TRCS00-24, Comp. Sci. Dept., UC, Santa Barbara(2000).
14. Smith, J. R. and Chang, S-F.:VisualSEEk:A Fully Automated Content-Based Image Query System, Proc. ACM Intl. Conf. on Multimedia, Boston (1996)87-98.
15. Pentland, A., Picard, R. W. and Schlaroff, S.: Photobook: Tools for Content-Based Manupulation of Image Databases, SPIE Storage and Retrieval for Image and Video Databases II, San Joes, (1994) 34-47.
16. Jacobs, C. E., Finkelstein, A. and Salesin, D. H.: Fast multiresolution Image Querying, Proc. SIGGRAPH95, Los Angeles, California, (1995)6-11.
17. Ma, W. Y. :NETRA: A Toolbox for Navigating Large Image Databases, Ph.D. Dissertation, Dept. of Electrical and Computer Engineering, University of California at Santa Barbara (1997).
18. Oh, K.-S., Kaneko, K., and Makinouchi, A.: Image Classification and Retrieval based on Wavelet-SOM, DANTE'99(1999)164-167.
19. Keim, H. A.:What is the Nearest Neighbor in High-Dimensional Spaces?, VLDB Conference(2000) 506-515.

Video Retrieval by Feature Learning
in Key Frames

Marcus J. Pickering[1], Stefan M. Rüger[1], and David Sinclair[2]

[1] Department of Computing, Technology and Medicine, Imperial College of Science
180 Queen's Gate, London, SW7 2BZ, UK
m.pickering@doc.ic.ac.uk
[2] AT&T Laboratories Cambridge
24a Trumpington Street, Cambridge, CB2 1QA, UK

Abstract. We evaluate the application of feature-vector based image retrieval methods to the problem of video retrieval. A vast number of primitive features is calculated for each of the key frames generated by a segmentation process, and we examine the use of three methods for retrieving video segments using the features — a vector space model, a learning method using the AdaBoost algorithm, and a k-nearest neighbour approach.

1 Introduction

The availability of cheap digital storage has led to a rapid expansion of digital video archives, and this has precipitated a need for effective search and retrieval systems. Initial attempts to meet this need involved the use of textual data obtained from subtitles [3,8] and speech recognition transcripts [6]. However, while these methods are effective for some needs, such as broadcast news retrieval, they take no account of the visual content of the video.

There are a number of advantages to being able to use the visual content as the basis for a search. It is often difficult to fully express a visual query in words, and yet a single image can completely describe what is being searched for. Describing an object such as an aeroplane simply in terms of its shapes and colours would be a demanding task, yet providing an example can give all the information that is required. Similarly, one can easily visualise a tropical beach with palm trees and a clear ocean, but describing it clearly to a retrieval system without the use of examples would be a challenge. Systems which employ subtitles or speech recognition are relying solely on the audio component of a broadcast, and while this provides a great deal of the necessary information in news material, there is still much information that can only be gleaned from the visual component. A news report from London, with Big Ben in full view, will probably not have the words "Big Ben" actually spoken. Many modern feature films rely far more on their visual effects than they do on any spoken material. The use of visual cues also allows the retrieval system to become language independent, an advantage where the database is to be made available internationally, such as on the Internet.

M. S. Lew, N. Sebe, and J. P. Eakins (Eds.): CIVR 2002, LNCS 2383, pp. 309–317, 2002.

A major problem with content-based retrieval of video material is the sheer volume of information which must be processed, and we address this problem by segmenting the video into shots and processing a representative frame for each shot. Our segmentation method is based on a scheme proposed by Pye et al [2], in which scene boundaries are declared where audio segmentation boundaries and video shot boundaries coincide. Their video shot boundary detection algorithm, which we use in our implementation [4] and describe in more detail in Section 2, is based on differences of colour histograms and uses a multi-timescale filter bank to detect cuts and gradual effects.

Having segmented the video and generated a key frame for each clip, we treat the key frames as a database of still images. For each image we generate a large feature vector, as described by Tieu and Viola [1] and outlined in Section 3. We have tested three retrieval algorithms on these feature vectors: (1) A vector space model, (2) a nearest neighbour approach, and (3) a variation of the AdaBoost algorithm described by Tieu and Viola [1], in which a hypothesis based on a few of the features is generated from a set of example images. Each of these methods is described in more detail in Section 4. Having retrieved a still image associated to a video shot, the video can then be played by the user.

To facilitate a detailed evaluation of the retrieval methods, we tested each of them on a database of 658 still images. Our results are given in Section 5. In order to demonstrate the effectiveness of the learning methods on video retrieval we also show graphical results for two specific video queries.

2   Key Frame Generation

In order to generate enough key frames to capture the salient moments from the video sequence it is necessary to use an algorithm which detects significant breaks in the video. Early video segmentation algorithms which simply looked at changes between consecutive frames were unreliable because they were defeated by gradual changes which take place across a number of frames, such as fades, or by transients, such as flashes, which cause a sudden change on a single frame where no shot change is actually taking place. In many cases shot boundaries are not the most logical divisions - for example in news broadcasts it is perhaps more useful to segment on story boundaries. Hauptmann and Witbrock [7] have suggested the integration of information from subtitles, speech recognition transcripts and the video stream itself to determine story boundaries. However, since our aim is to obtain enough salient key frames for effective retrieval, it is necessary for us to segment the video into its most basic components.

We use the video shot change detection algorithm proposed as part of a scene change detection scheme by Pye et al [2]. The algorithm makes use of colour histograms to characterise frames and looks at differences across time periods up to 16 frames either side of the current frame in order to determine if a shot boundary is occurring at that point. The multi-timescale approach facilitates detection of gradual changes and rejection of transients. Our own implementation of the algorithm [4] was shown to be effective through participation in the shot boundary detection task of the TREC video track [12] where, over a variety of cuts and gradual transitions, the system achieved an average of 83% recall and 80% precision.

Having segmented the video, a representative frame (key frame) is taken for each shot. If the shot began with a hard cut, the key frame is the first frame of the shot, otherwise the key frame is the tenth frame of the shot.

3   Feature Generation

Describing an object to a retrieval system typically involves the use of such characteristics as texture, colour and edge orientation, which may be common to a large number of images. Tieu and Viola's method [1] depends on the definition of highly selective features which are determined by the structure of the image, as well as capturing information about colour, texture and edges. By defining a vast set of features, each feature is such that it will only have a high value for a small proportion of images, and by discovering a number of features which distinguish the example set in question we are able to perform an effective search.

The feature generation process is based on a set of 25 primitive filters, which are applied to each of the three colour channels to generate 75 feature maps. Each of these feature maps is rectified and downsampled before being fed again to each of the 25 filters to give 1875 feature maps. The process is repeated a third time, and then each feature map is summed to give 46,875 feature values. The idea behind the three stage process is that each level 'discovers' arrangements of features in the previous level. The feature generation process is computationally quite costly, but only needs to be done once and then the feature values can be stored with the image in the database. With the features pre-computed, one of the query algorithms we use simply has to check the few features of a reduced feature space for each image.

4   Retrieval Methods

4.1   Vector Space Model

The vector space model [10] uses the entire feature vector, and images in the database are ranked according to the cosine of the angle between their feature vector and the sum of the feature vectors of the positive examples.

4.2   Boosting

The AdaBoost algorithm [5] aims to select the most relevant features by determining which features generate the minimum error when classifying the example set.

We deploy Tieu and Viola's method [1] in which a hypothesis is determined from the positive and negative images by building a strong classifier from a weighted combination of weak classifiers. The ranking of an image in the database is based on how closely it fits the hypothesis.

A user specifies a query by selecting some positive examples. Negative examples are also required, but in a sufficiently disparate database, these can be randomly selected without too much danger of accidentally picking positive results. When a categorised image database is used, as in our evaluation, it is possible to forcibly exclude from the negative set images which fall into the same category as the positive examples. Even when positive results are selected by chance, they can be removed in an interactive relevance feedback stage. The two example sets are then used to generate a strong classifier using the AdaBoost [5] algorithm. The strong classifier is generated from the combination of a series of weak classifiers, each of which is determined by one iteration of the algorithm. An initial weight is assigned to each image, such that the total weights of the positive and negative images are equal, and weights are evenly distributed within these two groups. A hypothesis is generated at each iteration. Weights are re-assigned on successive iterations according to whether the image was correctly classified at the previous stage, so that more attention is paid to images which have thus far been incorrectly classified, and the hypotheses generated are combined to give an overall hypothesis which can be used to rank the images in the database.

4.3   K-nearest Neighbour

We use a variant of the distance-weighted k-nearest neighbour approach [9]. As with the boosting method, a number of positive examples are supplied, and a number of negative examples are randomly selected from the database. The distances from the test image i to each of the k nearest positive or negative examples (where 'nearest' is defined by the Euclidean distance in feature space) are determined, and a relevance measure calculated as follows:

$$R(i) = \frac{\displaystyle\sum_{p \in P} \frac{1}{\mathrm{dist}(i, p) + \varepsilon}}{\displaystyle\sum_{n \in N} \frac{1}{\mathrm{dist}(i, n) + \varepsilon} + \varepsilon}$$

where P and N are the sets of positive and negative examples respectively amongst the k nearest neighbours, such that $|P| + |N| = k$. ε is a small positive number to avoid division by zero. Images are ranked according to $R(i)$.

5   Results

We tested all three methods on a database of 658 images, divided into 31 categories of conceptually (and often visually) similar images. Categories include planes, surfing, golf, bears, miscellaneous people, market stalls, horses, wolves, military and computers. Firstly, each image in the database was used as a single-image query, and returned results were judged to be relevant if they belonged to

Table 1. Mean average precision for retrieval for boosting, vector space model and k-nn, compared with a random ranking. The left column is the number of positive examples used in the query

Queries	Boost	VSM	k-nn	Random
1 image	0.1609	0.1946	0.1676	0.0536
2 images	0.1659	0.1343	0.1841	0.0408
3 images	0.1877	0.1397	0.2105	0.0387
4 images	0.1971	0.1247	0.2269	0.0379
5 images	0.2029	0.1299	0.2272	0.0361

the same category. For the boosting and k-nn methods, negative examples were randomly picked from the whole database, excluding the category to which the positive query image belonged. For k-nn, we used $k = 40$.

For each category we then generated 6 random queries containing 2 positive examples, and repeated the process for 3, 4 and 5 positive examples. The results are shown in Table 1. Average precision is computed using trec_eval [11]. For all queries the query images are removed from the returned list before calculating statistics. Paired t-tests performed on the data proved that all pairwise comparisons across the rows are statistically significant at an α level of 0.01, with the exception of the comparison between boosting and k-nn, for 1, 2 and 3 image queries.

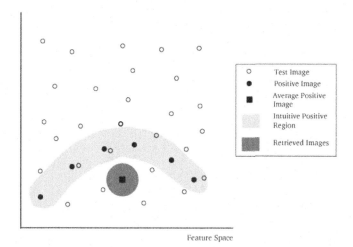

Fig. 1. Illustration of degradation of VSM performance with multiple image queries

The mean average precision values of all deployed methods are significantly better than random ranking of database images. The k-nn and boosting approaches returned particularly impressive results and, predictably, precision increased as more query images were used. The relative performance of the learning (boosting and k-nn) algorithms and the vector space model are interesting. The results show a clear trend of feature learning performing better with more image examples, while the vector space model deteriorates in performance. Figure 1 demonstrates the problem that we believe gives rise to these results. Since the positive feature vectors (represented by the solid black dots) are summed for the query in VSM, the resultant query (the black square) is effectively an average of the positive images. When the query images do not form a convex cluster the resultant query may be nowhere near any of the query images, so the returned images (represented by the dark grey shaded region) bear no relation to the images one would intuitively expect to have returned (represented by the lighter grey region). k-nn intuitively ranks highly images found close to the positive query images, hence its superior performance for the multiple image queries. It must be pointed out that we did no performance tuning with the parameters of the boosting algorithm. It must be expected that when the parameters of the AdaBoost algorithm are optimised, its performance will further increase.

We have also tested the boosting algorithm on a number of the queries set for the TREC video track [12]. Figure 5 shows the positive images supplied for the query "the planet Jupiter" and Figure 6 the results returned by our system, where the Jupiter known-item appears ninth.

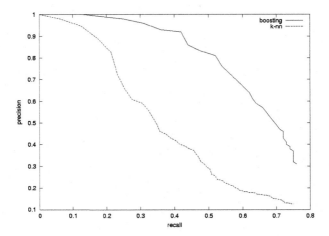

Fig. 2. Precision-recall graph for the anchor-person query, using the boosting and k-nn methods. The number of relevant images was estimated from a random sample

Fig. 3. The 14 positive examples of anchor-person shots

Fig. 4. The first 49 returned results for the anchor-person query, using the boosting method

The effectiveness of the boosting algorithm has been further demonstrated by an experiment in which we attempted to retrieve all anchor-person shots from a database of keyframes generated from recordings of daily news broadcasts over a period of three months. The database of 14590 images was queried using the 14 positive examples shown in Figure 3 and the result was a remarkable 81% precision in the first 500 returned shots. The first screen of results is shown in Figure 4, and the precision-recall graph is shown in Figure 2, where the results are compared to the same experiment carried out using the k-nn method.

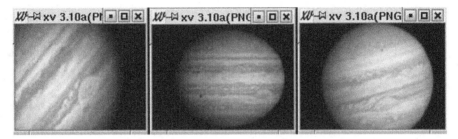

Fig. 5. Query images for the Jupiter query

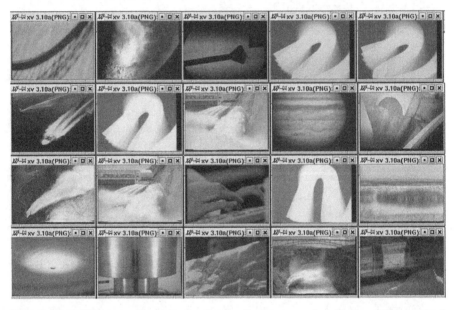

Fig. 6. First 20 returned results for the Jupiter query, using the boosting method

The superior performance of k-nn relative to boosting in the category experiments is probably due to the way in which the queries are constructed. k-nn copes well with visually disparate positive examples since positive results can be found in disjoint clusters. The boosting algorithm depends on discovering visual similarity in order to build a good retrieval hypothesis, hence results were much more impressive for the anchor-person experiment - where the positive examples were similar - than the category queries for which this was not necessarily the case.

6   Conclusions

This paper clearly demonstrates the benefits of feedback based selective learning techniques. The k-nn approach returned impressive results, even with visually disparate queries, and the AdaBoost algorithm performs well at retrieving key frames from video sharing similar visual composition. This demonstrates the potential for some of the skeletal structure of a video to be actively modelled, or at least familiar way points (such as anchor-person shots) to be automatically identified.

Acknowledgements

This work was partially supported by EPSRC, UK, and AT&T Laboratories, Cambridge, UK.

References

1. K. Tieu and P. Viola. Boosting Image Retrieval. In Proceedings of IEEE Conference on Computer Vision and Pattern Recognition, 2000. 310, 311
2. D. Pye, N. J. Hollinghurst, T. J. Mills and K. R. Wood. Audio-visual segmentation for content-based retrieval. 5th International Conference on Spoken Language Processing, December 1998. 310
3. M. J. Pickering. Video Search Engine. http://km.doc.ic.ac.uk/video-se/, June 2000. 309
4. M. J. Pickering and S. M. Rüger. Multi-timescale video shot change detection. In Proceedings of the Tenth Text REtrieval Conference, 2001. 310
5. Y. Freund and R. E. Schapire. A decision-theoretic generalization of on-line learning and an application to boosting. Journal of Computer and System Sciences, 55(1):119-139, 1997. 311, 312
6. H. D. Wactlar, A. G. Hauptmann and M. J. Witbrock. Informedia: News-on-demand experiments in speech recognition. In Proceedings of ARPA Speech Recognition Workshop, February 1996. 309
7. A. G. Hauptmann and M. J. Witbrock. Story segmentation and detection of commercials in broadcast news video. In ASL-98 Advances in Digital Libraries, April 1998. 310
8. M. G. Brown, J. T. Foote, G. Jones, K. Spärck-Jones and S. J. Young. Automatic content-based retrieval of broadcast news. ACM Multimedia, 1995. 309
9. T. M. Mitchell. Machine Learning. McGraw Hill, 1997. 312
10. G. Salton. Automatic Text Processing. Addison-Wesley, 1989. 311
11. E. M. Voorhees and D. Harman. Overview of the eighth text retrieval conference (TREC-8). In Proceedings of the Eighth Text REtrieval Conference, 1-33 and A.17-A.18, 1999. 313
12. TREC-10. Video track. http://www-nlpir.nist.gov/projects/t01v/   311, 314

Local Affine Frames for Image Retrieval[*]

Štěpán Obdržálek[1,2] and Jiří Matas[1,2]

[1] Center for Machine Perception
Czech Technical University, Prague, CZ
[2] Centre for Vision Speech and Signal Processing
University of Surrey, Guildford, UK

Abstract. A novel approach to content-based image retrieval is presented. The method supports recognition of objects under a very wide range of viewing and illumination conditions and is robust to occlusion and background clutter. Starting from robustly detected 'distinguished regions' of data dependent shape, local affine frames are established by affine-invariant constructions exploiting invariant properties of the second moment matrix and bi-tangent points. Direct comparison of photometrically normalised colour intensities in normalised frames facilitates robust, affine and illumination invariant, but still very selective matching. The potential of the proposed approach is experimentally verified on FOCUS — a publicly available image database — using a standard set of query images. The results obtained are superior to the state of the art. The method operates successfully on images with complex background, where the sought object covers only a fraction (around 2%) of the database image. Examples of precise localisation of the query objects in an image are shown too.

1 Introduction

In recent few years, the number of digital images that a user could search increased rapidly, fueling the need for content-based image retrieval systems. Many approaches addressing the problem of image retrieval were introduced, the most common being those using colour histograms [1,2,3], texture [4], shape [5,6,7], colour invariants [8,9] or graph representations of colour content [10]. For a comprehensive survey, see [11].

In this paper we focus on a class of retrieval problems where the query depicts (a part of) an object of interest. We assume that the query object may cover only a franctional part of the database image and that it may be viewed from a significantly different viewpoint and under different illumination.

The proposed approach is based on robust, affine and illumination invariant detection of local affine frames (local coordinate systems). Local correspondences between the query and database images are established by a direct comparison of normalised colour in image patches with shape normalised according to the affine

[*] The authors were supported by the EU project IST-2001-32184, the Czech Ministry of Education project MSM 210000012 and a CTU grant No. CTU0209613.

M. S. Lew, N. Sebe, and J. P. Eakins (Eds.): CIVR 2002, LNCS 2383, pp. 318–327, 2002.

frames. The method achieves the discriminative power of template matching while maintaining the invariance to illumination and object pose changes of techniques using more general feature descriptors. In addition, for every local correspondence obtained, the local inter-image transformation is known, making it possible to robustly localise the query in the database image.

The most closely related work is that of Tuytelaars and Gool [9], where local regions were also affine-invariantly found, but these regions were used to determine the image area over which affine moment invariants were computed. We argue here, that once image regions are found in a affine-invariant way, matches can be established by direct comparison of intensity profiles over these regions.

The main contribution of the paper is the utilisation of several affine-invariant constructions of local affine frames (LAFs) for determination of local image patches being put into correspondence. Robustness of the matching procedure is accomplished assigning multiple frames to each detected image region, and not requiring all of the frames to match. The outline of the proposed retrieval process is as follows:

1. For every database and query image compute distinguished regions, establish local affine frames, and generate intensity representation of local image patches normalised according to the local frames.
2. Establish correspondences between frames of query and database images, directly comparing the local image intensities. Estimate the match score based on the number and quality of established correspondences.
3. Combine affine transformations provided by every matched frame pair to establish an estimate of query location in the database image.

The paper is organised as follows. In Section 2 we briefly review the concept of distinguished regions. Section 3 gives a description of procedures giving local affine frames on distinguished regions of complex shapes. Section 4 details how correspondences between the local affine frames are established, and in Section 5 experimental results are presented.

2   Distinguished Regions

Distinguished Regions (DRs) are image elements (subsets of image pixels), that posses some distinguishing, singular property that allows their repeated and stable detection over a range of image formation conditions. In this work we exploit a new type of distinguished regions introduced in [12], the *Maximally Stable Extremal Regions* (MSERs). An extremal region is a connected component of pixels which are all brighter (MSER+) or darker (MSER-) than all the pixels on the region's boundary. This type of distinguished regions has a number of attractive properties: 1. invariance to affine and perspective transforms, 2. invariance to monotonic transformation of image intensity, 3. computational complexity almost linear in the number of pixels and consequently near real-time run time, and 4. since no smoothing is involved, both very fine and coarse image structures

Fig. 1. An example of detected distinguished regions of MSER type

are detected. We do not describe the MSERs here; the reader is referred to [12] which includes a formal definition of the MSERs and a detailed description of the extraction algorithm. The report is available online. Examples of detected MSERs are shown in Figure 1. Note that DRs do not form segmentation, since 1. DRs do not cover entire image area, and 2. DRs can be (and usually are) nested.

3   Local Frames of Reference

Local affine frames facilitate normalisation of image patches into a canonical frame and enable direct comparison of photometrically normalised intensity values, eliminating the need for invariants. It might not be possible to construct local affine frames for every distinguished region. Indeed, no dominant direction is defined for elliptical regions, since they may be viewed as affine transformations of circles, which are completely isotropic. On the other hand, for some distinguished regions of a complex shape, multiple local frames can be affine-invariantly constructed in a stable and thus repeatable way. Robustness of our approach is thus achieved by 1. selecting only stable frames and 2. employing multiple processes for frame computation.

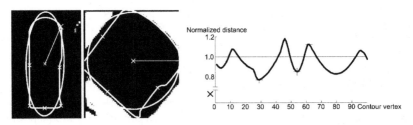

Fig. 2. Construction of affine frames. Original image, shape-normalised image and normalised distances between the center of gravity and contour

Frame constructions. Two main groups of affine-invariant constructions are proposed, one based on region normalisation by the covariance matrix (see Ap-

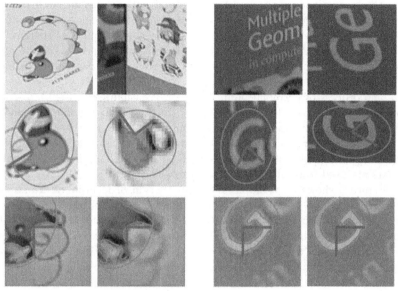

Fig. 3. Construction of affine frames. top row - original views, middle row - detected frames, bottom row - normalised frames

pendix A for definition and proof of affine invariance), second on detection of stable bi-tangents.

Transformation by the square root of inverse of the covariance matrix normalises the DR up to an unknown rotation. To complete an affine frame, a direction is needed to resolve the rotation ambiguity. The following directions are used: 1. Center of gravity (CG) to contour point of extremal (either minimal or maximal) distance from the CG 2. CG to contour point of maximal convex or concave curvature, 3. CG of the region to CG of a concavity, 4. direction of a bi-tangent of a concavity of the region.

In frame constructions derived from the bi-tangants (the line segments of convex hull bridging concavities), the two tangent points are combined with a third point to complete an affine frame. As the third point, either 1. the center of gravity of the distinguished region, 2. the center of gravity of the concavity, 3. the point of the distinguished region most distant from the bi-tangent, or 4. the point of the concavity most distant from the bi-tangent is used. Another type of frame construction is done by combining covariance matrix of a concavity, CG of the concavity and the bi-tangent's direction.

Frame constructions involving the center of gravity or the covariance matrix of a DR rely on the correct detection of the DR in its entirety, while constructions based solely on properties of the concavities depend only on a correct detection of a part of the DR (containing the concavity).

Affine covariance of the center of mass and of the covariance matrix is shown in Appendix A. The invariance of the bi-tangets is a consequence of the affine invariance (and even projective invariance) of the convex hull construction [13,14]. The invariance of the maximal-distance-from-a-line property is easily appreciated taking into account that affine transform maintains parallelism of lines and their ordering.

The process of extremal point detection is visualised in Figure 2. A region detected in an image (left) is transformed to the shape-normalised frame (middle). Normalised distances of contour points are plotted on the right. The ellipse defined by the covariance matrix of the region is transformed to the unit circle in the normalised frame. To complete affine frames, directions to the extremal points are used to resolve the unknown rotation.

Figure 3 shows an example of a construction of local affine frame for two objects. In the leftmost two columns, frames are build from bi-tangent points and the points most distant from the bi-tangents. The two rightmost columns show frame construction based on DR's covariance matrix and extremal distance points. The top row displays two different views of the same object, detected frames are shown in the middle row, and normalised frames are depicted in the bottom row.

4   Matching

As a final step of our method, a matching score is computed between a pair of images. Since local affine frames have been established on distinguished regions, (geometrically) invariant descriptors of local appearance are not needed. The similarity assessment can rely simply on correlating photometrically normalised regions defined intrinsically in terms of local coordinate frames.

Figure 4 shows an example of normalised frames for a pair of images being put into correspondence, demonstrating some desirable properties of the LAF method. The object of interest undergoes a full affine change (anisotropic scale, rotation, skew), it is partially occluded, and covers only about 5% of the database image. However, many of the detected LAFs cover the same part of the object surface, which is clearly seen in the right part of Figure 4.

Having two sets of local affine frames, set S_1 of frames computed on the query image, and set S_2 on a database image, the computation of the matching score can be outlined as follows:

1. For every frame $f_{1i} \in S_1$ find such a frame $f_{2i} \in S_2$ so that f_{1i} and f_{2i} are of the same frame type, and that the intensity-normalised distance $d_i = |f_{1i}, f_{2i}|$, $d_i \in \langle 0, 1 \rangle$ is minimal, ie. $d_i = min|f_{1i}, f_2|, \forall f_2 \in S_2$
2. matching score $m = \sum_i (1 - d_i)^2$

Considering only the best match for every query frame makes the matching score independent of the number of frames defined on individual database images.

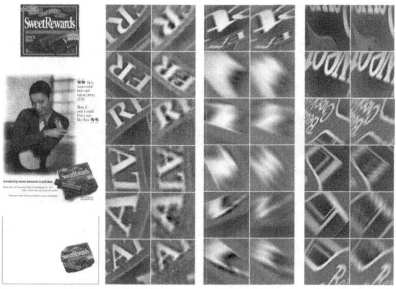

Fig. 4. Samples of correspondences established between frames of query (left columns) and database (right columns) images. On the left, the query, the database image, and the query localisation is shown

5   Experiments on the FOCUS Database

We tested the retrieval performance of the proposed method on the FOCUS image database, containing 360 colour high-resolution images of commercials scanned from miscellaneous magazines. For comparison purposes, we run an experiment with an identical setup as the SEDL system introduced by Cohen [15]. The quality of the retrieval is assessed by the same two quantities as defined by Cohen, the recall rate r_R and the precision ρ_R:

$$r_R = \frac{n}{N} \qquad \rho_R = \frac{\sum_{i=1}^{n}(R + 1 - r_i)}{\sum_{i=1}^{n}(R + 1 - i)} \qquad (1)$$

where n is the number of correct answers in the first R retrieved images, N the number of all correct answers contained in the database, and r_i the rank of the i-th correctly retrieved answer.

In Table 1, average recall rate r_{20} and average precision ρ_{20} are given for the number of retrieved images $R = 20$. For each of the 25 queries used by Cohen, the database images were sorted according to the matching score (similarity measure) m, and the recall r_{20} and the precision ρ_{20} were computed according to formula (1). Each of the 25 queries has 2 to 9 correct answers in the database, with the total number of all correct answers equal to 90. The proposed local affine frame (LAF) method achieves a 83% recall, which is approximately 5% better than results reported by Cohen. Note that the LAF method is not attempting

Table 1. Retrieval performance compared to the SEDL system

SEDL		LAFs	
recall r_{20}	avg precision ρ_{20}	recall r_{20}	avg precision ρ_{20}
$70/90 = 77.8\%$	88%	$75/90 = 83.3\%$	93.5%

Fig. 5. Examples of query (left) and database images (right) not retrieved in the FOCUS experiment

to generalise the query (i.e. to categorise). Most database images missed depict *objects different from the query*. Figure 5 shows three such examples. The 'failure' in such cases might be viewed as a strength, demonstrating the very high selectivity of the method, distinguishing items that superficially look identical, while being immune to severe affine deformations. Adapting the method for categorisation task is an open problem, possible approaches include adopting more flexible local representations, eg. local colour and texture distributions or flexible local templates.

Query localisation. Since the matching establishes explicit correspondences and mappings between parts of the query and the database image, it is possible to localise precisely the position of the query in the database image. The process is demonstrated in Figure 6. On the left, the query and database images are shown. Every individual correspondence of frames provides a single estimate of the query location in the database image. These estimates are displayed as white parallelograms in the central image. A voting scheme is applied to accumulate the estimates to form a clipping mask, shown in the fourth image. The clipped-out part of the database image, where the query was located, is shown on the right of Figure 6. Two other examples of query localisation are presented in Figure 7.

Fig. 6. Query localisation in the database image

Fig. 7. Sample query localisation results, query images, database images, and query localisations

6 Conclusions

In this paper, a novel procedure for image retrieval was introduced. Starting from robustly detected distinguished regions of data dependent shape, local affine frames were obtained. The constructions of affine frames was proved affine covariant, and experimentally shown to be stable. Direct comparison of photometrically normalised colour intensities in normalised frames allowed for robust and selective matching. Fully determined frame to frame correspondences made it possible to robustly localise the occurrence of the query in the retrieved database image.

Experimental results obtained on a publicly available image database, a recall of 83% and a precision of 93% were superior to the state of the art [15].

References

1. Swain, M., Ballard, D.: Color indexing. In: International Journal of Computer Vision, vol. 7, no. 1. (1991) 11–32   318
2. Finlayson, G. D., Chatterjee, S. S., Funt, B. V.: Color angular indexing. In: ECCV. (1996) 16–27   318
3. Funt, B., Finlayson, G., Color, C.: Color constant color indexing. IEEE Transactions on Pattern Analysis and Machine Intelligence **17** (1995) 522–529   318
4. Liu, F., Picard, R. W.: Periodicity, directionality, and randomness: Wold features for image modeling and retrieval. IEEE Transactions on Pattern Analysis and Machine Intelligence **18** (1996) 7–733   318
5. E., Gary, R., Mehrotra: Shape similarity-based retrieval in image database systems. SPIE **1662** (1992) 2–8   318
6. Mokhtarian, F., Abbasi, S., Kittler, J.: Robust and efficient shape indexing through curvature scale space. In: In Proceedings of British Machine Vision Conference, Edinburgh, UK. (1996) 53–6   318
7. Bimbo, A., Pala, P.: Effective image retrieval using deformable templates. In: ICPR96. (1996) 120–123   318
8. Mindru, F., Moons, T., Gool, L. V.: Recognizing color patterns irrespective of viewpoint and illumination. In: CVPR99. (1999) 368–373   318

9. Tuytelaars, T., Gool, L. V.: Content-based image retrieval based on local affinely invariant regions. In: Proc. Visual '99: Information and Information Systems. (1999) 493–500 318, 319

10. Park, K., Yun, I., Lee, S. U.: Color image retrieval using a hybrid graph representation. Journal of Image and Vision Computing **17(7)** (1999) 465–474 318

11. Smeulders, A. W. M., Worring, M., Santini, S., Gupta, A., Jain, R.: Content-based image retrieval at the end of the early years. IEEE Transactions on Pattern Analysis and Machine Intelligence **22** (2000) 1349–1380 318

12. Matas, J., Chum, O., Urban, M., Pajdla, T.: Distinguished regions for wide-baseline stereo. Research Report CTU–CMP–2001–33, Center for Machine Perception, K333 FEE Czech Technical University, Prague, Czech Republic (2001) ftp://cmp.felk.cvut.cz/pub/cmp/articles/matas/matas-tr-2001-33.ps.gz. 319, 320

13. Suk, T., Flusser, J.: Convex layers: A new tool for recognition of projectively deformed point sets. In Solina, F., Leonardis, A., eds.: Computer Analysis of Images and Patterns : 8th International Conference CAIP'99. Number 1689 in Lecture Notes in Computer Science, Berlin, Germany, Springer (1999) 454–461 322

14. Mundy, J. L., Zisserman, A., eds.: Geometric Invariance in Computer Vision. The MIT Press (1992) 322

15. Cohen, S.: Finding color and shape patterns in images. Technical Report STAN-CS-TR-99-1620, Stanford University (1999) 323, 325

A   Affine Invariance of Covariance Matrix Construction

An affine transformation is a map $F : \mathbb{R}^n \to \mathbb{R}^n$ of the form $F(\mathbf{x}) = A^T \mathbf{x} + \mathbf{t}$, for all $\mathbf{x} \in \mathbb{R}^n$, where A is a linear transformation of $\mathbb{R}^n$, assumed non-singular here. Let's consider a region Ω_1, and its transformes image $\Omega_2 = A\Omega_1$. Area of Ω_2 is given as

$$|\Omega_2| = \int_{\Omega_2} d\Omega_2 = \int_{\Omega_1} |A| \, d\Omega_1 = |A||\Omega_1|, \tag{2}$$

where $|A|$ is the determinant of A, and $|\Omega|$ is the area of region Ω. The center of gravity of region Ω is $\mu = \frac{1}{|\Omega|} \int_\Omega \mathbf{x} d\Omega$. The relation between the centers of gravity of transformed regions is:

$$\mu_2 = \frac{1}{|\Omega_2|} \int_{\Omega_2} \mathbf{x_2} \, d\Omega_2 = \frac{1}{|A||\Omega_1|} \int_{\Omega_1} (A^T \mathbf{x_1} + \mathbf{t})|A| \, d\Omega_1$$

$$= A^T \frac{1}{|\Omega_1|} \int_{\Omega_1} \mathbf{x_1} \, d\Omega_1 + \frac{1}{|\Omega_1|} \int_{\Omega_1} \mathbf{t} \, d\Omega_1 = A^T \mu_1 + \mathbf{t} \tag{3}$$

so the center of gravity changes covariantly with the affine transform. The covariance matrix Σ of a region Ω is a 2x2 matrix defined as $\Sigma = \frac{1}{|\Omega|} \int_\Omega (\mathbf{x} - \mu)$

$(\mathbf{x} - \mu)^T \, d\Omega$. Covariance matrix of a transformed region Ω_2 is then

$$
\begin{aligned}
\Sigma_2 &= \frac{1}{|\Omega_2|} \int_{\Omega_2} (\mathbf{x_2} - \mu_2)(\mathbf{x_2} - \mu_2)^T \, d\Omega_2 \\
&= \frac{1}{|A||\Omega_1|} \int_{\Omega_1} (A^T \mathbf{x_1} + \mathbf{t} - (A^T \mu_1 + \mathbf{t}))(A^T \mathbf{x_1} + \mathbf{t} - (A^T \mu_1 + \mathbf{t}))^T |A| \, d\Omega_1 \\
&= \frac{1}{|\Omega_1|} \int_{\Omega_1} (A^T(\mathbf{x_1} - \mu_1))(A^T(\mathbf{x_1} - \mu_1))^T \, d\Omega_1 \\
&= A^T \left(\frac{1}{|\Omega_1|} \int_{\Omega_1} (\mathbf{x_1} - \mu_1)(\mathbf{x_1} - \mu_1)^T \, d\Omega_1 \right) A = A^T \Sigma_1 A
\end{aligned}
\tag{4}
$$

Cholesky decomposition of a symmetric and positive-definite matrix Σ is a factorization $\Sigma = U^T U$, where U is an upper triangular matrix. Cholesky decomposition is defined up to a rotation, since $U^T U = U^T R^T R U$ for any rotation R. For the decomposition of covariance matrix of a transformed region we write

$$
\Sigma_2 = U_2^T R_2^T R_2 U_2 = A^T \Sigma_1 A = A^T U_1^T R_1^T R_1 U_1 A
\tag{5}
$$

thus

$$
R_2 U_2 = R_1 U_1 A \qquad\qquad U_2 = R_2^{-1} R_1 U_1 A = R U_1 A
\tag{6}
$$

Hence the triangular matrix U, obtained through the cholesky-decomposition of a covariance matrix Σ, is covariant, up to an arbitrary orthonormal matrix R, with the affine transform applied to the region.

A Ranking Algorithm Using Dynamic Clustering for Content-Based Image Retrieval*

Gunhan Park, Yunju Baek, and Heung-Kyu Lee

Division of Computer Science
Department of Electrical Engineering & Computer Science,
Korea Advanced Institute of Science and Technology
373-1 Kusung-Dong Yusong-Gu Taejon, 305-701, Republic of Korea
{gunhan,yunju,hklee}@rtlab.kaist.ac.kr

Abstract. In this paper, we propose a ranking algorithm using dynamic clustering for content-based image retrieval(CBIR). In conventional CBIR systems, it is often observed that visually dissimilar images to the query image are located at high ranking. To remedy this problem, we utilize similarity relationship of retrieved results via dynamic clustering. In the first step of our method, images are retrieved using visual feature such as color histogram, etc. Next, the retrieved images are analyzed using a HACM(Hierarchical Agglomerative Clustering Method) and the ranking of results is adjusted according to distance from a cluster representative to a query. We show the experimental results based on MPEG-7 color test images. According to our experiments, the proposed method achieves more than 10 % improvements of retrieval effectiveness in ANMRR(Average Normalized Modified Retrieval Rank) performance measure.

1 Introduction

According as multimedia data increases in recent years, effective and efficient methods for storing and retrieving multimedia data have been required. In particular, images are used as important information representation in a many variety of areas such as medicine, entertainment, education, trademark, fashion design, manufacturing, etc. Over the previous years, techniques for content-based image retrieval(CBIR) from image collection have been studied.

In conventional content-based image retrieval systems, images are represented by visual features, and the retrieval process is performed as calculating similarity between visual features of the query image and images from database. The retrieved results are shown as orders by similarity ranking algorithm in the practical CBIR system. Users evaluate the performance of the system by the ranked results. A ranking algorithm is an important component for CBIR systems. Unfortunately, it is often observed that visually dissimilar images have

---

* This work was supported by the Korea Science and Engineering Foundation (KOSEF) through the Advanced Information Technology Research Center(AITrc)

M. S. Lew, N. Sebe, and J. P. Eakins (Eds.): CIVR 2002, LNCS 2383, pp. 328–337, 2002.

higher ranking than visually similar images in the conventional CBIR systems[1]; we call it *ranking inversion*.

To remedy the ranking inversion problem, we re-calculate the similarity distance by grouping and analyzing retrieved results. Retrieval results can be classified into some sub-groups via dynamic clustering. The formed groups should have a high degree of association between members of the same group and a low degree between members of different groups.

In this paper, we use 2-step methods for improving retrieval performance. In the first step of our method, images are retrieved using visual features. Next, the retrieved images are analyzed using clustering, and adjusted similarity according to distance from a cluster representative to a query. In experiments, we show that the application of dynamic clustering over retrieved results can significantly improve retrieval performance in CBIR systems.

The rest of the paper is organized as follows. In section 2, we explain about an image retrieval model. In section 3, we define the clustering method for CBIR systems and describe steps of a ranking algorithm using dynamic clustering and its advantages in CBIR system. In section 4, we explain experimental environments, results, and performance evaluations. Finally, we conclude in section 5.

2   Image Retrieval Model

There are many visual features such as color, texture, shape, etc. for CBIR systems. Generally, visual features can be represented as a vector in a n-dimensional vector space. We denote the images as a feature vector as follows,

$$I = (f_1, f_2, f_3, \cdots, f_n) \tag{1}$$

where f_n is the element of visual features.

We can define similarity functions as one of vector space distance models, and the definition is as follows,

$$D(I, I') = \sum_{k=1}^{n} d(f_k, f'_k) \tag{2}$$

$d()$ function is one of that similarity measures such as absolute difference($L1$ norm)[2], square root($L2$ norm), quadratic distance($L2$-related norm)[3], and so on.

Using these definitions, the typical ranking method in image retrieval is defined as follows,

$$\text{Sorting}\{D(I, I')\} \text{ where } I = \text{ query image} , I' \in \text{ image DB.} \tag{3}$$

Our goal is to improve the $D()$ function in equation (2) via clustering analysis of retrieved results.

3   The Ranking Algorithm Using Dynamic Clustering

A typical CBIR system retrieves and ranks images according to a similarity function based on a feature vector distance model. In this paper, we define another properties in deciding ranking of results. As we use dynamic clustering methods about retrieved images, we make relevant groups that contain similar images. Using the groups, we analyze similarity relationship of retrieved results and the query image. The ranking and the similarity value of retrieved images are adjusted according to the cluster analysis.

The proposed method is depicted in Fig. 1. In the first step, we calculate difference value using vector distance between visual feature vectors. In the second step, we apply the dynamic clustering method to the retrieved results of the first step and the query image. We make a tree structure of hierarchical agglomerative cluster. We select a cluster representative to compute distance between query image and the cluster. After we investigate cluster analysis, we adjust the similarity distance from the query image.

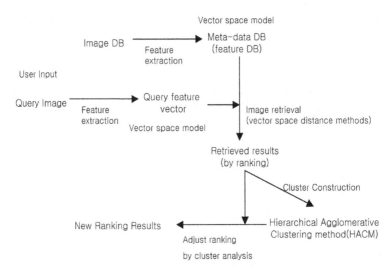

Fig. 1. Overview of the proposed method

There has been some research on how to employ clustering to improve retrieval performance in document information retrieval field. Hearst[6] shows that clustering method is effective in browsing retrieved results using document subgroups and summary text. Lee[7] shows that document clustering improve performance significantly. These results show that clustering is an effective method in document information retrieval. In case of CBIR systems, as images are represented as vectors, clustering can be an effective factor for retrieval performance improvement.

In general, there are two kinds of method in clustering methods. One is a hierarchical method, and the other is a non-hierarchical method. And the hierarchical method has two kinds of methods, agglomerative and divisive. In non-hierarchical clustering methods, the initial value and the number of clusters must be predefined. Non-hierarchical clustering method shows good performance for fixed number of clusters. But the flexible clustering method is needed for information retrieval. As these reasons, hierarchical agglomerative clustering methods(HACM) are frequently used in information retrieval field[6].

The cluster structure resulting from a HACM is made as a tree structure : *dendrogram*. There are several methods to make a dendrogram for HACM such as single link, complete link, Ward'd method, etc. In this paper, we use *Ward's method* for constructing dendrogram based on stored similarity matrix and Lance-Williams update formula[5]. In Ward's method, images join the cluster that minimizes the increase in the total within-group error sum of squares. It relatively makes an un-biased clustering tree.

After dynamic clustering using retrieved results, we modify the distance value of the results according to a equation (4).

$$D'(I, I') = \alpha D(I, I') + \beta Dc(I, I') \tag{4}$$

where $Dc(I, I')$ is distance from the query image to the cluster I'.

According to clustering hypothesis, more similar images are divided into same cluster and irrelevant images are divided into different cluster. Distance value is adjusted, as images in same cluster with query image have small distance value, otherwise images have large distance value as shown in equation (4).

In our method, clustering result is very important. There are some factors in considering. The factors are as follows,

- Cut-off size(N) : how much high rank images are clustered?
- Clustering construction threshold(T) : what is the best dividing value at agglomerative hierarchical clustering tree?
- Combining value : what is the more important element in distance computation?(α, β in equation (4))

Considering to these factors, we evaluate and investigate the effect of various methods and parameters in the next section.

4   Experimental Results

4.1   Experimental Environments

In our experiments, we use around 5,000 images from MPEG-7 experimental data set to form the image database. This test set includes a variety of still images which include stock photo galleries, screen shots of television programs, and animations etc. In our experiments, number of queries was about 1% of the number of images in the database. A set of 50 common color queries(query sets), each with specified ground truth images(manually predefined truth images), is used.

We use color histogram as the visual features for first retrieval step; 128 color bins of HSV color space is used. The distance measure in our experiments is $L1$ norm[2].

4.2 Retrieval Effectiveness Evaluation Measure: ANMRR

For performance evaluation, there is no standard measure like PSNR in image processing. There are several evaluation measures such as precision/recall graph, simple ranking method, precision/recall with scope, etc. for CBIR systems. In our experiments, we use a kind of ranking measure, ANMRR(Average Normalized Modified Retrieval Rank) that is defined from MPEG-7 research group. The ANMRR value is a normalized ranking method. This value is defined as follows[8,9].

First, we denote NG(q), K(q), R(k) as follows,

- $NG(q)$: the number of the ground truth images for a query q.
- $K(q) = min(4 * NG(q), 2 * GTM)$, Where GTM is max$\{NG(q)\}$ for all q's.
- $R(k)$ = rank of an image k in retrieval results.

Rank(k) is defined as follows,

$$Rank(k) = \begin{cases} R(k) & if\ R(k) \leq K(q) \\ (K+1)\ if\ R(k) > K(q) \end{cases} \tag{5}$$

Using equation (5), AVR(Average Rank) for query q is defined as follows:

$$AVR(q) = \sum_{k=1}^{NG(q)} \frac{Rank(k)}{NG(q)} \tag{6}$$

However, with ground truth sets of different size, the AVR value depends on $NG(q)$. To minimize the influence of variations in $NG(q)$, MRR(Modified Retrieval Rank) is defined as follows,

$$MRR(q) = AVR(q) - 0.5 - \frac{NG(q)}{2} \tag{7}$$

The upper bound of MRR depends on $NG(q)$. To normalize this value, NMRR(Normalized Modified Retrieval Rank) is defines as follows,

$$NMRR(q) = \frac{MRR(q)}{K + 0.5 - 0.5 * NG(q)} \tag{8}$$

NMRR(q) has values between 0(perfect retrieval) and 1(nothing found). And evaluation measure value for whole set over query sets, ANMRR(Average Normalized Modified Retrieval Rank) is defined as follows,

$$ANMRR(q) = \frac{1}{Q} \sum_{q=1}^{Q} NMRR(q) \tag{9}$$

4.3   Results

The goal of the experiments is to validate the proposed method. In order to evaluate the performance of the proposed method, we change the parameters of clustering, clustering threshold(T), distance weight(β), and cut-off size(N). We use a centroid of cluster as a cluster representative for calculating the distance from a cluser to a query. The results are shown in Table 1 - Table 3. The ANMRR value of an initial method, that is, the method without clustering is 0.904. The experimental results show that the ANMRR value is improved by more than 10% comparing to that of an initial method.

In table 1, the T value represents the same meaning of the number of clusters. The smaller T value has the larger number of clusters. As shown in table 1, too small T value or too large T value cannot influence the improvement of the overall performance of the system. In other words, too small number of clusters or too large number of clusters show same results of the method without clustering. Table 2 shows the influence and the weights of cluster analysis and first step retrieval method. In experiments, we first fix the α value, and then adjust β value. The ANMRR value is similar for any β values in table 2. But in case of large β value, the result is not improved comparing to the method without clustering. These results show the only cluster analysis cannot improve performance of systems. In table 3, we change the cut-off size. In case of small cut-off size, the performance is not improved because the system cannot perform the cluster analysis using small retrieved results. Also, too large cut-off size cannot improve performance of systems because the clustering results contain many irrelevant images.

Table 1. Comparison of the performance of different T values, where N=100, β=0.5

T	0.6	0.8	1.0	1.2	1.4	1.6	1.8
ANMRR	0.0858	0.0795	0.0816	0.0818	0.0823	0.0793	0.0801
improvement	+5%	+12%	+9%	+9.5%	+8.9%	+12.2%	+11.3%

Table 2. Comparison of the performance of different β values, where $T = 1.6$

β	1.0	0.75	0.35	0.25
ANMRR	0.0826	0.0799	0.0777	0.0814
improvement	+8.6%	+11.6%	+14%	+9.9%

Table 3. Comparison of the performance of different N values, where $T = 1.6$, $\beta = 0.35$

N	80	120	140	160	180	200
ANMRR	0.0810	0.0788	0.0782	0.0784	0.0777	0.0826
improvement	+10.3%	+12.8%	+13.4%	+13.2%	+14%	+8.6%

The retrieval examples without clustering and our method are shown in Fig. 2, Fig. 3. It is clear that relevant images to the query are located at higher rank in the proposed method than the method without clustering. See the image at rank 3, rank 10 in Fig. 2(b). and rank 7, rank 8 in Fig. 3(b). It shows visually performance improvement of the proposed method.

The results indicate significant performance improvement using the dynamic clustering mechanism in CBIR systems. As analyzing experimental results, we show evidence validating our method is effective in CBIR systems.

5   Conclusions

In this paper, we present an efficient ranking algorithm using dynamic clustering for image retrieval. Experimental results show that our method improves more than 10% retrieval performance in ANMRR measure. In the future work, we will use several visual features such as texture, shape, color layout features and different clustering strategy.

References

1. Yong Rui, Thomas S. Huang, and Shih-Fu Chang, "Image Retrieval : Current Techniques, Promising Directions, and Open Issues," *Journal of Visual Communication and Image Presentation*, Vol 10, pp. 39 62, March, 1999. 329
2. Michael J. Swain and Dana H. Ballad, "Color Indexing," *International journal of computer vision*, Vol. 7, No. 1, p.11 - p.32, 1991. 329, 332
3. Christos Faloutsos, Ron Barber, Myron Flickner, Jim Hafner, Wayne Niblack, Dragutin Petkovic and Will Equitz "Efficient and Effective Querying by Image Content," *Journal of Intelligent Information Systems*, 3, 3/4, pp. 231-262, July 1994. 329
4. J. R Smith, "Integrated Spatial and Feature Image Systems : Retrieval, Analysis and Compression," Doctoral Dissertations, Columbia University, 1997.
5. William B. Frakes, and Ricardo Baeza-Yates, "Information Retrieval : Data Structures & Algorithms," Prentice Hall, 1992. 331
6. Marti A. Hearst, and Jan O. Pederson, "Reexamining the Cluster Hypothesis : Scatter/Gather on Retrieval Results," *Proceedings of 19th ACM SIGIR International Conference on Research and Development in Information Retrieval*, pp. 76-84, 1996. 330, 331

7. Kyung-Soon Lee, Young-Chan Park, and Key-Sun Choi, "Re-ranking model based on document clusters," *Information Processing and Management*, vol. 37, pp.1-14, 2001. 330

8. B. S. Manjunath, Jens Rainer Ohm, Vinod V. Vasudevan, and Akio Yamada, "Color and texture descriptors," *IEEE Trans. On circuits and systems for video technology*, Vol. 11, No. 6, pp. 703 715, June 2001. 332

9. V. V. Vinod and B. S. Manjunath, *Report on AHG of color and texture*, ISO/IEC/JTC1/SC29/WG11, Doc. M5560, Maui, December 1999. 332

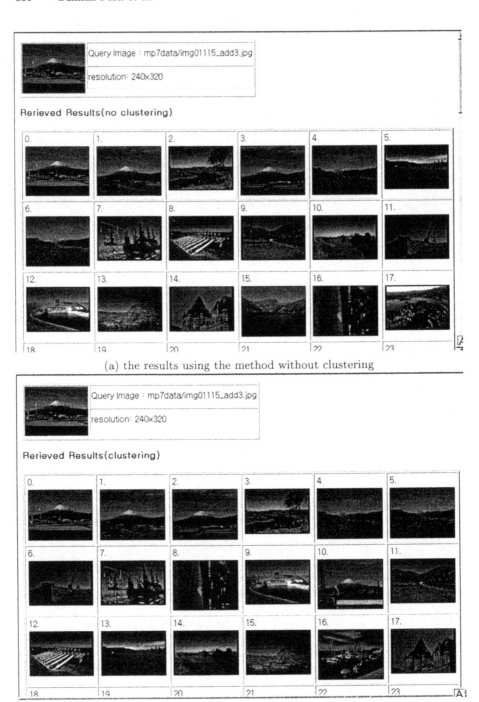

(a) the results using the method without clustering

(b) the results using the proposed method

Fig. 2. The retrieval example using query set no. 35

(a) the results using the method without clustering

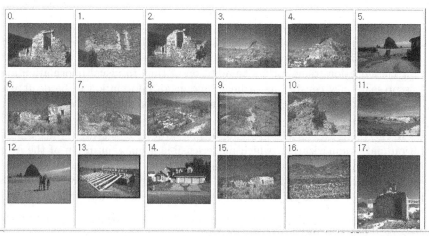

(b) the results using the proposed method

Fig. 3. The retrieval example using query set no. 50

Online Bayesian Video Summarization and Linking

Xavier Orriols and Xavier Binefa*

Centre de Visió per Computador (CVC), Dept. Informàtica
Universitat Autònoma de Barcelona
Barcelona, Spain, 08193 Bellaterra
{xevi,xavierb}@cvc.uab.es

Abstract. In this paper, an online Bayesian formulation is presented to detect and describe the most significant key-frames and shot boundaries of a video sequence. Visual information is encoded in terms of a reduced number of degrees of freedom in order to provide robustness to noise, gradual transitions, flashes, camera motion and illumination changes. We present an online algorithm where images are classified according to their appearance contents -pixel values plus shape information- in order to obtain a structured representation from sequential information. This structured representation is presented on a grid where nodes correspond to the location of the representative image for each cluster. Since the estimation process takes simultaneously into account clustering and nodes' locations in the representation space, key-frames are placed considering visual similarities among neighbors. This fact not only provides a powerful tool for video navigation but also offers an organization for posterior higher-level analysis such as identifying pieces of news, interviews, etc.

1 Introduction

In the last few years, video analysis has generated an enormous interest in Computer Vision and Pattern Recognition communities. It offers a novel and exciting challenge for applying and developing new techniques in the recognition/classification framework. The addition of time to visual information analysis presents new constraints -a huge amount of information to be dealt with- and specific demands (such as real-time analysis) on the formulation of feasible and reliable techniques.

In this paper, we focus on two important subjects in this area: video segmentation and summarization, which make feasible a quick intuition of the evolution, under a low streaming cost, of higher-level perceptual structures, such as stories, scenes or pieces of news. Shot partitioning is considered as the extraction of the basic units for video analysis. Usually, shot boundary detection has

* This work was supported by CICYT grant TEL99-1206-C02-02 and CERTAP Generalitat de Catalunya.

M. S. Lew, N. Sebe, and J. P. Eakins (Eds.): CIVR 2002, LNCS 2383, pp. 338–348, 2002.
© Springer-Verlag Berlin Heidelberg 2002

been analyzed through feature based techniques, such as: pairwise pixel comparison [15], which are very sensitive to noise, color or grayscale histogram comparison [10,13], which fails in distinguishing images with very different structures but similar color distributions, analysis of compressed streams [1,2], and local feature based techniques [14]. However, many problems arise when it comes to dealing with noise, shape information, camera motion, illumination change, fades, and flashes. In these cases, specific problem-oriented techniques are applied: camera compensation [15], illumination reduction [16]. This sort of peculiarities are also treated by combining different measures [3]: dissolve, cut and fade measures. On the other hand, appearance based methods [5] use a representation where visual changes are encoded with a reduced number of degrees of freedom, and which provide more flexibility to the system in order to tolerate camera motions or illumination changes. However, the method presented in [5] requires loading the whole sequence. Offline techniques are not practical when dealing with large pieces of video. In this sense, online solutions, combined with certain robustness to gradual transitions and a reasonable computational complexity, are rather preferable.

To this end, we present a novel algorithm that provides an online treatment of video analysis plus the mentioned advantages of working under a Bayesian appearance-based framework. We address the problems of key-frame extraction and shot partitioning relying on a feature space where not only pixel value distributions (grayscale or color) are encoded but also shape information is taken into account. The algorithm online classifies the different shots of a video sequence and automatically extracts the most significant key-frames. Often, due to post-production work (in commercials, movies, etc.), there are many sequences that contain the same shot in different time positions, that make standard algorithms to produce repeated key-frames and forcing posterior ad hoc merging/removing techniques in order to avoid unnecessary redundancies. Given that the algorithm is embedded in a Bayesian formulation, questions such as sufficient number of key-frames to represent a video sequences or avoiding extra key-frame detection due to flashes, are automatically solved.

The problems of key-frame extraction and shot partitioning are treated in terms of a probabilistic unsupervised learning approach. Each frame of a video sequence is assumed to belong to a cluster of images that are related in terms of their appearance contents, and, each cluster has a representative image that will be used for summarization purposes, i.e., a key-frame. The algorithm's process is controlled by a tuning parameter whose range embraces appearance representation [11,8,12] (such as PCA, FA or NNMF techniques) and hard clustering (competitive learning). The goal of the learning is to identify the latent variables (weights) and the unknown mapping parameters (key-frames). For this purpose, we present the estimation process in the context of the Expectation-Maximization algorithm.

The outline of the paper is as follows: first, we describe the basis of our approach in section 2. There, the model that connects a video sequence with a $2D$ graph representation is presented. We provide an EM algorithm for estimating the parameters of the model in two versions: offline and online. The experimental results analyze: (i) the effects of the tuning parameter that makes the algorithm embracing techniques compressed representation of appearance and clustering techniques, (ii) the effectiveness of dealing with prior information in order to automatically select the number of necessary key-frames to represent a video sequence.

2   The Model: Bayesian Framework

In order to describe our model, we first define the latent space were observations (images) are represented. We consider this latent space to be a grid of M nodes represented by a set of vectors $\{c_1, \ldots, c_M\}$. Each node represents a class of images that are similar in terms of their appearance contents. Consider a set of N images $\{y_1, \ldots, y_N\}$ in a vector form read in lexicographic order, and each image representation (location) on the $2D$ grid (latent variables) as $\{x_1, \ldots, x_N\}$. For each latent variable x_n we like to know the contribution to its generation for each node c_m in the grid. A measure that quantifies such a contribution can be expressed in terms of a Bayesian framework by the posterior probability $P(c_m \mid x_n)$. The model will provide the similarity measure that relates the ownership of an image under a specific class. Consider each node c_m has a representative image w_m that summarizes the images' appearance contents that belong to the m-class. In this sense, we can consider each image to be a weighed combination of the summarizing representative images w_m, where the weights correspond to the posterior probabilities $P(c_m \mid x_n)$. Therefore, we can construct a model where images and latent variables ($2D$ points on the grid) are related by the following mapping:

$$y_n = \sum_{m=1}^{M} w_m P(c_m \mid x_n) + e_n \qquad (1)$$

where e_n is gaussian independent identically distributed (idd) noise $\mathcal{N}(0, \sigma^2 I)$. Although equation (1) has the form of well known linear decompositions, PCA or FA, it is worth noting that the posterior probabilities $P(c_m \mid x_n)$. are restricted to be non-negative and the sum to be the unity. In the following sections, we show that the nature of these posterior probabilities determines whether the model can be used for performing hard clustering (key-frame extraction and shot partitioning applications) or a compressed representation of images (dimensionality reduction). The model has two main issues: on one hand, the estimation of the representative images w_m and the posterior probabilities $P(c_m \mid x_n)$, and on the other, the inference for the images' location in the latent space (grid).

Noise Model First, the noise model is expressed through a Gaussian distribution for the data: $\mathcal{N}(y_n - W h_n, \sigma^2)$, where W is a matrix whose columns are the vectors w_m and h_n is an array of the posterior probabilities for the image y_n: $h_n = (P(c_1 \mid x_n), \ldots, P(c_M \mid x_n))$. A Bayesian treatment of this model is obtained by introducing a prior distribution over the components $\{\mathbf{w}_1, \ldots, \mathbf{w}_M\}$. The key point is to control the effective number of sufficient parameters (number of classes). This is achieved by introducing a prior distribution $P(W \mid \alpha)$, where α is a M-dimensional vector of hyper-parameters $\{\alpha_1, \ldots, \alpha_M\}$. Each hyper-parameter α_m controls one of the cluster representative vector $\mathbf{w}_m$ by means of the following distribution: $P(W \mid \alpha) = \prod_{m=1}^{M} \mathcal{N}(\mathbf{w}_m, \alpha_m^{-1})$ Each hyper-parameter α_m corresponds to an inverse variance. For large values of α the corresponding $\mathbf{w}_m$ will tend to be small, and therefore, such component will be neglected. These hyper-parameters behave as switchers, activating or deactivating each component $\mathbf{w}_m$. Typically, the selection for the distribution of α_m corresponds to Laplace distributions or Gamma distributions due to their properties on "pruning". We select a gamma distribution for the hyper-parameters, since it offers a tractable analytical treatment for the estimation process: $P(\alpha) = \prod_{m=1}^{M} \Gamma(\alpha_m \mid a, b)$, where

$$\Gamma(\alpha_m \mid a, b) = \frac{b^a (\alpha_m)^{a-1} e^{-b\alpha_m}}{\Gamma(a)} \tag{2}$$

We select $a = 10^{-3}$ and $b = 10^{-3}$, which are the magnitude orders typically selected in this framework.

Modeling the Latent Space The distribution on the grid is modeled as a mixture of unimodal distributions. For instance, one might select a mixture of Gaussians, however, the algorithm that we present can be easily modified with another type of density functions (laplacians, gamma, etc.). The likelihood measure for a single point in the latent space is given by: $P(\mathbf{x}_n) = \sum_{m=1}^{M} P(\mathbf{c}_m) P(\mathbf{x}_n \mid \mathbf{c}_m)$, where we select the conditional distributions to have the form of Gaussian distributions: $P(\mathbf{x}_n \mid \mathbf{c}_m) = \mathcal{N}(x_n - \mathbf{c}_m, \tau_m^2)$.

2.1   Parameter Estimation

In this section, we present the framework where the parameters of this model are estimated. The estimation process has to take into account two issues at the same time: the noise model and the density distribution in the latent space. In the estimation procedure, there are two main steps communicated by a feedback process. Given a set of images, the cluster representative images $\mathbf{w}_m$ and the posterior probability vectors are computed in order to infer the location of each image on the grid. This is done through posterior expectation of the nodes, i.e.:

$$< x_n >= \sum_{m=1}^{M} \mathbf{c}_m P(\mathbf{c}_m \mid \mathbf{x}_n) = \sum_{m=1}^{M} \mathbf{c}_m h_n^m \tag{3}$$

These expected locations on the grid are used as data for estimating the grid's density distribution parameters, i.e., the variances τ_m^2, which play an important role as scale factors in the clusters distribution topography. In addition to this, the posterior probabilities are then recomputed using the Bayes' rule. Since these probabilities contribute to the re-estimation of the components $\mathbf{w}_m$, now, the estimation of the noise model parameters takes into account the topographical distribution of the clusters in the latent space. The following table summarizes the parameters to be estimated according to these two steps:

Model	Parameters	
Noise	$\mathbf{w}_m$	Cluster representative images
	$\mathbf{h}_n$	Posterior Probabilities
	σ^2	Noise variance
	α_m	Switchers
Latent Space	$< \mathbf{x}_n >$	Expected location
	τ_m	Node variance

We embed the estimation process in the framework of the Expectation-Maximization (EM) algorithm, which is useful to find maximum likelihood parameter estimates in problems where some variables are unobserved. In our case, posterior probabilities and posterior location points are unobserved. The M step maximizes w.r.t. the model parameters $(\mathbf{w}_m, \sigma^2, \alpha_m, \tau_m)$ and the E step maximizes it w.r.t. the distribution over the unobserved variables $(\mathbf{h}_n, < \mathbf{x}_n >)$. Typically, the algorithm consists of a set of fixed-point type equations that are iterated until convergence. In the following section, we show the procedure to estimate both sets of parameters and latent variables.

EM Algorithm The maximum likelihood estimation for the noise model parameters can be equivalently performed in terms of maximizing the logarithm of the joint distribution:

$$\mathcal{L} = -\frac{1}{2}\sum_{n=1}^{N} \mid \mathbf{y}_n - W\mathbf{h}_n \mid^2 -\frac{1}{2}\sum_{m=1}^{M}\left\{\alpha_m|\mathbf{w}_m|^2 + \frac{d}{2}\log\alpha_m - \log\Gamma(\alpha_m \mid a, b)\right\} \quad (4)$$

Given an initial guess for the parameters and unobserved variables iterate:

– **Expectation:** Find $\mathbf{h}_n$ that maximizes (4). Given the constraints for $\mathbf{h}_n$ –non-negativity and normalization– we need to apply a exponentiated gradient method [7] to ensure that the new estimates are always positive. This is done by introducing an auxiliary function as in [4]. In order to derive this update rule, we make use of an auxiliary function $G(\mathbf{h}_n, \mathbf{h^t}_n)$ such that $G(\mathbf{h}_n, \mathbf{h}_n) = \mathcal{L}(\mathbf{h}_n)$ and $G(\mathbf{h}_n, \mathbf{h^t}_n) \leq \mathcal{L}(\mathbf{h}_n)$ for all $\mathbf{h^t}_n$. For this auxiliary function, it can be seen that F is nondecreasing after the update $\mathbf{h}_n^{t+1} = \arg\max_{\mathbf{h}_n} G(\mathbf{h}_n, \mathbf{h^t}_n)$. So the update rule is given by making $\partial G(\mathbf{h}_n, \mathbf{h}_n^t)/\partial\mathbf{h}_n = 0$ on each step. An auxiliary function for $\mathcal{L}(\mathbf{h}_n)$ is constructed as,

$$G(\mathbf{h}_n, \mathbf{h}_n^{t+1}) = -\frac{1}{2}\sum_{n=1}^{N} \mid \mathbf{y}_n - W\mathbf{h}_n^{t+1} \mid^2 -\nu\sum_{k=1}^{M} h_{kn}^{t+1} \log\frac{h_{kn}}{h_{kn}^{t+1}} \quad (5)$$

which leads to the following update rule:

$$h_{kn}^{t+1} = h_{kn}\frac{\exp\left\{\frac{\nu}{2\sigma^2}[W'(\mathbf{y}_n - W\mathbf{h}_n)]_k\right\}}{\sum_{i=1}^{M} h_{in}\exp\left\{\frac{\nu}{2\sigma^2}[W'(\mathbf{y}_n - W\mathbf{h}_n)]_i\right\}} \quad (6)$$

Note that there has been introduced a scale parameter $\nu \in [0, \infty)$, which controls the degree of change from the old estimate to the new one. This parameter plays an important role, since its value determines whether the algorithm is performing vector coding (clustering when $\nu \to \infty$) or appearance encoding when $\nu \to 0$ (such as PCA, FA techniques). After estimating the posterior probabilities $\mathbf{h}_n$, the estimation of the images positions on the grid in done by means of eq. (3).

- **Maximization:** Given the inference of the positions on the grid $\mathbf{x}_n$, we can compute the node variances: $\tau_m = \frac{1}{2N} \sum_{n=1}^{N} h_{nm} \mid \mathbf{x}_n - \mathbf{c}_m \mid^2$. And therefore, the posterior probabilities under the grid model making use of the Bayes' rule. We define the new computed posterior probabilities as vectors $\varphi_n = [P(\mathbf{c}_1 \mid \mathbf{x}_n), \dots, \mathbf{c}_M \mid \mathbf{x}_n)]$. These are used for computing the cluster representative images:

$$W = \left[\sum_{n=1}^{N} \mathbf{y}_n \varphi_n' \right] \left[\sum_{n=1}^{N} \varphi_n \varphi_n' + diag(\alpha_1, \dots, \alpha_M) \right]^{-1} \qquad (7)$$

and the noise variance: $\sigma^2 = \frac{1}{ND} \sum_{n=1}^{N} \mid \mathbf{y}_n - W\varphi_n \mid^2$, where D is the images' dimension $(\mathbf{y}_n)$. Finally, the switchers are computed as follows:

$$\alpha_m = \frac{a + \frac{D}{2}}{b + \mid \mathbf{w}_m \mid^2} \qquad (8)$$

Incremental Learning The manner the algorithm has been formulated implies that all data is necessary in the M step. When dealing with large data sets, this sort of batch algorithms may incur in computational memory problems. It could be more useful updating the parameters incrementally using data points one at a time. To this end, it is shown in [9] that, under some specific conditions, an EM algorithm can be performed incrementally converging to a local maximum as well. First condition is that the joint probability distribution over the observed data factorizes, and, another condition is that the sufficient statistics can be expressed as a sum over the contribution of each single sufficient statistics' point. In our case, the problem is consistent with these conditions. Therefore the update rules for each new point are:

Expectation: Compute the posterior probabilities as in eq.(5) an (3). This step does not change.

Maximization:

$$\tau_m = \tau_m^{new} + \frac{1}{2N} h_{N+1,m}^{new} \mid \mathbf{x}_{N+1}^{new} - \mathbf{c}_m \mid^2 - \frac{1}{2N} h_{N+1,m} \mid \mathbf{x}_{N+1} - \mathbf{c}_m \mid^2$$

For a given $A = \left[\sum_{n=1}^{N} \mathbf{y}_n \varphi_n' \right]$ and $B = \left[\sum_{n=1}^{N} \varphi_n \varphi_n' \right]$ set:

$$A^{new} = A + \left[\mathbf{y}_{N+1} \varphi_{N+1}^{new'} - \mathbf{y}_{N+1} \varphi_{N+1}' \right]$$

$$B^{new} = B + \left[\varphi_{N+1}^{new'} \varphi_{N+1}^{new'} - \varphi_{N+1} \varphi'_{N+1} \right]$$
$$W^{new} = A^{new} [B^{new} + diag(\alpha_1, \ldots, \alpha_M)]^{-1}$$

and

$$[\sigma^2]^{new} = \sigma^2 + \frac{1}{ND} \mid \mathbf{y}_{N+1} - W^{new} \varphi_{N+1}^{new} \mid^2 - \frac{1}{ND} \mid \mathbf{y}_{N+1} - W \varphi_{N+1} \mid^2$$

Finally, the switchers α_m are computed as in eq.(8). These two steps are iterated for each new image $\mathbf{y}_{N+1}$.

3   Experiments

In this section, we analyze the introduced algorithm. We specially emphasize two facts: the effects of introducing switchers and the consequences of selecting a specific value for the tuning parameter ν. In the first case, we obtain from the switchers the necessary number of classes to represent a video sequence. With regard to the selection of the tuning parameter, we show how the algorithm embraces vector coding (clustering) and appearance learning encoding.

3.1   Features of the Tuning Parameter

The aim of this experiment is to show the effects of selecting a specific value for the tuning parameter. For this purpose, we chose the MNIST[6] data set of handwritten digits. In this case, we select 10 $\mathbf{w}_m$ components. Figure 1 shows two sets of images: left one corresponds to perform the learning process using a relatively high tuning parameter $\nu = 3.5$, and figure 1(b) is the result of using a low tuning parameter $\nu = 1$. Notice that, in the first case and according to equation 5, for each sample $\mathbf{y}_n$ the posterior probabilities $\mathbf{h}_n$ are forced to be close to zero except one which corresponds to the biggest one and its value is assigned to be close to the unity. This fact forces the learning process to perform hard clustering. On the other hand, when the tuning parameter is set to $\nu = 1$ the learning process is finding a compressed representation for the observed data through a sparse basis. Finally, the extreme case of $\nu = 0$ makes all the posterior probabilities to have the same value, which means that all the basis $\mathbf{w}_m$ components will be the same.

3.2   Applications to Video Analysis

The purpose of this experiment is twofold, we address the problem of characterizing key-frames basing partitions on appearance visual information criterion, and in addition to this, we show the use of the switchers that provide a measure for deciding how many components (key-frames) are necessary for a given video sequence. To this end, we chose an interview of 20 minutes (30000 frames), where there are mainly three camera shots. Note that a priori we are not assumed to

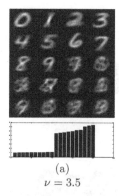

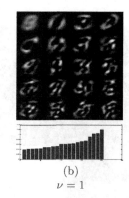

(a)   (b)

$\nu = 3.5$ $\nu = 1$

Fig. 1. Basis components $\mathbf{w}_m$ resulting of running our algorithm on the MNIST data set. (a) Hard clustering corresponds to perform the learning process using a relatively high tuning parameter $\nu = 3.5$ and (b) is the result of using a low tuning parameter $\nu = 1$

know how many shots will be on the interview. Therefore, we selected 5 possible classes. Since we are focused on clustering we selected a relatively high value for the tuning parameter $\nu = 3.5$. Figure 2 shows the results of the algorithm's performance under these conditions. First, the resulting key frames are shown in fig. 2(a), where one of them corresponds to a meaningless image. The switcher value corresponding this particular "key-frame" is the highest one (figure 2(b)); therefore as mentioned in section 2.2.1 it can be neglected. In this sense, we describe this interview in terms of four key-frames that corresponds to four different camera shots (fig. 2(a)). Adding time as a third dimension to the grid, we can have in figure 2(c) notion of the evolution of the sequence in terms of linking in time the four clusters. For visualization purposes we have skimmed the total contents of sample images in this last figure 2(c). It took 1 hour and 24 minutes to do the computation in MATLAB; however, we believe that the same algorithm under C++ will speed up the process at least in a factor of 3, providing a real time solution for video indexing and annotation. In figure 3, we analyze our algorithm when it comes to dealing with professionally produced material such as commercials. The one[1] we analyze contains flashes, illumination changes, and fast camera movements. Performing pixelwise comparison between consecutive frames flashes, many images are detected as key-frames because of rapid camera or objects movements, flashes and illumination changes. This makes the resulting storyboard to be redundant in these specific types of scenes. Adding flexibility to the system by means of codifying appearance in terms of a few degrees of freedom, we can detect representative images and determine their corresponding relevance (bottom of fig 3(a)).

---

[1] All material -videos and images- can be found at the following URL : http://www.cvc.uab.es/~xevi/videos/

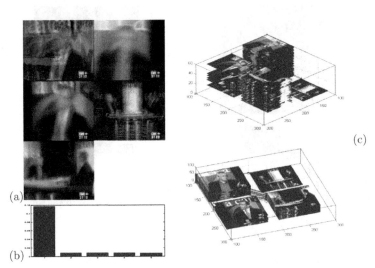

(a)

(b)

(c)

Fig. 2. (a) Cluster means, where key-frame selection can consist of picking up the closest image to each of them.(b) Switchers. (c) Adding time to the grid space

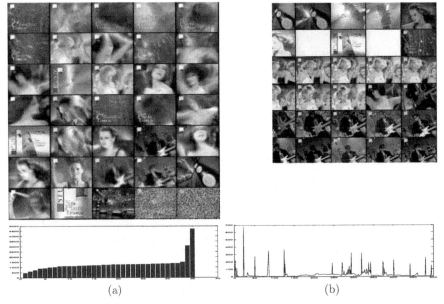

(a)   (b)

Fig. 3. (a) Cluster means extracted from a commercial using our algorithm and using pixelwise comparison (b). Note that in (a) key-frames are presented in this case by criteria of key-frames' relevance. The figure is read in lexicographic order

4 Conclusions

As an alternative to standard feature-based key-frames selection, in this paper we propose a Bayesian framework for video summarization. We address the problem of characterizing key-frames basing partitions on appearance visual information criterion. This fact, not only allows embedding in a more numerical tractable framework the video *retrieval*, but also yields a new approach to extract underlying information from temporal evolution of sequences. A suitable selection of these basic perceptual units allows the transformation of a continuous temporal data structure into a discrete meaningful one, where the intention is that the semantics remains preserved.

References

1. F. Arman, A. Hsu, and M.-Y. Chiu. Feature management for large video databases. In *Storage and Retrieval for Image and Video Databases (SPIE)*, pages 2–12, 1993. 339

2. F. Arman, A. Hsu, and M.-Y. Chiu. Image processing on compressed data for large video databases. In *Computer Graphics (Multimedia '93 Proceedings)*, pages 267–272. Addison-Wesley, 1993. 339

3. J. Corridoni and A. D. Bimbo. Structured representation and automatic indexing of movie information contents. *Pattern Recognition*, 31:2027–2045, 1998. 339

4. A. Dempster, N. Lair, and D. Rubin. Maximum likelihood from incomplete data via the EM algorithm. *Journal of the Royal Statistical Society Series B*, 39:1–38, 1977. 342

5. Y. Gong and X. Liu. Generating optimal video summaries. In *IEEE International Conference on Multimedia and Expo (III)*, pages 1559–1562, 2000. 339

6. G. E. Hinton, P. Dayan, and M. Revow. Modeling the manifolds of images of handwritten digits. *IEEE trans. on Neural Networks*, 8(1):65–74, 1997. 344

7. J. Kivinen and M. Warmuth. Additive versus exponentiated gradient updates for linear prediction. *Journal of Information and Computation*, 132:1–64, 1997. 342

8. B. Moghaddam and A. Pentland. Probabilistic visual learning for object representation. *IEEE Transactions on Pattern Analysis and Machine Intelligence*, 19(7):696–710, 1997. 339

9. R. Neal and G. Hinton. *A new view of the EM algorithm that justifies incremental and other variants*. Kluwer, 1998. 343

10. M. Smith and M. Christel. Automating video database indexing and retrieval. *ACM Int. Conf. on Multimedia*, pages 357–358, 1995. 339

11. K. Sung and T. Poggio. Example-based learning for view-based human face detection. *A.I. Memo 1521, C.B.C.L. Paper 112*, 1994. 339

12. M. Turk and A. Pentland. Eigenfaces for recognition. *Journal of Cognitive Neuroscience*, 3(1):71–86, 1991. 339

13. H. Yu and W. Wolf. A visual search system for video and image databases. *Int. Conf. on Multimedia Computing and Systems*, pages 517–524, 1997. 339

14. R. Zabih, J. Miller, and K. Mai. A feature-based algorithm for detecting and classifying scene breaks, 1995. 339

15. H. Zhang, A. Kankanhalli, and S. Smoliar. Automatic partitioning of video. *Multimedia Systems*, 1:10–28, 1993. 339

16. W. Zhao, J. Wang, D. Bhat, K. Sakiewicz, N. Nandhakumar, and W. Chang. Improving color based video shot detection. In *ICMCS, Vol. 2*, pages 752–756, 1999.

Face Detection for Video Summaries

Jean Emmanuel Viallet and Olivier Bernier

France Télécom Recherche & Développement
Technopole Anticipa, 2, avenue Pierre Marzin 22307 Lannion, France
{jeanemmanuel.viallet,olivier.bernier}@rd.francetelecom.com

Abstract. In an image, the faces of the persons are the first information looked for. Performing efficient face detection in a video with persons (excluding cartoons and nature videos) allows to classify shots, and to obtain automatically face summaries. Shot sampling greatly improves time processing. Scene layout (same number of person, similar face position and size) provides a criterion to establish a similarity measure between shots. Similar shots are gathered within shot clusters and all but one shot of a cluster are discarded from the summaries.

1   Introduction

Persons are the principal shooting operator, the major subject of shooting and also the primary concern of audience. This concern and aptitude for persons and specifically for person faces is illustrated every day in television magazines (paper or electronic) where, it is a well established convention, summaries of programs are quasi systematically illustrated with images of persons and most of the time with (cropped) close-up shots of face.

An alternative to summarizing a whole video to a unique image consists in segmenting the video in shots. Segmentation consists in finding the location and the nature of the transition between two adjacent shots and has led to numerous techniques, trimmed to the nature of the transition both in the compressed [1] and uncompressed domain [2]. Each identified shot is summarized by a keyframe [3]. Shot detection and key frame extraction rely on low level information (colour, movement) but nothing is known on the content of the shot or key frame (presence/absence of persons or of specific objects). We present a technique to summarize video using face information obtained by face detection. This technique is adequate for videos with persons but unsuitable for videos such as cartoons or nature documentaries (without faces of persons).

Sequences or scenes are narrative units, of level of abstraction greater than shots and thus scene segmentation may vary subjectively according to the director, editor or audience. Some key frames or shots can be viewed as carrying little information (intermediate shots) or similar information (alternate shots in a dialogue scene). We remove such key frames and decrease the size of the summary [4].

M. S. Lew, N. Sebe, and J. P. Eakins (Eds.): CIVR 2002, LNCS 2383, pp. 348-355, 2002.
© Springer-Verlag Berlin Heidelberg 2002

2   Face Detection

Since face represents high-level information to which humans are very sensitive, face/non face shot classification [5] and face-based summaries are relevant. Early work on face detection performed by Rowley, Pentland and others dealt with frontal face detection. Most of the work performed on face based video indexation deals with video news and face detection of anchors [6, 7]. Such videos are of particular interest since overlaid text and audio recognition efficiently contribute to indexation. As face detection is concerned, these videos are characterized by typical frontal face, waist high shot and central position when there is one anchor. The anchor usually looks at the camera, which eases the face detection process. Unfortunately, in most non-news videos, such as those dealt with in this paper, frontal views are not always available and side view face detection is needed [8, 9]; our own face detector achieves detection up to angles of 60° [10].

The performances of a face detector on a video can be evaluated by processing every image. Apart from being tedious, such an evaluation is biased since many images of a shot are highly similar. The performances are thus estimated on a shot basis (Table 1).

Since this work does not focuss on automatic shot detection, shot boundaries (cuts for the videos processed) are manually determined. A shot is manually labelled face, if at least one image of the shot exhibits a least a face; otherwise it is labelled non-face shot. A face shot is correctly classified if at least one face is detected on at least one frame of the shot. A face shot is incorrectly classified if no face is detected on any of the frames of the shot. A non-face shot is correctly (incorrectly) classified if there is no (at least one) detection is detected on any (at least one) frame of the shot.

Face detection rate, estimated on shots, if of 56% (Table I) and is below the 75% rate obtained on still images [10]. Face detection typically fails because of face orientation, size, occultation and colorimetry (when skin tone pre-filtering is implemented in order to accelerate the face detection process).

Only five non-face shots (less than 4%) are misclassified (Table 1). The false alarm rate on video is equivalent to the one obtained with still images [10] (one false alarm for every full 250 images processed). Although the obtained false alarms are highly correlated (they look alike), their temporal stability is low and no false alarm is obtained on more than two consecutive frames. This low temporal stability could be used to automatically filter the false alarms [11].

The overall rate of correctly classified shots is 85%.

2.1   Frame Sampling and Face Detection

An alternative to using a fast face detector such as the one described in [12] is to process a limited number of frames per shot. This is of particular interest for the non-face shots (the more numerous of the tested videos), which otherwise must be entirely scanned before being classified as a non-face shot. On the 185 shots of the seven videos we have processed, a 737 shot/hour rhythm is found, corresponding to an average shot length of 109 frames. Once the limits of a shot are known (obtained with an automatic shot segmentation for example), face detection is performed on frames sampled along the shot, until detection occurs; the corresponding frame represents the key frame of the shot. When no detection is found, the shot is classified as non-face and will be discarded from the video summary.

On the processed videos, the face shot detection yield only slightly increases for sampling rate greater than 3 to 4 samples per shot. A face shot is a shot where detection occurs (face or false alarm). The detection yield is the ratio of the number of face shot obtained for s samples over the number of face shot obtained when all the frames are processed. Sampling is equivalent to Group Of Picture processing for compressed video [11]. On average, only 3.65 frames are processed when a maximum of four samples per shot is selected.

Table 1. Face and non-face shot classification

	Correctly classified	Incorrectly classified
50 face shots	56%	44%
135 non face shots	96.3%	3.7%
Total 185 shots	85.4%	14.6%

3   Video Summaries

The shot summary, obtained from shot segmentation, has a number of key frames images equal to the number of detected shot (key frames are manually selected as the middle of the shots images) (Fig. 1 top). Each of the shots (and corresponding key frames) has a priori the same importance. The face shot summary (Fig. 1 bottom left), far smaller than the shot summary, only keeps the key frames where detection has occurred. The face summary collects the (cropped) images of the faces detected and discards similar faces (Fig. 1 bottom right).

A video could be summarized with the (cropped) image of the first face detected. A one face image summary limits processing time and provides a summary more interesting than the first image of a video (often a dark image or a video credits image) with which, until recently, video search engines used to summarize video before selecting a within video frame as summary [13].

From the face information knowledge, different processes may be thought off. For example, retaining the face key frame corresponding to the longest face shot is presumably preferable to selecting the key frame of the longest shot.

A more difficult process deals with selecting an image corresponding to the face detected in the greatest number of shots. Ascertaining that the face belongs to the same person [14] can be straightforward when the images have little difference (for example, top images in Fig. 1 bottom left) but is usually difficult (bottom images in Fig. 1 bottom left).

4   Scene Layout Similarity and Shot Clusters

A same person may be found in different scene corresponding either to a change of location, time or characters (for example the bottom images in Fig. 1 bottom left). A same person may also be encountered in different shots of a scene, owing to the editing technique of alternate shots or to the insertion of shots.

From one scene to another, the changes of pose, of facial expression, of light conditions and of background are among the major reasons which make face identification difficult, regardless of the fact that face recognition only succeeds with front view faces [15].

On the contrary, within a scene without camera change, the position of the face and the background do not change much.

We consider that two shots i, j have a similar scene layout and belong to the same shot cluster if the number of detected faces is the same in both shots and if the positions and scales of the detected faces have changed less than a predefined value between the two shots. Let us consider the relative variation of the horizontal position (Equation 1) and of the vertical position (Equation 2) with respect to the width w and the height h of the face, together with the relative variation of the size z (Equation 3) of the face; x and y are the image coordinates of the position of the face.

Equation (4) measures the scene layout similarity, according to our criteria, when only one person is found in the shots i and j. These shots are said to be similar, when their mutual similarity $S_{i,j}$ is greater than a given threshold. If similar, these shots are merged within a same shot cluster. Otherwise, if a shot cannot be merged to cluster, a new cluster is initiated from this shot. The value of the threshold used in the following experiment is set to 0.5 and corresponds to relative variation of lateral, vertical position and size of face of 25%.

$$X_{i,j} = 2 * \frac{|x_i - x_j|}{w_i + w_j} \tag{1}$$

$$Y_{i,j} = 2 * \frac{|y_i - y_j|}{h_i + h_j} \tag{2}$$

$$Z_{i,j} = 2 * \frac{|z_i - z_j|}{z_i + z_j} \tag{3}$$

$$S_{i,j} = \frac{1}{1 + X_{i,j}} * \frac{1}{1 + Y_{i,j}} * \frac{1}{1 + Z_{i,j}} \tag{4}$$

From the 28 shots for which faces have been detected, and the threshold value of 0.5, the shot clusters obtained are given in figure 3, and presented on a video per video basis.

The criteria used (Equations 5 to 8) to compare the effectiveness of shot segmentation techniques [16], are also used to measure the quality of the shot cluster obtained.

$$\text{Accuracy} = \frac{N_C - N_I}{N_T} = 0.77 \tag{5}$$

$$\text{Recall} = \frac{N_C}{N_T + N_D} = 0.70 \tag{6}$$

$$\text{Error rate} = \frac{N_D + N_I}{N_T + N_I} = 0.11 \tag{7}$$

$$\text{Precision} = \frac{N_C}{N_C + N_I} = 1 \tag{8}$$

The total number of cluster is estimated by the author to $N_T = 18$. This estimation is subjective and a different, although close, number of clusters could have been found by someone else. The situation is similar for shot segmentation for which there is no ground truth. For instance, in one of the videos, the first cluster is obtained in the same room as the second cluster, but with a greater field of view and a slightly different camera angle (Fig. 2 top left and Fig. 3). In another video, the first cluster corresponds to a location estimated to be different from the location of the last cluster (Fig. 2 top right and Fig. 3).

The number of correctly identified cluster is $N_C = 14$, the number of incorrectly inserted cluster is $N_I = 0$ and $N_D = 2$ is the number of incorrectly deleted clusters. The first deleted cluster corresponds to an obtained cluster, which incorrectly merges a same person in a similar position but in two different places (Fig. 2 bottom left and Fig. 3). The second deleted cluster (Fig. 2 bottom right and Fig. 3) corresponds to the cluster that incorrectly merges a man and a woman. If the colorimetry of the images had been taken into account, these two errors would have probably been dismissed as shown by keyframe clustering based on compressed chromaticity signature [17].

Keeping only one sample per cluster yields smaller video summaries. Summaries assembling (cropped) images of face (Fig. 1 bottom right) focalise on the person to the detriment of contextual information.

5   Conclusion

Face detection is a mean to obtain video summaries, which people are familiar with that is to say that focus on face information. The size of the obtained video summaries is far smaller than the standard shot summary, and even benefits from non-detected faces together with a low false alarm rate. Many of the face images are similar and can be gathered in shot clusters and discarded from the summary.

Fig. 1. Top: the standard key frame "shot" summary. Bottom left: the "shot-face" summary obtained by selecting shots where faces are detected. Bottom right: the "face" summary, keeping only facial parts of the images and discarding similar redundant faces

Fig. 2. Top left: same person and location, different frames. Top right: same person, different location and face position. Bottom left: same person, different locations and similar positions. Bottom right: different persons and locations, similar face position

Fig. 3. Face shot clusters. Each of the seven black frames corresponds to a video. Columns show the different clusters of a video and rows show the shots of a cluster, according to the chronology of the video. Only the information's on the number, size and position of faces are used and image colorimetry is not taken into account. Two clusters are incorrect: one merges a woman and a man and, in the second one, a man is first in front of a bookshelves than in front of a window. For the top video, two images of the second shot cluster, enclosed with hyphen lines, are positioned on the first row for convenience

References

1. Wang, H. L., Chang, S. F.: A Highly Efficient System for Automatic Face Region Detection in MPEG Video. CirSys Video, 7(4) (1997) 615-628
2. Demarty, C. H., Beucher, S.: Efficient morphological algorithms for video indexing. *Content-Based and Multimedia Indexing, CBMI'99* (1999)

3. Chen, J.-Y., Taskiran C., Albiol, A., Delp, E. J., Bouman, C. A.: ViBE: A Video Indexing and Browsing Environment. Proceedings of the SPIE Conference on Multimedia Storage and Archiving Systems IV, 20-22 septembre, Boston, vol. 3846 (1999) 148-164

4. Aoki, H., Shimotsuji, S., Hori, O.: A shot classification method of selecting effective keyframe for video browsing. In Proc. of ACM Int'l Conf. on Multimedia, pages 1--10, Boston, MA, November 1996

5. Chan, Y., Lin, S. H., Tan, Y. P., Kung, S. Y.: Video Shot Classification Using Human Faces. ICIP (1996) 843-846

6. Eickeler, S., Muller, S.: Content-Based Indexing of TV Broadcast News Using Hidden Markov Models. *IEEE Int. Conference on Acoustics, Speech, and Signal Processing (ICASSP)*, Phoenix, Arizona (1999)

7. Liu, Z., Wang, Y.: Face Detection and Tracking in Video Using Dynamic Programming, ICIP00 (2000) MA02.08

8. Schneidermann, H., Kanade, T.: Probabilistic Modeling of Local Appearance and Spatial Relationships for Object Recognition. IEEE Computer Vision and Pattern Recognition, Santa Barbara (1998) 45-51

9. Wei, G., Li, D., Sethi, I. K.: Detection of Side View Faces in Color Images. WACV00 (2000) 79-84

10. Féraud, R., Bernier, O., Viallet, J. E., Collobert, M.: A fast and accurate face detector based on neural networks. IEEE Trans. Pattern Analysis and Machine Intelligence, Vol. 23 (2001) 42-53

11. Wang, H., Stone, H. S., Chang, S.-F.: FaceTrack: Tracking and Summarizing Faces from Compressed Video. SPIE Multimedia Storage and Archiving Systems IV, Boston (1999)

12. Viola, P., Jones, M.: Robust Real-Time Face Detection. International Conference on Computer Vision 01 (2001) II:747

13. Altavista video search engine: http://www.altavista.com

14. Eickeler, S., Wallhoff, F., Iurgel, U., Rigoll, G.: Content-Based Indexing of Images and Video Using Face Detection and Recognition Methods. *IEEE Int. Conference on Acoustics, Speech, and Signal Processing (ICASSP)*, Salt Lake City, Utah (2001)

15. Satoh, S.: Comparative Evaluation of Face Sequence Matching for Content-based Video Access. Proc. of Int'l Conf. on Automatic Face and Gesture Recognition (FG2000) (2000) 163-168

16. Ruiloba, R., Joly, P., Marchand-Millet, S., Quenot, G.: Towards a standard protocol for the evaluation of video-to-shots segmentation algorithms. CMBI 1999 Proceedings of the European workshop on content-based-multimedia indexing, Toulouse, France (1999)

17. Drew, M. S. Au. J.: Video keyframe production by efficient clustering of compressed chromaticity signatures. ACM Multimedia '00, pp.365368, November 2000

A Method for Evaluating the Performance of Content-Based Image Retrieval Systems Based on Subjectively Determined Similarity between Images

John A. Black, Jr., Gamal Fahmy, and Sethuraman Panchanathan

Visual Computing and Communications Lab, Arizona State University
PO Box 5406, Tempe, AZ 85287-5406, USA
{john.black,fahmy,panch}@asu.edu

Abstract. In recent years multimedia researchers have attempted to design content-based image retrieval systems. However, despite the development of these systems, the term "content" has still remained rather ill defined, and this has made the evaluation of such systems problematic. This paper proposes a method for the creation of a reference image set in which the similarity of each image pair is estimated by two independent methods – by the subjective evaluation of human observers, and by the use of "visual content words" as basis vectors that allow the multidimensional content of each image to be represented with a content vector. The similarity measure computed with these content vectors is shown to correlate with the subjective judgment of human observers, and thus provides both a more objective method for evaluating and expressing image content, and a possible path to automating the process of content-based indexing in the future.

1   Introduction

The widespread availability of reasonably priced multimedia development tools has greatly increased the number of people who are creating multimedia content. The raw materials used by these people are often extracted from databases of images and video clips. While the volume of the visual material available in these databases has risen astronomically, the means of accessing this visual material has not kept pace. As a result, multimedia designers might spend as much time searching for the right images or video clips as they do in putting together their multimedia content, once they have found the right raw materials.

2   Background and Related Work

In order to provide a solution to this search problem, multimedia researchers have attempted to design *content-based image retrieval* (CBIR) systems [5], [11], [9], [4]. However, despite the development of these systems, the term "content" has still

M. S. Lew, N. Sebe, and J. P. Eakins (Eds.): CIVR 2002, LNCS 2383, pp. 356-366, 2002.

remained rather ill defined. Most researchers accept the assertion that there are multiple levels of content. For example, luminance, and color are regarded as low-level content, and physical objects (such as an automobile or a person) are regarded as high-level content. (Textures and patterns, which blend different types of content, might be regarded as mid-level content.) However, there is no broad agreement about how many levels of content can be perceived by a human, how many types of content there are in each level, or how the content of a particular image might be classified into types and levels.

Ideally, a user would like to present a query image to an image retrieval system, and ask the system to find images with "similar content." If the search is to be based on low level content, the user might specify search criteria such as color or spatial frequency. However, if the search is to be based on high level content, the query image provides the system with little information about what type (or level) of content the user considers to be important. Some researchers have emphasized the importance of prompting the user to indicate what types of content are important. MacArthur [9] describes a system that allows the user to iteratively indicate which of the retrieved images also contains the desired content, allowing the system to progressively determine the desired content by looking for similarities between the original query image and retrieved images that are validated by the user. Ma [7] takes a more direct approach by providing an algorithm that segments an image into homogeneous regions, and then allows the user to specify which regions are most relevant.

Manmatha [8] observes that objects can often be recognized by their characteristic colors, textures or luminance patterns, suggesting that low-level content can sometimes be useful for recognizing objects. He emphasizes the importance of allowing the user to indicate which portions of the query image contain the characteristic colors, textures, or luminance patterns. By indicating the most salient and distinctive portions of the query image (such as the eyes and mouth on a face, or the wheels on a car) the user provides "semantic information" to the system. Thus, while the system might not be able to recognize these high-level features, it is able to look for similar features in the image database – effectively providing a semantic search.

This implicit linkage between the low-level and high-level content in the query image is important, because most content-based image retrieval systems perform searches based on low-level (or mid-level) content, which can be extracted automatically from images or video clips. As Smith [12] observes (based on his own experience with the WebSEEK image search engine) users typically want to perform their searches based on higher-level content. If the CBIR system they are using performs its search based on low-level content, and if there is no correlation between this low-level content and the high-level content in their query image, they are likely to rate the performance of the system as less than satisfactory.

3   Limitations of Existing Techniques

Evaluating a CBIR system based on the retrieved images *after* the retrieval has been completed certainly invites subjective bias. Manmatha [8] says that the "objective of the query" should be stated by the user *before* the retrieval is done, and then the "retrieval instances are gauged against the stated objective." While this approach certainly reduces the uncertainty in the evaluator's mind about the intent of the user, the system is not provided with the same information. In a sense, it is expected to "read the user's mind" in order to know what the user is looking for. For example, a user who presents a query image with surging ocean waves might be looking for images with "water" or "inclement weather", or "foam" or "gray skies". When presenting a pastoral scene, the user might be looking for images with dirt roads, cows, tractors, fences, or partly cloudy skies.

The fact that a query image typically contains many different kinds of content makes it very difficult to objectively measure the performance of a content-based image retrieval system. The "correct" set of images might depend simultaneously on many different kinds of content, or on some particular content that the user's attention is currently focused on – and that might change over time. Should a system be judged "successful" if it finds images with similar color content, despite the fact that the user judges those images to be very dissimilar?

Smith [12] recommends that a standard set of database images and query images be provided to allow researchers to compare the performance of different CBIR systems. However, this still leaves open the question of how we decide which database images are the "correct" ones to retrieve, based on each query image. Jorgensen [3] suggests that the correct set might be determined by using the "best" results from several similarity metric algorithms, but does not specify how the performances of those algorithms will be evaluated, to determine which are best.

4   Theory

Given the fact that the performance of a CBIR system is ultimately evaluated by the human user, it seems logical to base its image indexing on the types of image content that are detected by the human visual system. Much has been learned about the early visual system in recent years, and several researchers have used this information to construct models that gauge the similarity of images [2], [13], [1]. Most of these models are based on low-level content (such as color, shape, or luminance distribution) because these types of content are extracted by the early visual system, and can be automatically extracted from images. Frese [2] (after presenting his own image similarity metric based on a multiscale model of the human visual system that employs channels for color, luminance contrast, color contrast, and orientation) concludes that his model performs "better than conventional methods".

4.1   Using Higher Level Content for CBIR

Texture is a higher level of content than luminance or color, and some distinctive textures are readily linked to higher level content – sometimes even allowing object recognition. (The dimpled surface of a golf ball or the textured surface of an orange are two examples.) However, texture is a multidimensional feature, and vision researchers have not yet been able to determine exactly how textures are analyzed in the human visual system. Thus there is only limited information for those trying to build a model. Rao [10] used 20 subjects to rate the similarity of 30 different textures, and found three features (which he called repetition, orientation, and complexity) that were useful in characterizing the similarity of textures. However, he concluded that there is still a "need for new or improved computational measures of texture." Given the multidimensional nature of textures, there is a need for a set of versatile basis functions that can be used to characterize them.

4.2 Using Characteristic Textures as Basis Functions

During the critical period of visual development, the feature detectors in the primary visual cortex are shaped largely by the visual environment. In fact, exposure to an impoverished environment during this period can lead to serious visual deficits, including the inability to perceive some low-level stimuli. There is evidence that the feature detectors in the human visual system are "tuned" to frequently-occurring visual stimuli by a process known as Hebbian learning. It therefore seems plausible that feature detectors would be particularly sensitized to the most commonly encountered textures in the visual environment. Perhaps these textures then serve as basis functions for the perception of textures later in life. In fact Leung [6] uses a clustering approach with commonly-encountered textures to derive a set of about 100 "3D textons", which he suggests can be used to characterize other textures.

By combining basis functions (such as Leung's 3D textons) it might be possible to characterize many complex textures in a manner similar to the human visual system. However, the human visual system typically characterizes and recognizes textures at the subconscious level. Because of this we often don't have adequate words to describe the textures that we perceive and recognize. Fortunately, some of the more important (and more salient) textures that we perceive become associated with higher level content, and words associated with them can be found by studying a lexicon. For example, English words that represent characteristic textures of particular objects include basket, braid, brickwork, canvas, cauliflower, concrete, feather, foam, and frost. These textures have potential for facilitating high level CBIR.

There are also adjectives that are associated with characteristics of scenes, such as alpine, aquatic, arctic, cloudy, cosmic, or dusty, and objects can be characterized as amoeboid, bleeding, colorful, or crumpled. The fact that these particular features have been given names seems to indicate that they are especially significant, and might be more perceptually salient. Thus, they might provide a means for evaluating the similarity of images.

5   An Overview of Our Approach

CBIR systems should employ similarity measures between images that parallel those of humans. One way to compare the similarity measure of a CBIR system to that of humans is to create a calibrated image database. This can be done by having humans compare all possible pairs of images within the database, assigning a similarity rating to each pair. As a practical matter, however, human evaluators would find it very tedious to compare and subjectively rate the similarity between hundreds (or even thousands) of image pairs in a large image database. Given the rather long time required to complete such a task, the results might not be very consistent from beginning to end – especially if performed by a single individual. An alternative would be to have a human evaluator first scan an array containing all the images in the database to select the images that are *most similar* to a particular query image. Once the evaluator selects the most similar images, he/she can then *rank* those images, based on their relative similarity to the query image. In doing so, the human evaluator is producing the same results that a CBIR system would be expected to produce – a rank-ordered set of the most similar images. These results would then provide a basis for evaluating the performance of automated CBIR systems. Systems that produce rank orderings similar to this human-calibrated set would hopefully be judged by human users to have a higher level of performance.

To generate a basis for characterizing visual content, we have reviewed a comprehensive lexicon of the English language and selected words that represent mid-level visual content, such as dimpled, dirty, frosty, hairy, rippled, striped, or wrinkled. Each of these visual content words can then be thought of as a kind of *basis function*, with a weighting factor that can be used to represent the degree of its presence in a particular natural image. All of the weighting factors for a given image can be collectively taken to represent a *content vector*. The content vectors representing two different images can be compared to determine the visual similarity of the two images. For example, the dot product of two content vectors could be used to provide a single scalar value representing the similarity of the two images.

The success of this approach depends upon whether the words chosen as basis functions adequately represent the multidimensional space of image content. More particularly, the question is whether those basis functions adequately represent the portion of the visual content space that can be perceived by humans. To answer this question, we conceived an experiment that allows us to compare the similarity of content represented by a set of basis functions to the similarity subjectively perceived by human observers.

6   The Experimental Procedure

The experiment consisted of two independent procedures that were performed by different participants. Both of these procedures used the same image set (called the **NaturePix** image set) which consists of 94 images of outdoor scenery – most of which were downloaded from the California Department of Water Resources website. These images contain a wide variety of surface textures, as well as pictorial textures (such as clouds and foliage) but no foreground objects except those used to "frame" the pictures.

Procedure 1 was performed by 53 different participants. Each of these participants was given a hardcopy of one of the 94 images from the **NaturePix** image set, and was then asked to (1) subjectively compare the visual content of that reference image to all 94 images in the set, (2) select the 20 most similar images (which included one image that was identical to the reference image) and (3) rank order those 20 images based on their relative similarity to the reference image. (Most of the participants repeated this procedure for 6 different reference images.)

Three trials were performed. During each trial Procedure 1 was iterated 94 times – in each iteration a different image from the **NaturePix** image set was used as a reference image. Thus, during each trial the participants chose 20 "similar" images for each of the 94 images in the **NaturePix** image set.

Procedure 2 was performed by a single participant, and employed a list of 178 words that had previously been selected by the investigators from an English lexicon containing over 230,000 words. These words were chosen based on their perceived usefulness in describing the content of natural scenery. The participant was given each of the 94 images from the **NaturePix** image set, along with an associated "check sheet" that contained these 178 descriptive words. The participant was then asked to indicate (with a check mark) which (if any) of these words were useful for describing each particular image in the image set. Thus, at the conclusion of Procedure 2 the investigators had a list of descriptive words for each of the 94 images in the **NaturePix** image set.

7   Results

Of the 178 words originally chosen for content evaluation, only 98 were checked on the check sheets for more than one image (See Table 1). These 98 words were then used to define a 98-element *binary vector* that represented the *content* of each of the 94 images.

Dot products of these vectors were then used to represent the *dot product similarity* between each pair of images. Fig 1 shows a histogram of the dot product similarity values obtained by applying the word list in Table 1 to the **NaturePix** image set.

Table 1. The 98 content words used to compute dot product similarities

agricultural	cirrus	desertlike	icy	panorama	sunset
alpine	clouded	desolate	jungly	peaceful	terrestrial
aquatic	cloudless	dry	landscape	pebbly	thorny
aqueous	cloudscape	dusky	leafy	rippled	treelined
arboreal	clouds	dusty	lush	rocky	uphill
autumnal	cloudy	eroded	man made	sandy	verdant
barren	cold	estuarial	misty	scrubby	waterless
beautiful	colorful	evergreen	moist	shady	waterscape
bloomy	colorless	flowered	mountainous	smoky	watery
blossomy	complex	foggy	muddy	snowy	wooded
blurry	coniferous	forestlike	murky	soggy	woodsy
bone dry	contrasty	frigid	natural	stony	meandering
brambly	cool	frosty	naturelike	streamy	plowed
bubbly	cumulus	grassy	nimbus	sunbaked	
cactuslike	darkish	hazy	orderly	sunlit	
chaotic	dark	humpy	outdoor	sunny	
chilly	denuded	ice cold	overcast	sunrise	

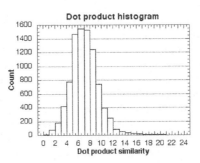

Fig. 1. A histogram of the dot product similarities generated using the word list in Table 1

These dot product similarity values were then arranged to form a 94 x 94 *similarity matrix*. That matrix is too large to present here, but Fig. 2 is a 94 x 94 grayscale image in which the value of each pixel represents a single similarity value. The pixel x coordinates (numbered left-to-right from 1 to 94, and y coordinates (numbered top-to-bottom from 1 to 94) represent the numbers of the two images being compared, and each pixel grayscale value represents the corresponding similarity value, with white representing the greatest similarity.

Unlike the dot product (which produces an *absolute* similarity value) the subjective similarity evaluation done in Procedure 1 (which produces a *rank ordering* of the 20 most similar images) provides only *relative* similarity values. Since each reference image was taken from the 94 image set, the most similar image in the set was *identical* to the reference image, and was assigned a similarity of 94. The other 19 images selected by the human evaluator were then assigned descending values of 93, 92, etc, with the least similar image being assigned a value of 75. The 74 images not chosen by the evaluator were each assigned a similarity value of zero. The resulting set of 94 values (most of which were zeros) were then used to form a single row in another 94 x 94 *subjective similarity matrix*. Figs. 2, 3, and 4 are 94 x 94 grayscale images in which the value of each pixel represents a single subjective similarity value. Fig 3 shows the matrix for Trial 1, Fig 4 shows the matrix for Trial 2, and Fig 5 shows the matrix for Trial 3.

Each of the subjective similarity matrices shown in Figs 3, 4, and 5 is *sparse*. The undefined values within each of these sparse matrices fall within a range from 1 to 74. However, we cannot be sure of the exact value of each element. (Note that we have arbitrarily assigned a value of zero to all of these undefined matrix elements for the purpose of rendering the gray scale images in Figs 3, 4 and 5.) This uncertainty makes the computation of a correlation coefficient between the dot product similarity matrix and each of the 3 subjective similarity matrices problematic.

One method for studying the correlation between the elements of these two matrices is to (1) treat the arbitrarily assigned values of zero as *actual* values, (2) pair each element of the 94 x 94 dot product similarity matrix of Fig 2 with its corresponding element in a chosen subjective similarity matrix (such as Fig 3, 4, or 5) to create 94 x 94 = 8836 pairs, (3) sort each of these pairs into a set of "bins," based on that pair's dot product similarity value, and (4) average the subjective similarity values in each bin and plot these averages against the dot product similarity of that bin. Figs 6a, 6b, and 6c show the results when this was done for Trials 1, 2 and 3, respectively.

Fig. 2.   Dot product similarity matrix

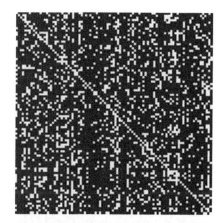

Fig. 3. Trial 1 Subjective similarity matrix

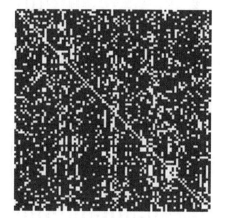

Fig. 4.   Trial 2 Subjective similarity matrix

Fig. 5. Trial 3 Subjective similarity matrix

An alternative method for studying the correlation between the elements of these two matrices is to simply *discard* all pairs of elements with an undefined subjective similarity value, before sorting the remaining pairs into the bins. Figs 6d, 6e, and 6f show the plots produced with this method.

8   Discussion

Figs 6a, 6b, and 6c all show a clear correlation between the Dot product similarity and the Subjective similarity for the **NaturePix** image set. The average Subjective similarity asymptotes to zero as the Dot similarity decreases. (This is due primarily to the predominance of the "zero" Subjective similarity values as the Dot product similarity approaches zero.) It could be argued that zero is not the best value to assign to the undefined elements in the Subjective similarity matrix, since the average value of those undefined elements is 37.5. When a value of 37.5 is used, the asymptote approaches that value instead of zero, but the overall correlation is still quite evident.

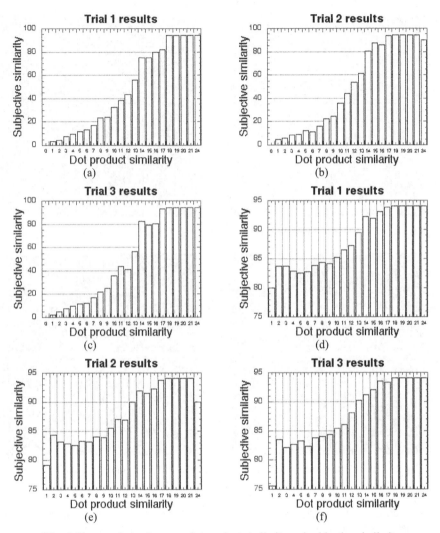

Fig. 6. The correlation between dot product similarity and subjective similarity

Figs 6d, 6e, and 6f show that the correlation persists even when the undefined subjective similarity elements are discarded. This indicates that the dot product similarity provides a useful measure of similarity not just for choosing the 20 most similar images, but also for ordering those 20 images, based on their similarity to the reference image.

Overall, these 6 plots indicate that this set of content words is useful for describing and comparing the content of the images in the **NaturePix** image set. If a content based image retrieval system could be designed to detect the content represented by these words, it would have a similarity measure for this image set that would parallel the judgment of the human participants in our experiment.

9   Conclusion

This paper has described research conducted to develop and validate an image database for evaluating the performance of content based image retrieval (CBIR) systems. This research has (1) defined a set of content types, as represented by a list of content words, (2) annotated a set of images, based on this list of content words, (3) documented an ordered set of the "most similar" images for each of the images within the set, based on subjective evaluations by human participants, and (4) measured the usefulness of the content words in gauging the similarity of the images within the set. It is anticipated that the **NaturePix** image set (along with the subjective similarity data collected in this experiment) will be useful to researchers who need to measure or compare the performance of CBIR systems. To permit the use of this resource, the entire **NaturePix** image set and the ordered subjective similarities of the images within that set have been placed in the public domain, and are available for download over the world wide web. [14]

References

1.   Chang, E. Y., Li Beitao, and Li Chen. "Toward Perception-Based Image Retrieval." *Proceedings IEEE Workshop on Content-Based Access of Image and Video Libraries*. IEEE Comput. Soc Los Alamitos CA USA, 2000. viii+119.

2.   Frese, T., C. A. Bouman, and J. P. Allebach. "Methodology for Designing Image Similarity Metrics Based on Human Visual System Models." *Proceedings of the SPIE The Intl Society for Optical Engineering* 3016 (1997): 472-83.

3.   Jorgensen, C., and R. Srihari. "Creating a Web-Based Image Database for Benchmarking Image Retrieval Systems." *Proceedings of the SPIE The International Society for Optical Engineering* 3644 (1999): 534-41.

4.   Kam, A. H., et al. "Content Based Image Retrieval through Object Extraction and Querying." *Proceedings IEEE Workshop on Content-Based Access of Image and Video Libraries*. IEEE Comput. Soc Los Alamitos CA USA, 2000. viii+119.

5.   La Cascia, M., S. Sethi, and S. Sclaroff. "Combining Textual and Visual Cues for Content-Based Image Retrieval on the World Wide Web." *Proceedings. IEEE Workshop on Content-Based Access of Image and Video Libraries* (Cat. No.98EX173). IEEE Comput. Soc Los Alamitos CA USA, 1998. viii+115.

6.   Leung, T., and J. Malik. "Recognizing Surfaces Using Three-Dimensional Textons." *Proceedings of the Seventh IEEE International Conference on Computer Vision*. IEEE Comput. Soc Los Alamitos CA USA, 1999. 2 vol.xxvii+1258.

7.   Ma, W. Y., Deng Yining, and B. S. Manjunath. "Tools for Texture/Color Based Search of Images." *Proceedings of the SPIE The International Society for Optical Engineering* 3016 (1997): 496-507.

8.   Manmatha, R., and S. S. Ravela. "Syntactic Characterization of Appearance and Its Application to Image Retrieval." *Proceedings of the SPIE The International Society for Optical Engineering* 3016 (1997): 484-95.

9. MacArthur, S. D., C. E. Brodley, and Shyu Chi Ren. "Relevance Feedback Decision Trees in Content-Based Image Retrieval." *Proceedings IEEE Workshop on Content-Based Access of Image and Video Libraries*. IEEE Comput. Soc Los Alamitos CA USA, 2000. viii+119.

10. Ravishankar Rao, A. "Identifying High Level Features of Texture Perception." *CVGIP: Graphical Models and Image Processing* 55.3 (1993): 218-33.

11. Shyu, C. R., et al. "Local Versus Global Features for Content-Based Image Retrieval." *Proceedings. IEEE Workshop on Content-Based Access of Image and Video Libraries* (Cat. No.98EX173). IEEE Comput. Soc Los Alamitos CA USA, 1998. viii+115.

12. Smith, J. R. "Image Retrieval Evaluation." *Proceedings. IEEE Workshop on Content-Based Access of Image and Video Libraries* (Cat. No.98EX173). IEEE Comput. Soc Los Alamitos CA USA, 1998. viii+115.

13. Yihong, Gong, G. Proietti, and C. Faloutsos. "Image Indexing and Retrieval Based on Human Perceptual Color Clustering." *Proceedings. 1998 IEEE Computer Society Conference on Computer Vision and Pattern Recognition* (Cat. No.98CB36231). IEEE Comput. Soc Los Alamitos CA USA, 1998. xvii+970.

14. The NaturePix reference image set is in the public domain, and may be downloaded at http://cubic.asu.edu/vccl/imagesets/naturepix.

Evaluation of Salient Point Techniques

N. Sebe[1], Q. Tian[2], E. Loupias[3], M. Lew[1], and T. Huang[2]

[1] LIACS, Leiden University
Leiden, The Netherlands
{nicu,mlew}@liacs.nl
[2] Beckman Institute, University of Illinois at Urbana-Champaign, USA
{qtian,huang}@ifp.uiuc.edu
[3] Laboratoire Reconnaissance de Formes et Vision
INSA-Lyon, France
loupias@rfv.insa-lyon.fr

Abstract. In image retrieval, global features related to color or texture are commonly used to describe the image content. The problem with this approach is that these global features cannot capture all parts of the image having different characteristics. Therefore, local computation of image information is necessary. By using salient points to represent local information, more discriminative features can be computed. In this paper we compare a wavelet-based salient point extraction algorithm with two corner detectors using the criteria: repeatability rate and information content. We also show that extracting color and texture information in the locations given by our salient points provides significantly improved results in terms of retrieval accuracy, computational complexity, and storage space of feature vectors as compared to global feature approaches.

1 Introduction

Haralick and Shapiro [1] consider a point in an image *interesting* if it has two main properties: distinctiveness and invariance. This means that a point should be distinguishable from its immediate neighbors and the position as well as the selection of the interesting point should be invariant with respect to the expected geometric and radiometric distortions. Considering these properties, Schmid and Mohr [2] proposed the use of corners as interest points in image retrieval. Different corner detectors are evaluated and compared in [3] and the authors show that the best results are provided by the Harris corner detector [4].

Corner detectors, however, were designated for robotics and shape recognition and they have drawbacks when are applied to natural images. Visual focus points do not need to be corners: when looking at a picture, we are attracted by some parts of the image, which are the most meaningful for us. We cannot assume them to be located only in corner points, as is mathematically defined in most corner detectors. For instance, a smoothed edge may have visual focus points and they are usually not detected by a corner detector. Corners also gather in textured regions. The problem is that due to efficiency reasons only a preset

M. S. Lew, N. Sebe, and J. P. Eakins (Eds.): CIVR 2002, LNCS 2383, pp. 367–377, 2002.

number of points per image can be used in the indexing process. Since in this case most of the detected points will be in a small region, the other parts of the image may not be described in the index at all.

We aim for a set of interesting points called *salient points* that are related to any visual interesting part of the image whether it is smoothed or corner-like. Moreover, to describe different parts of the image, the set of salient points should not be clustered in few regions. We believe multi-resolution representation is interesting to detect salient points. Our wavelet-based salient points [5] are detected for smoothed edges and are not gathered in texture regions. Hence, they lead to a more complete image representation than corner detectors.

We also compare our wavelet-based salient point detector with the Harris corner detectors used by Schmid and Mohr [3]. In order to evaluate the "interestingness" of the points obtained with these detectors (as was introduced by Haralick and Shapiro [1]) we compute the repeatability rate and the information content. We are also interested in using the salient points in a retrieval scenario. Therefore, in a small neighborhood around the location of each point we extract local color and texture features and use only this information in retrieval. It is quite easy to understand that using a small amount of such points instead of all image pixels reduces the amount of data to be processed. Moreover, local information extracted in the neighborhood of these particular points is assumed to be more robust to classic transformations (additive noise, affine transformations including translation, rotation, and scale effects, partial visibility).

2   Wavelet-Based Salient Points

A wavelet-based salient point detector has been presented in our previous work [6]. Here we briefly present the outline of the algorithm and we show some examples of detected salient points.

A wavelet is an oscillating and attenuated function with zero integral. We study the image f at the scales (or resolutions) $1/2$, $1/4$, ..., 2^j, $j \in Z$ and $j \leq -1$. The wavelet detail image $W_{2^j} f$ is obtained as the convolution of the image with the wavelet function dilated at different scales. We consider orthogonal wavelets with compact support. First, this assures that we have a complete and non-redundant representation of the image. Second, we know from which signal points each wavelet coefficient at the scale 2^j was computed. We can further study the wavelet coefficients for the same points at the finer scale 2^{j+1}. There is a set of coefficients at the scale 2^{j+1} computed with the same points as a coefficient $W_{2^j} f(n)$ at the scale 2^j. We call this set of coefficients the children $C(W_{2^j} f(n))$ of the coefficient $W_{2^j} f(n)$. The children set in one dimension is:

$$C(W_{2^j} f(n)) = \{W_{2^{j+1}} f(k), 2n \leq k \leq 2n + 2p - 1\} \tag{1}$$

where p is the wavelet regularity, $0 \leq n < 2^j N$, and N the length of the signal.

Each wavelet coefficient $W_{2^j} f(n)$ is computed with $2^{-j} p$ signal points. It represents their variation at the scale 2^j. Its children coefficients give the variations of some particular subsets of these points (with the number of subsets depending on the wavelet). The most salient subset is the one with the highest wavelet coefficient at the scale 2^{j+1}, that is the maximum in absolute value of

Fig. 1. Salient points extraction: spatial support of tracked coefficients

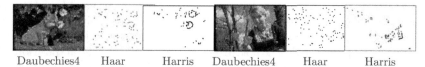

Daubechies4   Haar   Harris   Daubechies4   Haar   Harris

Fig. 2. Salient points examples. For Daubechies4 and Haar salient points are detected for smooth edges (fox image) and are not gathered in textured regions (girl image)

$C(W_{2^j} f(n))$. In our salient point extraction algorithm, we consider this maximum and look at his highest child. Applying recursively this process, we select a coefficient $W_{2^{-1}} f(n)$ at the finer resolution $1/2$. Hence, this coefficient represents $2p$ signal points. To select a salient point from this tracking, we choose among these $2p$ points the one with the highest gradient (Figure 1). We set its saliency value as the sum of the absolute value of the wavelet coefficients in the track:

$$saliency = \sum_{k=1}^{-j} |C^{(k)}(W_{2^j} f(n))|, \ -\log_2 N \leq j \leq -1 \qquad (2)$$

The tracked point and its saliency value are computed for every wavelet coefficient. A point related to a global variation has a high saliency value, since the coarse wavelet coefficients contribute to it. A finer variation also leads to an extracted point, but with a lower saliency value. We then need to threshold the saliency value, in relation to the desired number of salient points. We first obtain the points related to global variations; local variations also appear if enough salient points are requested.

The salient points extracted by this process depend on the wavelet we use. Haar is the simplest wavelet function, so is the fastest for execution. The larger the spatial support of the wavelet, the more the number of computations. Nevertheless, some localization drawbacks can appear with Haar due to its non-overlapping wavelets at a given scale. This can be avoided with the simplest overlapping wavelet, Daubechies4. Examples of salient points extracted using Daubechies4, Haar, and Harris detectors are shown in Figure 2. Note that while for Harris the salient points lead to an incomplete image representation, for the other two detectors the salient points are detected for smooth edges (as can be seen in the fox image) and are not gathered in texture regions (as can be seen in the girl image). Hence, they lead to a more complete image representation.

3   Repeatability and Information Content

Repeatability is defined by the image geometry. Given a 3D point P and two projection matrices M_1 and M_2, the projections of P into two images I_1 and I_2 are $p_1 = M_1 P$ and $p_2 = M_2 P$. The point p_1 detected in image I_1 is repeated in image I_2 if the corresponding point p_2 is detected in image I_2. To measure the repeatability, a unique relation between p_1 and p_2 has to be established. In the case of a planar scene this relation is defined by an homography: $p_2 = H_{21}p_1$.

The percentage of detected points which are repeated is the *repeatability rate*. A repeated point is not in general detected exactly at position p_2, but rather in some neighborhood of p_2. The size of this neighborhood is denoted by ε and repeatability within this neighborhood is called ε-*repeatability*. The set of point pairs (p_2, p_1) which correspond within an ε-neighborhood is $P(\varepsilon) = \{(p_2, p_1)|dist(p_2, H_{21}p_1) < \varepsilon\}$. Considering N, the total number of points detected, the repeatability rate is:

$$r(\varepsilon) = \frac{|D(\varepsilon)|}{N}, \; 0 \le r(\varepsilon) \le 1 \qquad (3)$$

We would also like to know how much average information content a salient point "has" as measured by its greylevel pattern. The more distinctive the greylevel patterns are, the larger the entropy is. In order to have rotation invariant descriptors for the patterns, we chose to characterize salient points by local greyvalue rotation invariants which are combinations of derivatives. We computed the "local jet" [7] which is consisted of the set of derivatives up to N^{th} order. These derivatives describe the intensity function locally and are computed stably by convolution with Gaussian derivatives. The local jet of order N at a point $\mathbf{x} = (x, y)$ for an image I and a scale σ is defined by: $J^N[I](\mathbf{x}, \sigma) = \{L_{i_1 \ldots i_n}(\mathbf{x}, \sigma)|(\mathbf{x}, \sigma) \in I \times R^+\}$, where $L_{i_1 \ldots i_n}(\mathbf{x}, \sigma)$ is the convolution of image I with the Gaussian derivatives $G_{i_1 \ldots i_n}(\mathbf{x}, \sigma)$, $i_k \in \{x, y\}$.

In order to obtain invariance under the group $SO(2)$ (2D image rotation), Koenderink and van Doorn [7] computed differential invariants from the local jet:

$$\boldsymbol{\nu}[0 \ldots 3] = \begin{bmatrix} L_x L_x + L_y L_y \\ L_{xx} L_x L_x + 2 L_{xy} L_x L_y + L_{yy} L_y L_y \\ L_{xx} + L_{yy} \\ L_{xx} L_{xx} + 2 L_{xy} L_{xy} + 2 L_{yy} L_{yy} \end{bmatrix} \qquad (4)$$

The computation of entropy requires a partitioning of the space of $\boldsymbol{\nu}$. Partitioning is dependent on the distance measure between descriptors and we consider the approach described by Schmid, et al. [3]. The distance we used is the Mahalanobis distance given by: $d_M(\boldsymbol{\nu}_1, \boldsymbol{\nu}_2) = \sqrt{(\boldsymbol{\nu}_1 - \boldsymbol{\nu}_2)^T \Lambda^{-1}(\boldsymbol{\nu}_1 - \boldsymbol{\nu}_2)}$, where $\boldsymbol{\nu}_1$ and $\boldsymbol{\nu}_2$ are two descriptors and Λ is the covariance of $\boldsymbol{\nu}$. The covariance matrix Λ is symmetric and positive definite. Its inverse can be decomposed into $\Lambda^{-1} = P^T D P$

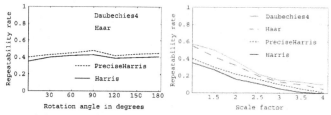

Fig. 3. Repeatability rate for image rotation (left) and scale change (right) ($\varepsilon{=}1$)

where D is diagonal and P an orthogonal matrix. Furthermore, we can define the square root of Λ^{-1} as $\Lambda^{-1/2} = D^{1/2}P$ where $D^{1/2}$ is a diagonal matrix whose coefficients are the square roots of the coefficients of D. The Mahalanobis distance can then be rewritten as: $d_M(\boldsymbol{\nu}_1, \boldsymbol{\nu}_2) = \|D^{1/2}P(\boldsymbol{\nu}_1 - \boldsymbol{\nu}_2)\|$. The distance d_M is the norm of difference of the normalized vectors: $\boldsymbol{\nu}_{norm} = D^{1/2}P\boldsymbol{\nu}$. This normalization allows us to use equally sized cells in all dimensions. This is important since the entropy is directly dependent on the partition used. The probability of each cell of this partition is used to compute the entropy of a set of vectors $\boldsymbol{\nu}$.

In the experiments we used a set of 1000 images taken from the Corel database and we compared 4 salient point detectors. In Section 2 we introduced two salient point detectors using wavelets: Haar and Daubechies4. For benchmarking purposes we also considered the Harris corner detector [4] and a variant of it called PreciseHarris, introduced by Schmid, et al. [3].

3.1   Results for Repeatability

Before we can measure the repeatability of a particular detector we first had to consider typical image alterations such as image rotation and image scaling. In both cases, for each image we extracted the salient points and then we computed the average repeatability rate over the database for each detector.

In the case of image rotation, the rotation angle varied between 0° and 180°. The repeatability rate in a $\varepsilon{=}1$ neighborhood for the rotation sequence is displayed in Figure 3.

The detectors using wavelet transform (Haar and Daubechies4) give better results compared with the other ones. Note that the results for all detectors are not very dependent on image rotation. The best results are provided by Daubechies4 detector.

In the case of scale changes, for each image we considered a sequence of images obtained from the original image by reducing the image size so that the image was aspect-ratio preserved. The largest scale factor used was 4.

The repeatability rate for scale change is presented in Figure 3. All detectors are very sensitive to scale changes. The repeatability is low for a scale factor above 2 especially for Harris and PreciseHarris detectors. The detectors based on wavelet transform provide better results compared with the other ones.

Table 1. The information content for different detectors

Detector	Entropy
Haar	6.0653
Daubechies4	6.1956
Harris	5.4337
PreciseHarris	5.6975
Random	3.124

3.2   Results for Information Content

For each detector we computed the salient points for the set of images and characterized each point by a vector of local greyvalue invariants (cf. Eq. (4)). The invariants were normalized and the entropy of the distribution was computed. The cell size in the partitioning was the same in all dimensions and it was set to 20. The σ used for computing the greylevel invariants was 3. We also considered random points in our comparison. For each image in the database we computed the mean number m of salient points extracted by different detectors and then we selected m random points using a uniform distribution.

The results are given in Table 1. The detector using the Daubechies4 wavelet transform has the highest entropy and thus the salient points obtained are the most distinctive. The results obtained for Haar wavelet transform are almost as good. The results obtained with PreciseHarris detector are better than the ones obtained with Harris but worse than the ones obtained using the wavelet transform. Moreover, the results obtained for all of the salient points detectors are significantly better than those obtained for random points. The difference between the results of Daubechies4 and random points is about a factor of two.

In summary, the most "interesting" salient points were detected using the Daubechies4 detector. These points have the highest information content and proved to be the most robust to rotation and scale changes. Therefore, in our next experiments we will consider this detector and as benchmark the PreciseHarris corner detector.

4   Content-Based Retrieval

Our next goal is to use the salient points in a content-based retrieval scenario. We consider a modular approach: the salient points are first detected for each image in the database and then feature vectors are extracted from a small neighborhood around each salient point. This approach assures the independence of the salient point extraction techniques and the feature extraction procedure and gives the user the liberty to use any features he wants for a specific application [8]. In our experiments in constructing the feature vectors we used color moments because they provide a compact characterization of color information and they are more robust and efficient in content-based retrieval than the well-known color histograms [9] and Gabor texture features because they are extensively used for texture characterization [10,11]. Of course the wavelet coefficients used during the salient point detection can be also used in constructing the feature vectors.

The number of salient points extracted will clearly influence the retrieval results. We performed experiments (not presented here due to space limitation) in which the number of salient points varied from 10 to several hundreds and found out that when using more than 50 points, the improvement in accuracy we obtained did not justify the computational effort involved. Therefore, in the experiments, 50 salient points were extracted for each image.

For feature extraction, we considered the set of pixels in a small neighborhood around each salient point. In this neighborhood we computed the color moments (in a 3×3 neighborhood) and the Gabor moments (in a 9×9 neighborhood). For convenience, this approach is denoted as the Salient W (wavelet) approach when Daubechies4 detector is used and as the Salient C (corner) approach when the PreciseHarris corner detector is used. For benchmarking purposes we also considered the results obtained using the color moments and the wavelet moments [10] extracted over the entire image (denoted as Global CW approach) and the results obtained using the color moments and the Gabor moments [11] extracted over the entire image (denoted as Global CG approach).

The overall similarity distance D_j for the jth image in the database is obtained by linearly combining the similarity distance of each individual feature:

$$D_j = \sum_i W_i S_j(f_i), \text{ with } S_j(f_i) = (\mathbf{x}_i - \mathbf{q}_i)^T (\mathbf{x}_i - \mathbf{q}_i) \text{ and } j = 1, \dots, N \quad (5)$$

where N is the total number of images in the database and $\mathbf{x}_i$ and $\mathbf{q}_i$ are the ith feature (e.g. $i = 1$ for color and $i = 2$ for texture) vector of the jth image in the database and the query, respectively. The low-level feature weights W_i for color and texture in Eq. (5) are set to be equal.

4.1   Results

In the first experiments we considered a database of 479 images (256×256 pixels in size) of color objects such as domestic objects, tools, toys, food cans, etc [12]. As ground truth we used 48 images of 8 objects taken from different camera viewpoints (6 images for a single object). Both color and texture information were used. The Salient approaches, the Global CW approach, and the Global CG approach were compared. Color moments were extracted either globally (the Global CW and Global CG) or locally (the Salient approaches). For wavelet texture representation of the Global CW approach, each input image was first fed into a wavelet filter bank and was decomposed into three wavelet levels, thus 10 de-correlated subbands. For each subband, the mean and standard deviation of the wavelet coefficients were extracted. The total number of wavelet texture features was 20. For the Salient approaches, we extracted Gabor texture features from the 9×9 neighborhood of each salient point. The dimension of the Gabor filter was 7×7 and we used 2 scales and 6 orientations/scale. The first 12 features represented the averages over the filter outputs and the last 12 features were the

Query

Salient W	1	2	4	15	21
Salient C	1	3	7	18	25
Global CW	2	7	12	25	33
Global CG	2	5	9	20	27

Fig. 4. Example of images of one object taken from different camera viewpoints and the corresponding ranks of each individual image using different approaches

corresponding variances. Note that these features were independent so that they had different ranges. Therefore, each feature was then Gaussian normalized over the entire image database. For the Global CG approach, the global Gabor texture features were extracted. The dimension of the global Gabor filter was 61×61. We extracted 36 Gabor features using 3 scales and 6 orientations/scale. The first 18 features were the averages over the filters outputs and the last 18 features were the corresponding variances.

In Figure 4 we show an example of a query image and the similar images from the database retrieved with various ranks. The Salient point approaches outperform both the Global CW approach and the Global CG approach. Even when the image was taken from a very different viewpoint, the salient points captured the object details enough so the similar image was retrieved with a good rank. The Salient W approach shows better retrieval performance than the Salient C approach. The Global CG approach provides better performance than the Global CW approach. This fact demonstrates that Gabor feature is a very good feature for texture characterization. Moreover, it should also be noted that: (1) the Salient point approaches only use the information from a very small part of the image, but still achieve a good representation of the image. For example, in our object database $9 \times 9 \times 50$ pixels were used to represent the image. Compared to the Global approaches (all 256×256 pixels were used), the Salient approaches only use less than 1/16 of the whole image pixels. (2) Compared to the Global CG approach, the Salient approaches have much less computational complexity.

Table 2 shows the retrieval accuracy for the object database. Each of the 6 images from the 8 classes was considered as query image and the average retrieval accuracy was calculated.

Table 2. Retrieval accuracy (%) using 48 images from 8 classes for object database

Top	6	10	20
Salient W	61.2	75.2	85.7
Salient C	58.9	73.8	83.2
Global CW	47.3	62.4	71.7
Global CG	58.3	73.4	82.8

Results in Table 2 show that using the salient point information the retrieval results are significantly improved (>10%) compared to the Global CW approach. When compared to the Global CG approach, the retrieval accuracy of the Salient W approach is 2.9%, 2.8%, and 2.9% higher in the top 6, 10, and 20 images, respectively. The Salient C approach has approximatively 2.5% lower retrieval accuracy comparing with the Salient W approach. Additionally, the Salient approaches have much lower computational complexity and 33.3% less storage space of feature vectors than the Global CG approach. Although the global wavelet texture features are fast to compute, their retrieval performance is much worse than the other methods. Therefore, in terms of overall retrieval accuracy, computational complexity, and storage space of feature vectors, the Salient W approach is best among all the approaches.

In our second experiments we considered a database consisted of 4013 various images covering a wide range of natural scenes such as animals, buildings, paintings, mountains, lakes, and roads. In order to perform quantitative analysis, we randomly chose 15 images from a few categories, e.g., building, flower, tiger, lion, road, forest, mountain, sunset and use each of them as queries. For each category, we measured how many hits, i.e. how many similar images to the query were returned in the top 20 retrieved images.

Figure 5 shows the average number of hits for each category using the Global CW approach, the Global CG approach, and the Salient W approach. Clearly the Salient approach has similar performance comparing with the Global CG approach and outperforms the Global CW approach when the first five categories are considered. For the last three categories, which are forest, mountain, and sunset, the global approaches (both Global CW and Global CG) perform better than the Salient approach because now the images exhibit more global characteristics and therefore, the global approaches can capture better the image content.

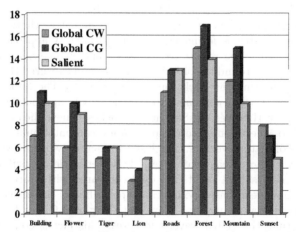

Fig. 5. The average number of hits for each category using the global color and wavelet moments (Global CW), the global color and Gabor moments (Global CG) and the Salient W approach (Salient)

As noted before, the Salient approach uses only a very small part of the image to extract the features. Therefore, comparing with the global approaches the Salient approach has much less computational complexity. Regarding the storage space of feature vectors, the number of Gabor texture features used in the Salient approach and the Global approach were 24 and 36, respectively. This does not have a big effect for small database. However, for very large image databases, the storage space used for these texture features will surely make big difference. As to the color features, both approaches have the same number of features.

5   Discussion

In this paper we compared a wavelet-based salient point extraction algorithm with two corner detectors using the criteria: repeatability rate and information content. Our points have more information content and better repeatability rate than the Harris corner detector. Moreover, the detectors have significantly more information content than randomly selected points.

We also show that extracting color and texture information in the locations given by our salient points provides significantly improved results in terms of retrieval accuracy, computational complexity, and storage space of feature vectors as compared to global feature approaches. Our salient points are interesting for image retrieval because they are located in visual focus points and therefore, they capture the local image information.

For content-based retrieval, a fixed number of salient points (50 points in this paper) were extracted for each image. Color moments and Gabor moments were extracted from the 3×3 and the 9×9 neighborhood of the salient points, respectively. For benchmark purpose, the Salient point approaches were compared to the global color and wavelet moment (Global CW) approach and the global color and Gabor moments (Global CG) approach.

Two experiments were conducted and the results show that: (1) the Salient approaches have better performance than the Global CW approach. The Salient approaches proved to be robust to the viewpoint change because the salient points were located around the object boundaries and captured the details inside the objects, neglecting the background influence; (2) The Salient approaches have similar performance compared to the Global CG approach in terms of the retrieval accuracy. However, the Salient approaches achieve the best performance in the overall considerations of retrieval accuracy, computational complexity, and storage space of feature vectors. The last two factors will have very important influence for very large image databases; (3) Better retrieval results are obtained when Daubechies4 salient points are used compared with Harris corners. This shows that our wavelet-based points can capture better the image content.

Our experimental results also show that the global Gabor features perform much better than the global wavelet features. This fact is consistent with the results of the other researchers in the field proving that Gabor features are very good candidates for texture characterization.

In conclusion, the novel contribution of this paper is in showing that a wavelet-based salient point technique beats the current leading method which uses the PreciseHarris corner detector [3] with respect to the area of content-based retrieval. In addition, we show that the wavelet-based salient point technique outperforms global feature methods, because the salient points are able to capture the local feature information and therefore, they provide a better characterization for the scene content. Moreover, the salient points are more "interesting" (as defined by Haralick and Shapiro [1]) than the Harris corner points since they are more distinctive and invariant.

In our future work, we plan to explore salient point extraction techniques which mimic the way the humans extract information in an image. This will hopefully lead to more semantically meaningful results. Moreover, we plan to extract shape information in the location of the salient points making the retrieval more accurate. We also intend to automatically determine the optimal number of the salient points needed to be extracted for each image.

References

1. Haralick, R., Shapiro, L.: Computer and Robot Vision II. Addison-Wesley (1993) 367, 368, 377
2. Schmid, C., Mohr, R.: Local grayvalue invariants for image retrieval. IEEE Trans on Patt Anal and Mach Intell **19** (1997) 530–535   367
3. Schmid, C., Mohr, R., Bauckhage, C.: Evaluation of interst point detectors. I. J. Comp. Vis. **37** (2000) 151–172   367, 368, 370, 371, 377
4. Harris, C., Stephens, M.: A combined corner and edge detector. Alvey. Vis. Conf. (1993) 147–151   367, 371
5. Loupias, E., Sebe, N.: Wavelet-based salient points: Applications to image retrieval using color and texture features. In: Visual'00. (2000) 223–232   368
6. Tian, Q., Sebe, N., Lew, M., Loupias, E., Huang, T.: Image retrieval using wavelet-based salient points. Journal of Electronic Imaging **10** (2001) 835–849   368
7. Koenderink, J., van Doorn, A.: Representation of local geometry in the visual system. Biological Cybernetics **55** (1987) 367–375   370
8. Sebe, N., Tian, Q., Loupias, E., Lew, M., Huang, T.: Color indexing using wavelet-based salient points. In: IEEE Workshop on Content-based Access of Image and Video Libraries. (2000) 15–19   372
9. Stricker, A., Orengo, M.: Similarity of color images. SPIE - Storage and Retrieval for Image and Video Databases III **2420** (1995) 381–392   373
10. Smith, J., Chang, S.F.: Transform features for texture classification and discrimination in large image databases. Int Conf on Imag Process **3** (1994) 407–411 373
11. Ma, W., Manjunath, B.: A comparison of wavelet transform features for texture image annotation. Int Conf on Imag Process **2** (1995) 256–259   373
12. Gevers, T., Smeulders, A.: PicToSeek: Combining color and shape invariant features for image retrieval. IEEE Trans Imag Process **20** (2000) 102–119   373

Personal Construct Theory as a Research Tool for Analysing User Perceptions of Photographs

Mary A. Burke

Department of Library and Information Studies, University College Dublin
Belfield, Dublin 4, Ireland
Mary.Burke@ucd.ie

Abstract. This paper describes a preliminary research project which applies Personal Construct Theory to individual user perceptions of photographs. The background to Personal Construct Theory and to the use of Repertory Grids is presented. The analysis provided by repertory grids is compared with that from various theoretical and practical approaches to indexing visual images, including 'ofness' and 'aboutness', facet analysis, iconology and MPEG7. A high level of consistency is found among the personal constructs which participants use to distinguish between photographs. The research concludes that repertory grids provide a useful method of collecting unbiased data about what users see in visual images and for comparing user perceptions with alternative retrieval vocabularies and methods. Incorporation of a participant's constructs in automatic classification systems for visual images remains a major challenge.

1   Introduction

There is considerable evidence that the description of subject content for visual images is a subjective process in which individual user characteristics play an important role [1], [2]. There are two main categories of solutions to this problem:

- intellectual analysis of content by human indexers with searchers using some combination of free-text, thesaurus terms, or classification codes.
- automatic analysis of content by computer software as exemplified by QBIC (IBM) [3].

However, neither of these reflect the variety of interpretation of subject content which may occur for a diverse user population, especially in a web-based environment with remote users of digitised collections.

This paper describes preliminary research on the application of Personal Construct Theory to the subject content analysis of photographs, using a test collection from the Irish Folklore Archive and participants from the Departments of Irish Folklore and Library and Information Studies in University College Dublin (UCD).

M. S. Lew, N. Sebe, and J. P. Eakins (Eds.): CIVR 2002, LNCS 2383, pp. 378–385, 2002.

2   Personal Construct Theory and Repertory Grids

Personal construct theory was developed by George Kelly in the mid 1950s as a contribution to Psychotherapy [4], [5], [6]. There are two main components of Personal Construct Theory: elements and constructs. Elements represent the area to be investigated. They are objects of people's thoughts, e.g., people, books, events, experiences. Constructs are qualities people attribute to those objects, e.g., kindness, level of interest, type of memory or effect on one's mood.

Kelly developed Repertory Grids as a technique for exploring the relationship between constructs and elements in a given context. They present a two-way classification of data and also facilitate the analysis of patterns and the presentation of relationships. They are created from a structured interview situation between the researcher and the participant, in which the participant generates constructs ands rates the elements on a scale between the poles of each construct.

The construct system is then analysed manually and by computer software. There are a wide range of software packages for this purpose, including WebGrid, a web-based version of RepGrid, which allows for multimedia elements and multiple options for data storage and analysis on the Knowledge Science Institute web-site at the University of Calgary [7], [8].

3   Research Project on User Perceptions of Photographs

3.1   Research Objectives

The aims of this research were both theoretical and practical. For the former, the researcher wanted to test the applicability of Personal Construct Theory (Kelly's theory and corollaries) for subject content analysis of a collection of photographs. For the latter, the aim was to use Personal Construct Theory and Repertory Grids to enhance retrieval of photographs.

The research intended to answer the following questions:

1. How similar are the constructs of the participants?
2. Is the level of similarity greater within a professional sub-group, e.g., Library and Information Scientists (LIS) and Irish Folklorists (IF)?
3. Will subject experts generate a greater number of constructs and how will these differ from those generated by generalists for the same collection?
4. How similar are the constructs of the participants to those which form the basis of the classification scheme currently used for the collection?
5. Do the constructs reflect Panofsky's primary and secondary levels of meaning [9], Shatford's 'ofness' and 'aboutness'[1], Ranganathan's facets[10], or other theoretical approaches to subject indexing?
6. Does the MPEG-7 Multilevel Indexing Pyramid accommodate these constructs [11]?

3.2   Experimental Methodology

The sample collection consisted of a set of nine digitised photographs from the Irish Folklore Archive at University College Dublin (UCD).   The photographs were selected as a stratified sample in order to represent five of the main subject categories in the in-house classification scheme.   The photographs were selected randomly within each subject category. The participants came from two sub-groups:

Students on professional post-graduate programmes in Library and Information Studies (23 participants);

Students and staff from the Department of Irish Folklore in UCD (11 participants).

The photographs were classified using the in-house classification scheme of the Folklore Archive and this categorisation was used as a baseline for comparison with each participant.

Various types of analysis were carried out on the participant's data, including a basic display of the Repertory Grid, a FOCUS cluster analysis, and a PrinCom principal components analysis using software on the WebGrid site [7].

4   Results

4.1   Summary of Constructs

Overall 264 constructs were identified, with huge diversity in emphasis and structure. These constructs were analysed for semantic and syntactic similarity and grouped under general headings. Semantic similarity was based on equivalent word meanings at both poles of the construct, e.g., 'people - no people' was deemed equivalent to 'humans - no humans'. Syntactic similarity was based on categorising the elicited constructs according to Ranganathan's PMEST formula [10]. Constructs were deemed to be equivalent if they had the same semantic and syntactic meaning; this process resulted in 73 unique constructs. It would be possible to reduce this number of unique constructs by applying less rigorous criteria for similarity but this step seemed inappropriate at this stage of the research. The most frequently occurring constructs are presented in Table 1 and the unique constructs in Table 2.

4.2   Consistency of Constructs

The researcher's initial reaction was that there was huge diversity of constructs, with very little overlap among participants. However, on analysing these constructs more closely, certain patterns emerged. The twelve most frequently used constructs, with eight or more occurrences, as shown in Table 1, account for 157 constructs from a total of 264 (59%), while the 18 most frequent constructs account for 187 individual constructs from a total of 264 (71%).

Table 1. Most popular constructs (8 or more occurences)

Construct		All (34)		LIS (23)		IF (11)	
		#	%	#	%	#	%
People	no people	24	71%	18	78%	6	55%
work	no work	16	47%	10	43%	6	55%
rural	urban	16	47%	11	48%	5	45%
photo composition (combined)		15	44%	12	52%	3	27%
land	sea	14	41%	6	26%	7	64%
activity	no activity	13	38%	7	30%	6	55%
leisure	work	13	38%	9	39%	4	36%
work type		11	32%	7	30%	4	36%
youth	age	10	29%	5	22%	5	45%
buildings	no buildings	9	26%	7	30%	2	18%
modern	historical	8	24%	6	26%	2	18%
children	no children	8	24%	7	30%	1	9%

4.3   Comparison of Constructs for Professional Sub-groups

There is a slightly greater use of frequent terms by the LIS participants than by the IF participants. Consideration of infrequent terms shows that nine of these terms came from the eleven IF participants (82%), with fourteen from the twenty-three LIS participants (61%) (Table 2). There is a clear pattern of IF participants seeing unique features in photographs, which presumably is due to their more detailed knowledge of the subject area. No difference was observed in the number of constructs generated by members of the IF and LIS sub-groups. All participants were encouraged to generate the same number of constructs, but there was no observable difference between the two sub-groups in terms of their ability or desire to continue generating constructs.

4.4   Comparison with Cataloguers and Indexers

The in-house classification scheme of the Folklore Archive was used to generate a baseline set of constructs for categorisation of the photographs. All but one of the categories used in the in-house classification scheme emerged as constructs in the experiment. However none of the clusters of photographs which the participants generated were remotely similar to the baseline. Extensive discussion with the curators of the collection indicated that the reason for this lack of similarity was the distinction between the needs of researchers in Irish Folklore and those of generalists looking at photographs. The classification scheme is designed to facilitate research by Folklorists using general subject categories.

Asking people to classify photographs in the context of specific user needs is not the same exercise as asking people to distinguish between them, dissociated from any context. Thus it is understandable that the classification scheme used in the archive does not correspond to what users from either participant group see in photographs.

Table 2. Unique constructs and users

Construct		User Sub-group
element of risk	element of security / safety	IF
emotional, engages one	less emotional	IF
homely feel	open / anonymous	IF
human	manufactured / artificial	IF
human life	nature	IF
more comfort	less comfort	IF
nature	less nature	IF
objects being made or already made	nothing being made	IF
work process	work product	IF
architectural style 1	architectural style 5	LI
concrete ground	earth	LI
couple	group	LI
exuberance	lack of exuberance	LI
female scene	male scene	LI
machine made	hand made	LI
more man-made influences	more natural	LI
nets	no nets	LI
non domestic	domestic scene	LI
non-human	human	LI
representational	evocative	LI
social conditions	working conditions	LI
value 1	value 5	LI
warmth	coldness	LI

4.5   Comparison of Constructs with Theoretical Approaches to Subject Retrieval

There is complete predominance of Panofsky's first level of analysis, corresponding to identification of primary subject matter which does not require any specialist subject knowledge.   There was no identification of specific people or places but the photographs did not lend themselves to this.

When one turns to Shatford's criteria of 'ofness' (generic and specific) and 'aboutness', there is a preponderance of 'specific of' terms (55 from 73, 75%), with 18 'aboutness' examples (25%), and no 'generic of' terms.   The 'aboutness' examples reflected the subjective meaning or role of some photographs, including constructs for mood, financial and social conditions, etc. If one takes the frequency of occurrence of each construct into consideration, there was even greater emphasis on the 'generic of' category (242 from 264, 92%).

Ranganathan's PMEST approach to facet analysis provided a sound basis for grouping the constructs, with all five types of facet represented. There was a small number of constructs which corresponded to combinations of terms from two or more facets, e.g., 'young, poor, female - none of these', or 'children playing - no children playing'. These results support the view that facet analysis has much to offer as an aid in classifying and indexing visual images.

4.6   Comparison with MPEG-7 Multilevel Pyramid

Jaimes et al. (2000) [11] describe experiments they performed for the MPEG-7 standard for describing multimedia objects. They used a ten-level pyramid to classify attributes obtained from images, video and audio. The top four levels of the pyramid deal with the syntax of the object, namely: 1, Type / Technique; 2, Global Distribution; 3, Local Structure; and 4, Global Composition. The bottom six levels of the pyramid deal with semantics, namely: 5, Generic Objects; 6, Specific Objects; 7, Abstract Objects; 8, Generic Scene; 9, Specific Scene; and 10, Abstract Scene. All of the constructs in the Irish Folklore experiment were accommodated by the MPEG-7 multilevel indexing pyramid. Levels 5, 8, and 10 occur most frequently, level 4 is also fairly common, and level 7 occurs twice. When frequency of occurrence of each construct was included, Generic Scene (116 from 264, 44%) and Generic Objects (91 from 264, 34%) become even more dominant.

5   Overall Conclusions and Recommendations

This research had both theoretical and practical objectives. It showed that Personal Construct Theory and Repertory Grids are applicable to research on what participants see in visual images. The nature of the difference between asking people to classify objects in the context of specific user needs and merely asking people to distinguish between objects out of context merits further research.

Considerable analysis has yet to be done on the data collected in this experiment. Analysis so far has focussed on the constructs; the next stage will do cross comparisons of the 34 subjects by comparing the clustering of photographs using shared constructs.

A new phase of the research would generate a core set of constructs from this experiment and use these as the basis for a new experiment with a different set of photographs and a different group of participants. This would enable testing of the independence of the constructs of a group of users and of a set of photographs and also provide a better basis for comparison of clustering by participants. The researcher already has a parallel collection of nine photographs with the same distribution across subject categories.

Alternatively, one could also allow complete freedom of constructs; it is even possible for a participant to avoid naming a construct at all and to rank photographs based on her unarticulated understanding of it. This facility proved useful already when one participant had difficulty articulating a construct, which was recorded as 'value 1 - value 5' on the grid.

A longitudinal study is needed to determine how the constructs used by an individual change as she develops expertise in her subject. It would be interesting to execute a similar experiment for a different collection, e.g., a general photographic collection without the Irish Folklore specialisation. There is tremendous potential for cross-cultural comparison, e.g., with ethnic minorities in Ireland or with international partners and for demographic and social studies.

An upgraded version of WebGrid III which enables the image of each photograph to be shown on screen during the ranking will be used in subsequent research [8].

It is difficult to assess the possible effects of the results of the research for operational visual image databases and search engines. This would require a user interface which would enable users to filter large image collections by selecting a construct value from a range of options. The range of options would be created from personal construct experiments with previous users of the collection. These would be likely to include generic and specific people, animals and objects, activities and events, and time and place.

An important next stage of the research will be to test the correspondence between the WebGrid categorisation and that of an automatic indexing system, e.g., Query by information content (QBIC) [1]. QBIC, an IBM search and access product, allows users to search a collection's media objects by colour, texture, shape and size. A query may be expressed graphically, for example, by specifying colours from a colour wheel or selecting textures from a list of sample images. Once a user finds one fairly useful item, he or she can then request other 'similar' items. The constructs could be included as a basis for determining similarity.

Personal Construct Theory and Repertory Grids have potential for enhancement of retrieval for other media, e.g., video and film material, leisure books, Web resources, etc. indeed for any materials which lend themselves to subjective interpretation. They have much to offer other areas of Library and Information Science, e.g., user studies, human resources management, etc. Many surveys of user needs and of library and information service performance would be enriched by using Repertory Grids to identify constructs which are important to users and to measure users' rankings of relevant elements for these constructs. McKnight's research on personal construction of information space is an encouraging move in this direction [12].

Personal Construct Theory provides a neutral framework for comparison of subject vocabularies and for many areas of study of human behaviour in an information society. It is encouraging that many diverse strands of research subject retrieval are converging on new applications of traditional tools to a multimedia web-based environment. The real challenge however will be to implement user interfaces which exploit this greater awareness of how users perceive photographs and other visual images.

Acknowledgements

The author would like to express her gratitude to all participants in the experimental part of this study and to colleagues in the Departments of Irish Folklore, Library and Information Studies, and Psychology in UCD who facilitated the research in many ways. Particular recognition and thanks go to Professor Brian Gaines and his colleagues at the Knowledge Science Institute, University of Calgary, for providing WebGrid software on the Web and for their gracious technical support and advice. Any errors or weaknesses in the paper are the author's sole responsibility.

References

1. Shatford, S.: Analyzing the Subject of a Picture: a Theoretical Approach. Cataloguing and Classification Quarterly **6** (3) (1986) 39-62
2. Markey, K.: Computer-Assisted Construction of a Thematic Catalog of Primary and Secondary Subject Matter. Visual Resources, **III**. (1983) 16-49
3. QBIC (IBM): [http://wwwqbic.almaden.ibm.com/]. (2001) Accessed 06/27/01
4. Kelly, G. A.: The psychology of personal constructs (Vols. 1 and 2). New York: Norton. 2nd printing (1991), Routledge, London (1955)
5. Bannister, D.: Personal Construct Theory: A Summary and Experimental Paradigm. Acta Psychologica **20**: (1968) 104-120
6. Burke, M.: Personal Construct Theory as a Research Tool in Library and Information Science. Case Study: Development of a User-Driven Classification of Photographs.   Proceedings of the Subject Retrieval in a Networked Environment IFLA Satellite Meeting. OCLC, Dublin, Ohio, 14-16 August 2001, In Print (2002)
7. Gaines, B. R. and M. L. G. Shaw: WebGrid II. [http://repgrid.com/]. (2001) Accessed 06/27/01
8. Gaines, B. R. and M. L. G. Shaw: WebGrid III. [http://gigi.cpsc.ucalgary.ca/]. (2001) Accessed 05/11/01
9. Panofsky, E.: Studies in Iconology: Humanistic Themes in the Art of the Renaissance. Harper and Row, New York (1962)
10. Ranganathan, S. R.: Elements of Library Classification. 2nd edn. Association of Assistant Librarians, London (1959)
11. Jaimes, A., Benitez, A., Jörgensen, C., & Chang, S.: Experiments in Indexing Multimedia Data at Multiple Levels. Classification for User Support and Learning. Proceedings of the 11th ASIS&T SIG/CR Classification Research Workshop. Held at the 62nd ASIS&T Annual Meeting, Chicago, IL, Nov. 12, 2000. American Society for Information Science and Technology, Silver Spring, MD (2000).163-168
12. McKnight, C.: The Personal Construction of Information Space. Journal of the American Society for Information Science, **51**(8): (2000) 730-733

Author Index

Alarcón, P. A. de 225

Baek, Y. 328
Bakker, E. M. 262, 271
Berens, J. 245
Bernier, O. 348
Binefa, X. 338
Black, Jr., J. A. 356
Bosson, A. 50
Brucale, A. 235
Bruijn, W. de 138
Burgard, W. 108
Burke, M. A. 378
Burkhardt, H. 108
Byun, H. 81

Carazo, J. M. 225
Cawley, G. C. 50
Chan, K.-L. 165
Chan, Y. 50
Choi, J. 278
Cox, G. 61

d'Amico, M. 235
Denman, H. 198

Eakins, J. P. 1, 147, 253
Enser, P. G. B. 206

Fahmy, G. 356
Fauzi, M. F. A. 91
Feng, G. 120
Ferri, M. 235
Finlayson, G. D. 245

Gualandri, L. 235

Harvey, R. 50
Howell, A. J. 129
Huang, J. 289
Huang, T. 367
Huang, T. S. 7
Hussain, M. 147

Jager, G. de 61
Jiang, J. 120

Kim, K. 278
Kim, N. 278
Kim, P. K. 299
Kim, P. 278
Ko, B. 81
Kokaram, A. 198

Lam, K.-M. 100
Lee, H.-K. 328
Leung, M.-W. 165
Lew, M. S. 1, 17, 138, 271, 367
Lewis, P. H. 91
Loupias, E. 367
Lovato, A. 235

Marchand-Maillet, S. 38
Matas, J. 318
Moghaddam, B. 7
Müller, H. 38

Navarrete, P. 157

Obdržálek, Š. 318
Oh, K. S. 299
Orriols, X. 338

Palakal, M. 289
Panchanathan, S. 356
Park, G. 328
Pascual-Montano, A. D. 225
Pickering, M. J. 309
Plataniotis, K. N. 70
Pun, T. 38

Qiu, G. 100

Rea, N. 198
Riley, K. J. 253
Rüger, S. M. 309
Ruiz-del-Solar, J. 157

Sandom, C. J. 206
Schaffalitzky, F. 186
Sebe, N. 1, 17, 262, 367
Sexton, G. 147
Sinclair, D. 309
Smeaton, A. F. 215
Smolka, B. 70
Song, Y. 29

Tian, Q. 7, 367

Umamaheswaran, D. 289

Vendrig, J. 175
Viallet, J. E. 348
Visser, R. 262

Wang, W. 29
Wolf, J. 108
Worring, M. 175

Young, D. S. 129

Zaher, A. 299
Zhang, A. 29
Zisserman, A. 186

Lecture Notes in Computer Science

For information about Vols. 1–2302
please contact your bookseller or Springer-Verlag

Vol. 2303: M. Nielsen, U. Engberg (Eds.), Foundations of Software Science and Computation Structures. Proceedings, 2002. XIII, 435 pages. 2002.

Vol. 2304: R.N. Horspool (Ed.), Compiler Construction. Proceedings, 2002. XI, 343 pages. 2002.

Vol. 2305: D. Le Métayer (Ed.), Programming Languages and Systems. Proceedings, 2002. XII, 331 pages. 2002.

Vol. 2306: R.-D. Kutsche, H. Weber (Eds.), Fundamental Approaches to Software Engineering. Proceedings, 2002. XIII, 341 pages. 2002.

Vol. 2307: C. Zhang, S. Zhang, Association Rule Mining. XII, 238 pages. 2002. (Subseries LNAI).

Vol. 2308: I.P. Vlahavas, C.D. Spyropoulos (Eds.), Methods and Applications of Artificial Intelligence. Proceedings, 2002. XIV, 514 pages. 2002. (Subseries LNAI).

Vol. 2309: A. Armando (Ed.), Frontiers of Combining Systems. Proceedings, 2002. VIII, 255 pages. 2002. (Subseries LNAI).

Vol. 2310: P. Collet, C. Fonlupt, J.-K. Hao, E. Lutton, M. Schoenauer (Eds.), Artificial Evolution. Proceedings, 2001. XI, 375 pages. 2002.

Vol. 2311: D. Bustard, W. Liu, R. Sterritt (Eds.), Soft-Ware 2002: Computing in an Imperfect World. Proceedings, 2002. XI, 359 pages. 2002.

Vol. 2312: T. Arts, M. Mohnen (Eds.), Implementation of Functional Languages. Proceedings, 2001. VII, 187 pages. 2002.

Vol. 2313: C.A. Coello Coello, A. de Albornoz, L.E. Sucar, O.Cairó Battistutti (Eds.), MICAI 2002: Advances in Artificial Intelligence. Proceedings, 2002. XIII, 548 pages. 2002. (Subseries LNAI).

Vol. 2314: S.-K. Chang, Z. Chen, S.-Y. Lee (Eds.), Recent Advances in Visual Information Systems. Proceedings, 2002. XI, 323 pages. 2002.

Vol. 2315: F. Arhab, C. Talcott (Eds.), Coordination Models and Languages. Proceedings, 2002. XI, 406 pages. 2002.

Vol. 2316: J. Domingo-Ferrer (Ed.), Inference Control in Statistical Databases. VIII, 231 pages. 2002.

Vol. 2317: M. Hegarty, B. Meyer, N. Hari Narayanan (Eds.), Diagrammatic Representation and Inference. Proceedings, 2002. XIV, 362 pages. 2002. (Subseries LNAI).

Vol. 2318: D. Bošnački, S. Leue (Eds.), Model Checking Software. Proceedings, 2002. X, 259 pages. 2002.

Vol. 2319: C. Gacek (Ed.), Software Reuse: Methods, Techniques, and Tools. Proceedings, 2002. XI, 353 pages. 2002.

Vol.2320: T. Sander (Ed.), Security and Privacy in Digital Rights Management. Proceedings, 2001. X, 245 pages. 2002.

Vol. 2321: P.L. Lanzi, W. Stolzmann, S.W. Wilson (Eds.), Advances in Learning Classifier Systems. Proceedings, 2002. VIII, 231 pages. 2002. (Subseries LNAI).

Vol. 2322: V. Mařík, O. Štěpánková, H. Krautwurmová, M. Luck (Eds.), Multi-Agent Systems and Applications II. Proceedings, 2001. XII, 377 pages. 2002. (Subseries LNAI).

Vol. 2323: À. Frohner (Ed.), Object-Oriented Technology. Proceedings, 2001. IX, 225 pages. 2002.

Vol. 2324: T. Field, P.G. Harrison, J. Bradley, U. Harder (Eds.), Computer Performance Evaluation. Proceedings, 2002. XI, 349 pages. 2002.

Vol 2326: D. Grigoras, A. Nicolau, B. Toursel, B. Folliot (Eds.), Advanced Environments, Tools, and Applications for Cluster Computing. Proceedings, 2001. XIII, 321 pages. 2002.

Vol. 2327: H.P. Zima, K. Joe, M. Sato, Y. Seo, M. Shimasaki (Eds.), High Performance Computing. Proceedings, 2002. XV, 564 pages. 2002.

Vol. 2328: R. Wyrzykowski, J. Dongarra, M. Paprzycki, J. Waśniewski (Eds.), Parallel Processing and Applied Mathematics. Proceedings, 2001. XIX, 915 pages. 2002.

Vol. 2329: P.M.A. Sloot, C.J.K. Tan, J.J. Dongarra, A.G. Hoekstra (Eds.), Computational Science – ICCS 2002. Proceedings, Part I. XLI, 1095 pages. 2002.

Vol. 2330: P.M.A. Sloot, C.J.K. Tan, J.J. Dongarra, A.G. Hoekstra (Eds.), Computational Science – ICCS 2002. Proceedings, Part II. XLI, 1115 pages. 2002.

Vol. 2331: P.M.A. Sloot, C.J.K. Tan, J.J. Dongarra, A.G. Hoekstra (Eds.), Computational Science – ICCS 2002. Proceedings, Part III. XLI, 1227 pages. 2002.

Vol. 2332: L. Knudsen (Ed.), Advances in Cryptology – EUROCRYPT 2002. Proceedings, 2002. XII, 547 pages. 2002.

Vol. 2333: J.-J.Ch. Meyer, M. Tambe (Eds.), Intelligent Agents VIII. Revised Papers, 2001. XI, 461 pages. 2001. (Subseries LNAI).

Vol. 2334: G. Carle, M. Zitterbart (Eds.), Protocols for High Speed Networks. Proceedings, 2002. X, 267 pages. 2002.

Vol. 2335: M. Butler, L. Petre, K. Sere (Eds.), Integrated Formal Methods. Proceedings, 2002. X, 401 pages. 2002.

Vol. 2336: M.-S. Chen, P.S. Yu, B. Liu (Eds.), Advances in Knowledge Discovery and Data Mining. Proceedings, 2002. XIII, 568 pages. 2002. (Subseries LNAI).

Vol. 2337: W.J. Cook, A.S. Schulz (Eds.), Integer Programming and Combinatorial Optimization. Proceedings, 2002. XI, 487 pages. 2002.

Vol. 2338: R. Cohen, B. Spencer (Eds.), Advances in Artificial Intelligence. Proceedings, 2002. XII, 373 pages. 2002. (Subseries LNAI).

Vol. 2340: N. Jonoska, N.C. Seeman (Eds.), DNA Computing. Proceedings, 2001. XI, 392 pages. 2002.

Vol. 2342: I. Horrocks, J. Hendler (Eds.), The Semantic Web – ISCW 2002. Proceedings, 2002. XVI, 476 pages. 2002.

Vol. 2345: E. Gregori, M. Conti, A.T. Campbell, G. Omidyar, M. Zukerman (Eds.), NETWORKING 2002. Proceedings, 2002. XXVI, 1256 pages. 2002.

Vol. 2346: H. Unger, T. Böhme, A. Mikler (Eds.), Innovative Internet Computing Systems. Proceedings, 2002. VIII, 251 pages. 2002.

Vol. 2347: P. De Bra, P. Brusilovsky, R. Conejo (Eds.), Adaptive Hypermedia and Adaptive Web-Based Systems. Proceedings, 2002. XV, 615 pages. 2002.

Vol. 2348: A. Banks Pidduck, J. Mylopoulos, C.C. Woo, M. Tamer Ozsu (Eds.), Advanced Information Systems Engineering. Proceedings, 2002. XIV, 799 pages. 2002.

Vol. 2349: J. Kontio, R. Conradi (Eds.), Software Quality – ECSQ 2002. Proceedings, 2002. XIV, 363 pages. 2002.

Vol. 2350: A. Heyden, G. Sparr, M. Nielsen, P. Johansen (Eds.), Computer Vision – ECCV 2002. Proceedings, Part I. XXVIII, 817 pages. 2002.

Vol. 2351: A. Heyden, G. Sparr, M. Nielsen, P. Johansen (Eds.), Computer Vision – ECCV 2002. Proceedings, Part II. XXVIII, 903 pages. 2002.

Vol. 2352: A. Heyden, G. Sparr, M. Nielsen, P. Johansen (Eds.), Computer Vision – ECCV 2002. Proceedings, Part III. XXVIII, 919 pages. 2002.

Vol. 2353: A. Heyden, G. Sparr, M. Nielsen, P. Johansen (Eds.), Computer Vision – ECCV 2002. Proceedings, Part IV. XXVIII, 841 pages. 2002.

Vol. 2355: M. Matsui (Ed.), Fast Software Encryption. Proceedings, 2001. VIII, 169 pages. 2001.

Vol. 2358: T. Hendtlass, M. Ali (Eds.), Developments in Applied Artificial Intelligence. Proceedings, 2002 XIII, 833 pages. 2002. (Subseries LNAI).

Vol. 2359: M. Tistarelli, J. Bigun, A.K. Jain (Eds.), Biometric Authentication. Proceedings, 2002. X, 197 pages. 2002.

Vol. 2360: J. Esparza, C. Lakos (Eds.), Application and Theory of Petri Nets 2002. Proceedings, 2002. X, 445 pages. 2002.

Vol. 2361: J. Blieberger, A. Strohmeier (Eds.), Reliable Software Technologies – Ada-Europe 2002. Proceedings, 2002 XIII, 367 pages. 2002.

Vol. 2363: S.A. Cerri, G. Gouardères, F. Paraguaçu (Eds.), Intelligent Tutoring Systems. Proceedings, 2002. XXVIII, 1016 pages. 2002.

Vol. 2364: F. Roli, J. Kittler (Eds.), Multiple Classifier Systems. Proceedings, 2002. XI, 337 pages. 2002.

Vol. 2366: M.-S. Hacid, Z.W. Raś, D.A. Zighed, Y. Kodratoff (Eds.), Foundations of Intelligent Systems. Proceedings, 2002. XII, 614 pages. 2002. (Subseries LNAI).

Vol. 2367: J. Fagerholm, J. Haataja, J. Järvinen, M. Lyly. P. Råback, V. Savolainen (Eds.), Applied Parallel Computing. Proceedings, 2002. XIV, 612 pages. 2002.

Vol. 2368: M. Penttonen, E. Meineche Schmidt (Eds.), Algorithm Theory – SWAT 2002. Proceedings, 2002. XIV, 450 pages. 2002.

Vol. 2369: C. Fieker, D.R. Kohel (Eds.), Algebraic Number Theory. Proceedings, 2002. IX, 517 pages. 2002.

Vol. 2370: J. Bishop (Ed.), Component Deployment. Proceedings, 2002. XII, 269 pages. 2002.

Vol. 2372: A. Pettorossi (Ed.), Logic Based Program Synthesis and Transformation. Proceedings, 2001. VIII, 267 pages. 2002.

Vol. 2373: A. Apostolico, M. Takeda (Eds.), Combinatorial Pattern Matching. Proceedings, 2002. VIII, 289 pages. 2002.

Vol. 2374: B. Magnusson (Ed.), ECOOP 2002 – Object-Oriented Programming. XI, 637 pages. 2002.

Vol. 2375: J. Kivinen, R.H. Sloan (Eds.), Computational Learning Theory. Proceedings, 2002. XI, 397 pages. 2002. (Subseries LNAI).

Vol. 2378: S. Tison (Ed.), Rewriting Techniques and Applications. Proceedings, 2002. XI, 387 pages. 2002.

Vol. 2380: P. Widmayer, F. Triguero, R. Morales, M. Hennessy, S. Eidenbenz, R. Conejo (Eds.), Automata, Languages and Programming. Proceedings, 2002. XXI, 1069 pages. 2002.

Vol. 2382: A. Halevy, A. Gal (Eds.), Next Generation Information Technologies and Systems. Proceedings, 2002. VIII, 169 pages. 2002.

Vol. 2383: M.S. Lew, N. Sebe, J.P. Eakins (Eds.), Image and Video Retrieval. Proceedings, 2002. XII, 388 pages. 2002.

Vol. 2384: L. Batten, J. Seberry (Eds.), Information Security and Privacy. Proceedings, 2002. XII, 514 pages. 2002.

Vol. 2385: J. Calmet, B. Benhamou, O. Caprotti, L. Henocque, V. Sorge (Eds.), Artificial Intelligence, Automated Reasoning, and Symbolic Computation. Proceedings, 2002. XI, 343 pages. 2002. (Subseries LNAI).

Vol. 2386: E.A. Boiten, B. Möller (Eds.), Mathematics of Program Construction. Proceedings, 2002. X, 263 pages. 2002.

Vol. 2389: E. Ranchhod, N.J. Mamede (Eds.), Advances in Natural Language Processing. Proceedings, 2002. XII, 275 pages. 2002. (Subseries LNAI).

Vol. 2391: L.-H. Eriksson, P.A. Lindsay (Eds.), FME 2002: Formal Methods – Getting IT Right. Proceedings, 2002. XI, 625 pages. 2002.

Vol. 2392: A. Voronkov (Ed.), Automated Deduction – CADE-18. Proceedings, 2002. XII, 534 pages. 2002. (Subseries LNAI).

Vol. 2393: U. Priss, D. Corbett, G. Angelova (Eds.), Conceptual Structures: Integration and Interfaces. Proceedings, 2002. XI, 397 pages. 2002. (Subseries LNAI).

Vol. 2398: K. Miesenberger, J. Klaus, W. Zagler (Eds.), Computers Helping People with Special Needs. Proceedings, 2002. XXII, 794 pages. 2002.

Vol. 2399: H. Hermanns, R. Segala (Eds.), Process Algebra and Probabilistic Methods. Proceedings, 2002. X, 215 pages. 2002.

Vol. 2405: B. Eaglestone, S. North, A. Poulovassilis (Eds.), Advances in Databases. Proceedings, 2002. XII, 199 pages. 2002.